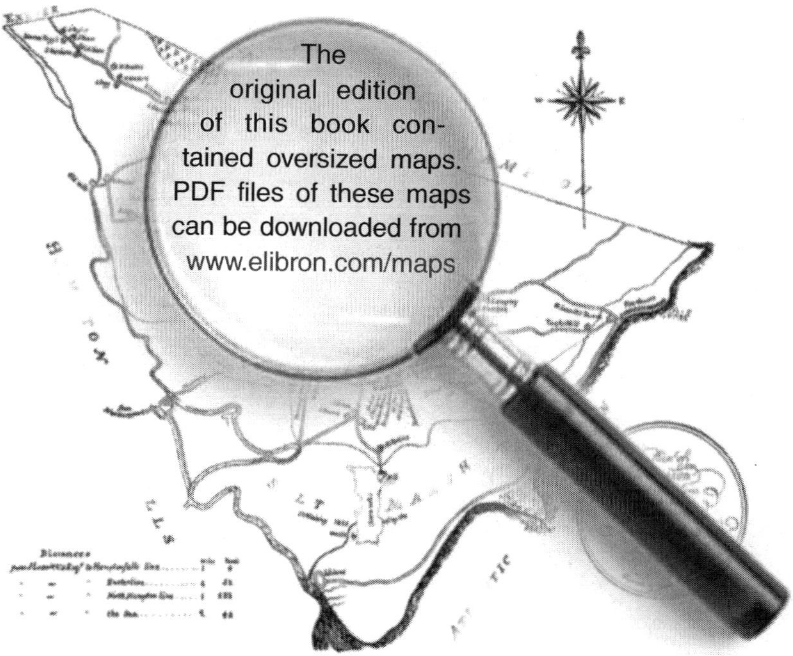
The
original edition
of this book con-
tained oversized maps.
PDF files of these maps
can be downloaded from
www.elibron.com/maps

CHARLES WILKES

NARRATIVE
OF THE
UNITED STATES EXPLORING EXPEDITION

DURING THE YEARS

1838, 1839, 1840, 1841, 1842

VOLUME II

Elibron Classics
www.elibron.com

Elibron Classics series.

ISBN 1-4212-2115-2 (paperback)
ISBN 1-4212-2114-4 (hardcover)

This Elibron Classics Replica Edition is an unabridged facsimile of the edition published in 1856 by G. P. Putnam & Co., New York.

UNITED STATES

EXPLORING EXPEDITION.

Wm L. HUDSON

COMMANDER, U S N

NARRATIVE

OF THE

UNITED STATES

EXPLORING EXPEDITION.

DURING THE YEARS

1838, 1839, 1840, 1841, 1842.

BY

CHARLES WILKES, U. S. N.

COMMANDER OF THE EXPEDITION,

MEMBER OF THE AMERICAN PHILOSOPHICAL SOCIETY, ETC.

IN FIVE VOLUMES, WITH THIRTEEN MAPS.

VOL. II.

NEW YORK:

G. P. PUTNAM & CO., 321 BROADWAY.

1856.

STEREOTYPED BY J. FAGAN.

PRINTED BY C. SHERMAN.

CONTENTS OF VOL. II.

CHAPTER I.

CHAPTER II.

CHAPTER III.

CHAPTER IV.

CHAPTER V.

CHAPTER VI.

CHAPTER VII.

CHAPTER VIII.

CHAPTER IX.

CHAPTER X.

CHAPTER XI.

CHAPTER XII.

LIST OF ILLUSTRATIONS.

VOLUME II.

PLATES.

VIGNETTES.

WOOD-CUTS.

MAPS.

Those marked with a †, were drawn on the wood by J. H. Manning; those marked with a ‡, by W. G. Armstrong. Those not marked, by the Artists of the Expedition.

CHAPTER I.

CONTENTS.

GENERAL ASPECT OF TAHITI—ARRIVAL AND RECEPTION—GENERAL FIGURE OF THE ISLAND—ITS GEOLOGICAL STRUCTURE—OVERHAUL OF THE VESSELS—AN OBSERVATORY ESTABLISHED—SURVEY OF HARBOURS—LONGITUDE OF POINT VENUS—HEIGHT OF MOUNTAINS—VISITERS—TAUA, THE GOVERNOR OF MATAVAI—HIS HOSPITALITY—CHURCH AT MATAVAI—CONCOURSE OF NATIVES—CHARACTER OF THE NATIVES—SCHOOL AT PAPIETI—ADDRESS TO THE SCHOLARS—REPLY OF PAOFAI—GENERAL DIFFUSION OF EDUCATION—COMPLAINTS OF THE AMERICAN CONSUL—COUNCIL OF THE CHIEFS—INFLUENCE OF THE MISSIONARIES—CHARACTER OF A PART OF THE FOREIGN RESIDENTS—CHANGE OF THE NATIONAL DRESS—GENERAL VIEW OF THE LABOURS OF THE MISSIONARIES—IMPROVEMENT IN THE NATIVE CHARACTER—CAUSES OF WANT OF INDUSTRY—CONSTITUTION OF TAHITI—COURTS OF JUSTICE—QUEEN AND ROYAL FAMILY—JUDGES OF THE SUPREME COURT—STATE OF PARTIES—CASE OF THE CATHOLIC PRIESTS—DWELLINGS OF THE NATIVES—APPEARANCE OF THE FEMALES—DRESS OF THE TWO SEXES—TASTE OF THE NATIVES FOR FLOWERS—COOKERY AND MODE OF EATING—MUSIC OF THE ISLANDERS—THEIR DANCES—EXPEDITION TO LAKE WAIHEREA—ABORTIVE ATTEMPT TO ASCEND OROHENA—DESCENT TO THE COAST—RETURN ALONG THE SOUTHERN AND WESTERN COAST TO PAPIETI—NAVAL ARCHITECTURE AND FOREIGN TRADE.

NARRATIVE

OF

THE EXPLORING EXPEDITION

CHAPTER I.

TAHITI.

1839.

The beauty of the distant view of Tahiti has been celebrated by all navigators, but I must confess that it disappointed me. The entire outline of the island was visible for too short a time and at too great a distance to permit its boasted features to be distinctly seen. Upon a second and nearer view, its jagged peaks and rugged inaccessible mountains were visible, but we looked in vain for the verdant groves which are said by all writers to clothe it. These indeed exist, but are confined to a narrow belt of low land, lying between the mountains and the shore, and being unseen at a distance, the general aspect of the island is that of a land recently thrown up by volcanic action.

When, however, Tahiti is approached so near as to make separate objects visible, the contrast between it and the barren coast of Peru becomes striking. Even upon the steep surface of its cliffs, vegetation abounds; the belt of low land is covered with the tropical trees peculiar to Polynesia; while the high peaks and wall-faced mountains in the rear are covered with vines and creeping plants. This verdure is seen to rise from a quiet girdle of water, which is again surrounded by a line of breakers, dashing in snow-white foam on the encircling reefs of coral. Such objects are sufficient to form a beautiful landscape, and my disappointment probably arose in part from finding every thing more diminutive than I had been led to imagine from the highly-wrought descriptions I had been perusing only a few days before.

We were surrounded, even before we anchored, by canoes of all shapes and sizes, whose crews made a prodigious clamour. I at once interdicted any one who was not a chief from coming on board; but upon this being announced, every one claimed to be a chief of some description or other. Only the great chiefs, therefore, were admitted. These came off in whale-boats, which are now superseding the canoe, and brought with them trifling presents of fruit. It was soon found that their errand was not one of mere ceremony, but was intended to solicit the washing of our dirty linen, a business which is among the prerogatives of the queen and chiefs. I was informed that the queen, being *enceinte*, was residing on the opposite side of the island, which would prevent her from paying us a visit. I was, therefore, at liberty to choose a less distinguished laundress, and spared the pain of resisting her royal solicitations for soap, an article much needed and in great request at Tahiti.

I was glad when the night closed in, to be rid of our numerous visiters. The pilot, who goes by the name of "English Jim," was equally so, for he chose to be considered as the only privileged person, and, besides, was looking somewhat to his own profit in the line of clothes-washing, a business which the presence of the chiefs threatened to interfere with. Jim is quite a respectable-looking man, dresses in the European fashion, and speaks English, which he has acquired on board of whale-ships, tolerably well. Although a good pilot, so far as a knowledge of the shoals goes, he does not understand what to do with a vessel, in case of difficulty. He told me that he had been looking out for vessels for some days, for it had thundered.

Although the shape and extent of Tahiti are well known, I venture to give a map of it, which has been made as correct as our opportunities would permit. The two peninsulas, if they may be so termed, of which it is made up, are of very different characters. The smaller one, called Tairaboo, and usually spoken of as "the small island," is said to be the most fertile: it possesses some harbours, but they are little better known than they were half a century ago. Both peninsulas possess twenty-four harbours, including the good and bad. Tahiti proper contains the best, and therefore engrosses all the commerce. It has in consequence been for many years the seat of government.

The whole island is of volcanic formation, but there is no longer any active igneous action, nor is there any well-defined crater to be seen. Coral reefs, with occasional openings, are attached to the shores, and the larger island (Tahiti) has also a sea reef. Between the two reefs is an almost continuous channel for boat navigation, and on the northern side they enclose many safe and commodious harbours for shipping. On this side also vessels may pass from harbour to harbour, within the outer reef. This reef varies in breadth from a few yards to fifty, or even a hundred. The shore that adjoins the coral reef is formed of black volcanic sand, occasionally mixed with comminuted shells, which give it a grayish hue. Basaltic ridges reach the sea at intervals, and form projecting points of moderate elevation.

We began without delay to overhaul the vessels, and the few sick persons we had on board were sent on shore to a shed hired for the purpose on Point Venus.

An observatory was established at the same place, and furnished with both astronomic and magnetic instruments; and as soon as the repairs of the vessels had made such progress as to permit it, parties were formed for the survey of the four principal harbours and the channels between them. These harbours, Matavai, Papaoa, Toanoa, and Papieti, are so important to the many whale-ships which visit this island, that I felt it an imperative duty to obtain accurate charts of them all. At the same time, a large party of officers and naturalists was ordered to cross the island, to reach, if possible, Orohena, one of the highest peaks, and to visit Lake Waiherea.

I had been in hopes of obtaining a full series of moon culminating stars on Point Venus; but I was disappointed, for it rained almost every night. I was, therefore, compelled to rely for the longitude on the chronometers alone, and restricted even in that method to observations of the sun. I was, however, well pleased to find that my results differed from the best preceding authorities no more than 1′ 33″ of space. These authorities give 149° 29′ 43″ W., for the longitude of Point Venus.

The mountains were obscured by clouds during the whole time of my stay, and no angles could be taken for the measurement of their heights, nor could the party I detached for the purpose reach their summits; but the Peacock remained for some days after my departure, and Captain Hudson, with his officers, succeeded in measuring the height of Aorai, the peak which is next in height to Orohena. This he found to be six thousand nine hundred and seventy-nine feet; and as Orohena appeared to be about one thousand five hundred feet higher, the height of the latter peak may be set down as about eight thousand five hundred feet above the level of the sea. From these two peaks, ridges diverge to all parts of the coast, throwing off spurs as they descend. These ridges are precipitous, and for the most part narrow. In many instances their summit is a mere edge, making walking upon them not only dangerous, but often impossible.

Soon after my arrival, I was visited by S. R. Blackler, Esq., our consul, who resides at Papieti, distant eight miles from Matavai. The resident missionaries, with Mr. Pritchard, Her Britannic Majesty's acting consul, and Mr. Morenhout, the French consul, also favoured me with a visit, as did several other persons, who all expressed their desire of being serviceable to us.

The governor of the district of Matavai, Taua, was the first acquaintance of any distinction that we made. He had already visited the Vincennes on her anchoring. He is a fine-looking man, of huge proportions, and has a large establishment near Point Venus, where he monopolized nearly all the washing, which was performed by his numerous dependants. By this business he derives some remuneration for the cost of feeding and clothing them, putting the gains of their labour into his own pocket. Such, at least, is his own account of the transaction.

Taua's usual dress was a striped cotton shirt, nankeen pantaloons that had once been yellow, and a round jacket of blue cloth. Both shirt and pantaloons were too tight, and he had neither suspenders nor stockings, although he wore shoes. In this guise he had an awkward look, which he probably would not have exhibited in a native costume.

He was profuse in offers of hospitality at his own house, and many of the officers were induced to accept his invitations. His entertainments appear to have been of the same general character with that to which I was treated, and which will, therefore, serve as a specimen of the mode in which such things are done by the "good society" of Tahiti.

We reached his dwelling in time to see the preparations for the feast. These were entrusted to his man of all work, Stephen, or as he called him, "Stiffin." This useful personage exhibited his dexterity, not only in cooking, but in killing the poultry. The bird selected was a cock, for the Tahitians well understand the difference in value between it and the hens; and Stephen exhibited much adroitness in the slaying, plucking, and dressing. While this was going on, the stones for the Tahitian oven, so often described by voyagers, were heating, and when they had acquired the proper temperature, the ashes were carefully swept off,—bread-fruit, taro, and plantains, wrapped in leaves, were then laid on the stones, with the fowl in the centre, and the whole covered up. In about an hour the oven was carefully opened, the contents exposed, and found to be thoroughly cooked. The dinner was then served in an earthen dish, with a knife and fork, when, although the fowl was somewhat tough, it was greatly relished. The dinner hour was one o'clock.

Taua, according to the universal opinion of the squadron, did not improve upon a closer acquaintance. His intrusive and greedy disposition, not to mention his fondness for the bottle, rendered him daily a less welcome visiter than at first. I must, however, do him the justice to say, that if he were wanting in other traits of character that ought to distinguish a chief, he did the honours of his house admirably, and that he must be seen in the capacity of a host, if a favourable opinion is to be formed of his character.

On the invitation of the Rev. Mr. Wilson, I visited him at the mission-house, and was kindly received. This gentleman is seventy-two years of age, and is the oldest missionary on the island. In spite of his advanced age, he still performs all the duties of his cure. The church and the parsonage are both frame houses. The former, which is neatly built, is capable of containing a large congregation. The Sabbath occurred on Saturday, by our reckoning, and all labour was suspended. I thought the attendance on worship small, compared with what I had been led to anticipate. There were less than two hundred persons present, and they did not appear to be as attentive as they had been represented. The women were more numerous than the other sex, and were dressed in a most unbecoming manner. They wore high flaring chip bonnets of their own manufacture, loose gay-coloured silk frocks, with showy kerchiefs tied around their necks. Nothing can appear more *outré* than they do in these habiliments, and I was at a loss to conceive how they could, in particular, have been induced to adopt a covering for the head, which affords no protection from the sun, and is in consequence so ill-adapted to the climate.

On Sunday, 15th September, as many of the officers and crews as could be spared from the vessels, attended divine service in the Mission church. Our chaplain performed the service, with the aid of the Rev. Mr. Pratt. This exercise attracted great crowds of the natives, of whom an unusual number had collected at Matavai. They flock, on the arrival of vessels, and particularly of ships of war, to the port in which they lie, partly from curiosity and a desire of amusement, but more from a hope of gain. A Tahitian changes his residence without difficulty or inconvenience; food is every where to be had in abundance, and lodgings never enter into his calculation. While the squadron was at Matavai Bay, the number of those who appeared to inhabit its shores would have given a very erroneous estimate of the usual population. They were assembled from every part of the island, and the right of occupying each spare nook in the houses of the permanent settlers, seems to be universally admitted. When this resource fails, they are to be seen beneath trees, or upon the beach, within a few feet of the water's edge, sleeping as soundly, although without any covering, as if they were beneath their own roofs.

Saturday is a holiday, and on it the children may be seen engaged in innocent plays and amusements, among which is swinging with a single rope from a tall cocoa-nut tree. This pastime is picturesque, and in good keeping with the landscape.

SWINGING, TAHITI

This people exhibits great curiosity, but does not manifest it by intrusive inquisitiveness. In our whole intercourse with them, we did not hear of a single act of theft, although there were innumerable opportunities for its commission, without the possibility of immediate detection. They seemed always in a good humour, gay, happy, and cheerful; nor did I witness a single quarrel among all the crowds that were assembled at Point Venus, during our stay. They are, however, inveterate beggars.

At the invitation of Mr. Pritchard, I visited the school under his direction at Papieti. This gentleman was, a few years since, a missionary, but now holds the station of Her Britannic Majesty's Consul. He has not, however, abandoned all his missionary duties. We were also invited to visit the school at Matavai Bay, under the direction of the Rev. Mr. Wilson. There was so much similarity in the two schools, that I shall content myself with a description of the former, and a circumstance or two that occurred at the latter.

The school at Papieti is held in the church, a large frame building, much like a New England meeting-house. It has numerous windows, a large gallery, and pews capable of containing a great number of people. All who were present were well dressed, and the assemblage, except from the colour of their skins, could have been, with difficulty, distinguished from a Sunday school in the United States.

The exercises were opened with prayer. The children then sang the A B C song, and went through the *a-b ab*, at the word of command from Mr. Pritchard. To this succeeded an examination of the larger children, and an exhibition of some of their performances. The number of scholars was from three hundred and fifty to four hundred, between eight and sixteen years of age. A large proportion of them read, sang, and wrote on a slate, manifesting a fair proficiency in each, and answered intelligently the questions that were put to them.

To test the character of the instruction given in this school, I felt desirous of putting some questions on subjects foreign to the ordinary routine, and particularly in relation to natural phenomena. For this purpose I called the attention of the scholars to the eclipse of the sun which had happened a few days before. This was received with a variety of expression of countenance by different scholars, but among them I could only remark stupid wonder, indifference, or listlessness which showed too clearly that no attempt had been made to awaken their attention to such subjects.

When the examination was concluded, I made a trifling present to the native teachers in the name of the American government, and desired Mr. Pritchard to express the satisfaction it had given myself and officers to witness the exhibition of a progress, which in the elementary branches of education had exceeded our expectations; and to state that I was assured it would be gratifying to our country to hear of their improvement. I concluded by expressing my hope that they would persevere in their attention to the missionaries, who were so zealously endeavouring to instruct them in knowledge and religion.

This short address was immediately answered by Paofai, a chief who holds the office of chief judge, and who is generally considered as the ablest and most clear-headed man in the nation. His reply was delivered in an earnest and animated manner, and contained many expressions of gratitude for the attention and kindness shown them by the people of the United States.

The exhibition of the schools did not surprise me so much as the fact that few natives are to be met with who cannot both read and write. This was not confined to the younger part of the population, but was true even of those advanced in years. I also learned that they had schools among themselves, and that parents were well aware of the advantages attendant on sending their children to them. In these schools great pains are taken to inculcate cleanly and industrious habits, with sound moral and religious principles.

The hours of attendance are confined to the forenoon, and during these the schools are crowded. The parents are unwilling that their children should be confined for a longer time.

Our consul, Mr. Blackler, had made complaints to me, as soon as I arrived, of the conduct of the queen and government, and asked my interference. The charges consisted in the following items:

1. The seizure of an American whale-boat and ill treatment of the crew.

2. That fines had been unjustly imposed on American seamen.

3. The refusal to apprehend deserters from American ships, or to provide a place for their safe keeping.

4. The evasion of a promise to provide a place for the transaction of the consular business.

In consequence of these complaints, I had immediately requested that a council of the chiefs might be held, and the 17th September was appointed for the purpose. On this day I ordered all the officers that could be spared from the vessels to attend. Captain Hudson and myself set out at an early hour, accompanied by several boats. We passed down through the reefs, and reached Papieti at ten o'clock, where we were joined by our consul, and in his company proceeded to the building which has been mentioned as the scene of the exhibition of the schools. Here we were received by Mr. Pritchard, who politely showed us to the seats we were to occupy. He then called the names of the chiefs, and each answering in his turn, took his seat on the side of the building opposite to us.

The meeting being ready for business, I read from a paper a list of the grievances complained of. This was translated sentence by sentence by a Mr. Darling. When I had finished, Paofai again appeared to make a reply. He began by apologizing for the absence of the queen, caused by her approaching confinement, and then requested a copy of the paper which had been read, in order that it might be considered and answered. He stated that it included too many points to be decided upon and answered at once, but promised that the matter should be examined, and the business concluded as speedily as possible.

PAOFAI.

This request was so reasonable that I at once assented to it. I thought the proposed mode far better, and it was more agreeable to me than a public discussion would have been, in which confusion could hardly be avoided. I therefore broke up the meeting, after stating that I should look for a satisfactory reply on my coming in the Vincennes to Papieti.

Many of the chiefs seemed disposed to act correctly and do justice. at least they repeatedly expressed their good intentions. It was also evident to me, that their minds were greatly relieved by the moderation of the demands, for they had feared that these were to be of some extraordinary kind, and might perhaps include a claim for heavy damages. Indeed, since the large contribution levied on this island by the French, the government has entertained apprehensions, and dreads the arrival of men-of-war. These fears are taken advantage of by many ill-disposed residents, who omit no opportunity to practise upon their alarms, and to threaten them with foreign interference.

Much complaint has been made of the influence which the missionaries, and Mr. Pritchard in particular, exercise over the government of Tahiti. They have, unquestionably, great influence; but I am satisfied that they are justly entitled to it. Indeed I cannot but consider it as part of their duty, nay, the great object of their mission, to acquire and exercise a salutary control over their converts, both of high and low degree. My own observations satisfied me that this control is exerted solely for the purpose of fulfilling the laudable object for which they were sent. It is possible that their views of the proper method of instructing an ignorant people are not at all times, or in every respect, the most enlightened; but no one can with propriety question their pious zeal, or the honesty of their intentions. We may perhaps lament their intolerance towards other sects, but no one can visit the island without perceiving on every side the most positive evidence of the great benefits they have already bestowed, and are daily conferring upon the inhabitants.

All this good has been done in the face of many and great difficulties. The most serious of these is the evil influence of a large portion of the other foreign residents. Although among these are some who are truly respectable, the majority is made up of runaways from the English convict settlements, and deserters from vessels. These men, the outcasts and refuse of every maritime nation, are addicted to every description of vice, and would be a pest even in a civilized community. It may easily be conceived what an injurious influence such a band of vagabonds, without trade or occupation by which they can support themselves, guilty of every species of profanity and crime, must exert upon the morals of the natives, and what a barrier they must oppose to their improvement in morals and religion.

Tahiti, when first visited, was proverbial for its licentiousness, and it would be asking too much, to require that after so short an enjoyment of the means of instruction, and in the face of such obstacles, its inhabitants should as a body have become patterns of good morals. Licentiousness does still exist among them, but the foreign residents and visiters are in a great degree the cause of its continuance, and an unbridled intercourse with them serves to perpetuate it. Severe laws have been enacted, but they cannot be put in force in cases where one of the parties is a foreigner. I see no reason, however, why this island should be pointed out as conspicuous for licentiousness. When compared with many parts of the world that arrogate a superior civilization, it appears almost in an advantageous light. Vice, at any rate, does not stalk abroad in the open day, as it did in some places we had lately visited upon the American continent. It would be unfair to judge of these natives, before they had received instruction, by our rules of propriety; and now many of those who bear testimony to the laxity of their morals, visit their shores for the very purpose of enticing them into guilt, and of rioting without fear or hindrance in debauchery. Coming with such intentions, and finding themselves checked by the influence of the missionaries, they rail against them because they have put an end to the obscene dances and games of the natives, and procured the enactment of laws forbidding illicit intercourse.

The missionaries are far from overrating their own success in effecting an improvement in morals, and inculcating the obligations of religion. So far from this, I found that they generally complained that sincere piety was rarely to be found among the natives. However this may be, the external signs of moral and religious improvement are conspicuous. Many of the natives are scrupulous in their attention to Christian duties, and members in communion of the church. All are strict observers of the Sabbath; indeed, nowhere is its institution more religiously attended to than in those Polynesian islands which are under missionary influence. On that day no canoe is launched upon the waters, and no person is seen abroad except while on his way to or return from church. When thus seen, they are neatly and decently clothed, although in very bad taste. At church they form a respectable-looking congregation, and listen with attention to the preacher.

The success of the missionaries in introducing this strict observance of a Sabbath is ascribed by themselves in a great degree to its analogy to the taboo-days of heathen times, and the continuance of its sanctity is now insured by the penalties which await an infraction of it. The punishment for Sabbath-breaking consists in the offender being compelled to make a certain number of fathoms of road, and upon a repetition of the offence, the number of fathoms is much increased.

I cannot pass without notice the untiring efforts of many of the foreign residents to disparage the missionaries and vilify the natives. They endeavour on all occasions to prepossess the minds of visiters against both. These efforts, however, generally fail of success; for no reflecting mind can fail to perceive how devoid they are of any foundation, nor avoid noticing the baneful effects these residents are themselves producing, by inculcating principles for which many of them have been compelled to fly their own countries, or teaching the practice of crimes from whose penalty they have made their escape.

There are about a hundred characters of this description on the island, and to give a better idea of them I shall divide them into three classes.

The first class comprises merchants, if they can be so called. The sole object of these is to make money. I regret to say that, as far as my observation went, this purpose is not accomplished without injury to the welfare of the natives. This class endeavours to place both the persons who compose it and the premises they occupy beyond the reach of the local laws.

The second class is composed of the children born upon the island of missionary parents. Of these many seem to have forgotten utterly the principles instilled into them in their infancy.

The third class is much the most numerous, and those I include in it appear destitute of all moral or religious principle. They stand out openly and boldly in defiance of all law and decency. Among them continual complaints are to be heard against the missionaries, the government, and the people. On being asked to state the ground of their complaints, most of them fail in presenting any other charges than that the missionaries are endeavouring to make the natives too good; that they deprive them of their innocent luxury of intoxicating liquors; that they interdict promiscuous intercourse, and have ruined the trade of the island by preventing the women from going on board ship; that they have interfered with their amusements by abolishing lascivious dances and songs, and requiring from them instead, prayers and hymns; that they have introduced too strict an observance of a Sabbath, translated the Scriptures, and taught the natives to read them. Others argue seriously, that this mild and amiable people had no need of instruction in divine revelation; that they would have been much happier had they been left to follow their own inclinations; and that they have been rendered miserable by being taught their responsibility as accountable beings.

The missionaries, however, receive countenance and support from a more respectable portion of the foreign residents. These, although they do not approve of the whole of the course the missionaries have pursued, are united in upholding the moral and religious principles which they endeavour to inculcate.

Although much has been done for the improvement of the natives, still it appears evident that much more might have been done if the missionaries had not confined themselves so exclusively to teaching from the Scriptures. The natives, by all accounts, are extremely fond of story-telling, and marvellous tales of their ancestors and ancient gods, are even now a source of amusement. The missionaries, as I am told, possess much information in relation to the history and mythology of the island, embodied in the superstitious tales still occasionally current among its inhabitants. It is to be hoped that they will preserve a record of these, before they are obliterated by their exertions to destroy the ancient superstition. But they would have succeeded sooner in eradicating the practice of reciting these legends, had they provided a substitute in works of fiction, inculcating moral and religious lessons, or teaching useful knowledge. So also, while it was indispensable to put down those amusements which were the means or incentives to debauchery, this measure ought to have been accompanied by the introduction of innocent modes of recreation. For want of the first resource, much time is now spent in unmeaning gossip, and the necessity for the other is often shown in a listless idleness.

No attempt has been made by the missionaries to introduce the mechanic arts, or improvements in agriculture, yet it cannot be doubted, that to have taught them even the simplest of these, would have materially aided the progress of civilization, and reacted favourably upon that of religion. The failure of a cotton manufactory, with expensive machinery, which was erected on the island of Eimeo, affords no argument against the probable success of less complex arts. The natives were not prepared to pass at once from habits of desultory exertion, to the regular and stated occupation of the mill. But the spinning-wheel, the hand-loom, and the plough, would not have required such a decided change, in the number of hours of labour, and would have served as a preparation for more continuous industry. The two former implements have at length been introduced by other hands, and have already been adopted with eagerness by some of the natives.

The change of dress which has been introduced by the missionaries and other foreigners, has, on the contrary, had an injurious effect on the industry of this people. While they wore their native tapa, the fabric, though of little value, gave employment to numbers of women; and this change of dress, intended as an advance in civilization, has had the effect of superseding employments which formerly engaged their attention, and occupied their time. The idleness hence arising, and the artificial wants thus created, have no little influence in perpetuating licentiousness among the females, to whom foreign finery is a great temptation. The European dress, at least as worn by them, is neither as becoming, nor as well adapted to the climate as that which it has almost superseded. Many of the missionaries now see these things in their true light, and informed me that they were endeavouring to pursue a more enlightened course.

Upon the whole, although the missionaries may be chargeable with misjudging zeal, and have exhibited a want of practical knowledge of human nature in their efforts, and in the solution of the difficult problem of bringing barbarians to civilization, they ought to receive due credit for what they have actually accomplished. I am decidedly of opinion, that in spite of all the drawbacks I have mentioned, as much would not have been done by any other class of persons. It has demanded a sense of religious duty, to enable them to persevere in a constant devotion to the cause in which they have embarked, to enable them to undergo the privations and trials to which they have been subjected, while continually at the mercy of uncivilized men. No desire of pecuniary emolument has been evinced by them, nor are they sustained by any expectation of temporal reward; and I can testify, from personal observation, that their position in a worldly sense, is not to be envied.

To judge of the amount of good they have accomplished, it is necessary to turn back to the records of early voyages, and compare the present with the former condition of these islanders. Now they are seen enjoying peace, possessing a written instead of a mere oral language, living under wholesome laws, and receiving the advantages of school education and church discipline. In former times, we read of perpetual intestine broils, of the worship of idols propitiated by human sacrifice, of the depraved association of the Ariore, and its accompanying crime of infanticide. In making this comparison, we cannot but acknowledge that the persons who have effected these changes, are both Christians and philanthropists, and that they have been reasonably successful in implanting the principles of civilization.

As a proof of the value of their labours, my experience warrants me in saying that the natives of Tahiti are honest, well-behaved, and obliging; that no drunkenness or rioting is to be seen, except when provoked by their white visiters and inmates, and that they are obedient to the laws and to their rulers. That they should be comparatively indolent is natural, in a climate where the fruits of the earth almost spontaneously supply the wants of nature, and where a mere animal existence may be maintained without labour. No People are, in truth, so independent of the aid even of their fellows as the Tahitians. A native may in the morning be wholly destitute even of implements wherewith to work, and before nightfall he may be found clothed, lodged, and have all the necessaries of life around him in abundance. These he derives from the cocoa-nut, the poorou (Hibiscus tiliaceus), banana, bread-fruit, and bamboo. That he does not find it necessary to call upon others for assistance, does not make him forget the duties of hospitality, but it does produce a thoughtlessness about his own wants, and takes away that incitement to labour, which is so powerful an aid in the promotion of civilization. Still, I am satisfied that the Tahitians do not avoid labour, when they can work with profit to themselves. Those who were employed on board the squadron, where their pay was liberal and regular, performed their tasks faithfully and well; and they bear the same character for fidelity in the whale-ships, on board of which they are much employed. Some of them are now engaged in the culture of the sugar-cane; and a single native plantation was mentioned to me, of which the preceding year's crop had amounted to five tons. Coffee has also been planted, and succeeds remarkably well. Much more, too, would have been done in these productions had their industry been encouraged by the missionaries, as a body; but, while some of them have done their utmost to stimulate the natives to exertion, others have altogether discountenanced any attempts to introduce new articles of culture.

One of the most important consequences of the introduction of civilization has been the establishment of a settled constitution. This was framed by the missionaries in 1823, upon the model of that of England, and was revised in 1826. The royal authority includes the power of the veto, the nomination of the supreme judges, and of all officers connected with the person of the sovereign. The crown is hereditary, descending either to males or females. The legislative power is lodged in an assembly, composed of two members from each district, chosen triennially by the people. This assembly is convened annually for the purpose of remodelling existing laws, or enacting new ones. It has also semi-annual meetings, and may be convened more frequently, if necessary, for the discussion of questions of importance. All enactments of the legislature, before they become laws, are laid before the queen for her approbation and signature. When this is affixed, they are carried into effect by the judges and the officers of the crown. Should she refuse her signature, they are revised and remodified, or laid aside altogether.

The island is divided into seven districts, each of which has an inferior court for the trial of ordinary cases. This consists of two judges, who are not unfrequently also members of the legislature. The decision of these courts must be founded upon evidence, and appeal lies to the supreme tribunal.

This supreme court is composed of seven judges, two of whom are residents of the island of Eimeo. The judges are also executive officers, and nearly all are chiefs. This double capacity gives them great influence, and their power is sufficient to supply, in part, the queen's want of energy, but at the same time serves as a check against any encroachment upon the prerogatives of the sovereign.

The powers of this court even extend to an impeachment of the royal ruler.

The mode of trial, both of civil and criminal cases, is by a jury, and free argument is allowed. The testimony is not given upon oath, but the penalty for giving false evidence is severe. The jury is composed of six persons; and every one has the right of being tried by his peers.

The reigning queen is named Aimata, but is more usually known as Pomare IV. She is the sister of the late king, and grand-daughter to that Pomare I. who acquired the sovereignty of Tahiti, soon after its discovery. She is now (1839) about twenty-seven years of age, and has been twice married: the first time to Pomare, a young chief of Tahaa, from whom she was divorced; the second, to a young chief of the island of Huaheine, by whom she has one son, the heir of the throne. The general appellation he goes by is Pomare *taue*, equivalent to king-consort.

Next in rank to the queen, is her aunt, Ariapaca, the eldest sister of her mother, and at one time queen-regent. She still possesses great influence.

In case of failure of the queen's posterity, the next heirs to the throne are the princesses Ninito and Taii, who are the queen's cousins, and nieces to Pomare II.

Uata, the godfather of the queen, although not a chief by birth, has from this connexion obtained great influence in the queen's councils, and may be termed prime minister.

The seven judges of the supreme tribunal are nominated by the queen, but the nomination must be confirmed by the legislature. Those who at present hold the office are all large landholders, and men of the highest character and intelligence to be found in the population. They are in fact the rulers of the kingdom. Five of them, viz.: Paofai, Mare, Utami, Taati, and Tanoni, reside on the island of Tahiti; the other two, Ruetone and Mahine, at Eimeo.

In spite of the small extent of the kingdom, it is not without subjects to distract its councils. There are two distinct parties: the one led by the queen and the missionaries; the other, by some of the chiefs. The leaders of the latter are Paofai, Hitoti, and Taua, who are descended from the ancient kings dethroned by Pomare I. These chiefs have large domains, and many of the raatiras (landholders) take part with them. They are, besides, distinguished by qualities which give them consideration among the islanders. Paofai, who has more than once been spoken of, holds the office of chief judge, and is considered as the best statesman on the island. Hitoti is distinguished for a dignity, uprightness, and good sense, which command universal respect. Taua possesses a high reputation as a brave and skilful warrior.

Of these three leaders, Hitoti alone is wholly free from reproach. Paofai is accused of covetousness, and a propensity to intrigue; and Taua, of a fondness for intoxicating drinks.

The queen, however, contrives to rule in all matters that rightfully belong to her; and, by the aid of the missionaries, maintains her ground against this strong opposition, although its leaders have generally the power to determine the course of policy to be pursued, and entire authority over the execution of the laws. They are much opposed to foreigners, and have made several attempts to have them banished from the island. They are supposed to entertain the design of setting aside the queen, on account of her irregular behaviour and vices; but this plan is not likely to succeed, because of the personal popularity she enjoys, and the number of adherents she possesses among the people. In conformity with such a design, these chiefs are said to be continually watching for opportunities to increase their own power and diminish the royal authority. Among the occasions of which they endeavoured to avail themselves, was the celebrated affair of the Roman Catholic priests, the circumstances of which, as nearly as I could learn from the statements of both parties, are as follows:

Two priests of this denomination, who had been stationed at the Manga Reva, or Gambier Group, landed on the southern side of the island, and travelled towards Papieti, preaching the doctrines of their church. They, however, found none willing to listen, and it is said, that no native would receive them into his house. On their arrival at Papieti, however, Paofai, Hitoti, and some other chiefs, gave them countenance, and they were hospitably received by Mr. Morenhout, the acting American consul, who, however, did not lodge them under his own roof, but in an adjacent building. The people, however, excited by the preaching of the English missionaries, broke into the building, and compelled the priests to embark on board a small vessel, which carried them to Uea, or Wallis Island, about two thousand miles to the west of Tahiti.

In considering this question calmly, and stripping it of the exaggerations with which both parties have loaded it, it is difficult to say which was most in the wrong. The Protestant religion was established by law upon the island, to the exclusion of all others, and this the priests well knew; nor can any but zealots, who think that those whom they style heretics are worse than infidels, excuse their intrusion upon missionary ground already fully and successfully occupied. On the other hand, their precipitate expulsion, under circumstances of great hardship, exhibited an unchristian spirit, for which the resident missionaries may justly be held responsible, as they unquestionably had it in their power to prevent any positive ill treatment on the part of the natives.

The consequences of this expulsion of the priests remain to be related. In due course of time the French frigate Venus, commanded by M. Du Petit Thouars, arrived at the island, and anchored in the harbour of Papieti. The commander immediately demanded satisfaction for the outrage committed on his countrymen the priests, and threatened that unless two thousand dollars were paid him within twenty-four hours, he would fire upon and burn the town of Papieti. The queen had no money, and was inclined, as I was told, to let the French do their worst; but as in this case the loss would have fallen wholly on the foreign residents, the required sum was collected from them by Mr. Pritchard, and paid to M. Du Petit Thouars. A treaty was also forced upon the government, allowing all Frenchmen to visit the island freely, to erect churches, and to practise their religion. Thus the local laws were abrogated under the threats of an irresistible force, and the national independence virtually surrendered.

This was a high-handed measure on the part of the French commander, and one that hardly admits of justification, particularly the demand for money; for he had himself been received with great hospitality, and not long before another of his sovereign's frigates, the Artemise, (I think), had been saved from wreck by the unrecompensed exertions of the Tahitians. The amount demanded also was at least four times as great as the pecuniary damage incurred by the priests would be reasonably valued at. The French commander, therefore, appears, in thus bullying a defenceless people into the payment of an exorbitant indemnity, and into a relinquishment of the right of admitting or excluding foreigners and strange religious creeds, by municipal regulation, in a light far from advantageous.

We have seen that Paofai and his party at first countenanced the French priests. This they no doubt did in the hope of introducing an influence which might be opposed to that of the English missionaries. Subsequently to these transactions, and after an attempt by two foreigners to murder Mrs. Morenhout, they have endeavoured to obtain the passage of a law for the expulsion of all foreigners whatsoever.

The aversion to the permanent residence of foreigners is general, and although there is no law forbidding the sale of land to them, yet no offers have hitherto been found sufficient to induce the chiefs to dispose of any portion of their soil. They find in its possession an acknowledged right to rank and respectability, and it spontaneously yields them and their followers the means of subsistence. So powerful is this repugnance to the admission of foreigners to any of the privileges arising from a possession of land, that those who are attempting to cultivate sugar, &c., hold their leases by so uncertain a tenure as to prevent their making any permanent improvement.

The canoes at Tahiti show an evident improvement on those of the Disappointment Islanders; the bottom is constructed of one piece. They use an out-rigger, and have a projection over the stern for landing. The paddle is also different.

COMMON TAHITIAN CANOE.

The fertile portion of the island of Tahiti lies in the valleys, which are of small extent, and in the plain which extends from the sea-shore to the spurs of the mountains. These produce tropical plants in great abundance and luxuriance, and are probably not exceeded in fertility by any portion of the earth's surface. The climate of this region is warm but not enervating, and is well adapted for the enjoyment of all the pleasures of life. To this climate the habits and pursuits of the natives are well adapted, or rather they are its necessary results. Their disposition leads them to the quiet enjoyment of the beautiful scenes around them. Their cottages are to be found in retired and lovely spots, and are usually surrounded by neatly-fenced enclosures. In these, which are often of considerable extent, are to be seen growing the bread-fruit, vi-apple, and orange, and sometimes extensive groves of tall cocoa-nut trees. In one corner are the patches of taro and sweet-potatoes.

The cottages are of an oval form, usually about fifty or sixty feet in length, and twenty in breadth. The walls are formed of bamboos set in the ground, with intervals of about an inch between them, for the admission of light and air. To the top of these a plate-piece of the hibiscus, a light and strong wood, is lashed with sinnet. From this the rafters rise on all sides, and meet in a ridge, which is about half the length of the building. The rafters touch each other, and are covered with small mats made of the pandanus-leaf. These are closely fitted together, and lapped over each other, forming an impervious and durable roof. The floor is the natural earth; there are no partitions, but tapa or matting is employed as an occasional screen. A building of this description may be erected for about fifty dollars.

NATIVE HOUSE, TAHITI.

The Tahitians use neither tables nor chairs. Their bedsteads are formed of a framework of cane, raised a short distance from the ground, upon which a few mats are laid. A pillow stuffed with aromatic herbs is in general use among the better class.

The natives are generally to be found in their houses, in a circle, chatting, reading, and singing, or smoking, unless they be, as is often the case, asleep. They are seldom to be seen engaged in manual labour, except on the sides of the streams where they are employed in washing, or at the residences of the chiefs.

I hesitate to speak of the females of this island, for I differ from all who have gone before me in relation to their vaunted beauty. I did not see among them a single woman whom I could call handsome. They have, indeed, a soft sleepiness about the eyes, which may be fascinating to some, but I should rather ascribe the celebrity their charms have obtained among navigators, to their cheerfulness and gaiety. Their figures are bad, and the greater part of them are parrot-toed. They are exceedingly prone to prattling, or may rather be said to have a tattling disposition, for they cannot keep even their own secrets.

This want of reserve is not confined to the women: the men are also incapable of keeping a secret. A crime is divulged almost as soon as committed, and for a small reward they will become informers against their nearest relatives and friends.

I have spoken of the incongruous character of the dress of the females. Among the men this is not as strongly marked as it is said formerly to have been, and they are no longer content with cast-off clothing. Those who can obtain it are dressed in sailors' garb. Others wear around their bodies, a wrapper called pareu, which extends to the calf of the leg. This is now usually made of blue cotton cloth, and with it some wear a cotton shirt of gaudy colours. Others luxuriate in a pair of duck trousers, and carry the pareu upon their shoulders.

The appearance of the dress of the women while at church, has already been spoken of. On ordinary occasions, they wear the pareu alone, but when dressed, put over it a loose dress, resembling a night-gown, buttoned at the wrists, and confined in no other place. Relics of their ancient dress may still occasionally be seen in wreaths of flowers around the head, and in the hair. The *hau* is a sort of rim made of pandanus, and when it has flowers beneath, it gives

TAHITIAN GIRL, WITH THE HAU.

a pleasing and rural look to the women, to whom it also affords a convenient and easily-procured protection from the sun. The wreaths are usually composed of the Cape Jasmine and Rosa Sinensis, the latter of which is often stuck through the lobes of their ears, and in their glossy black hair.

The natives of both sexes seem passionately fond of flowers, but the use of these in dress has been discouraged by their teachers, who have taught them that such vanities are unbecoming to Christians. I am at a loss to understand why so innocent a pleasure should not have been encouraged rather than discountenanced. In conformity with this opinion, the absence of flowers around the missionaries' dwellings is universal, and cannot fail to be remarked in a climate where the plants most admired in their own country, as exotics, are of almost spontaneous growth.

Cooking and eating occupy but a small portion of their time. The latter indeed is performed with more of the air of a business which requires despatch, than any thing else they do. Their food consists principally of bread-fruit, taro, banana, vi-apple (Spondias), oranges, cocoa-nuts, sugar-cane, fowls, and fish. They eat no salt, but employ instead of it a sort of sop, made of sea-water, cocoa-nut milk, and the root of the ti. Their mode of eating is somewhat disagreeable, for the bread-fruit or taro is dipped in the sop, and then sucked into the mouth with a smacking sound, that may be heard at some distance. The vessel most commonly used is a cocoa-nut shell. The children are fed upon poe, which is made of bread-fruit and taro, pounded together with a little sugar. The child is laid on its back, and is crammed with balls of poe of the size of a walnut, at which it shows its delight by flapping its arms, kicking, and chirping like a young bird.

At Tahiti the mode of carrying burdens is the same as we found prevailing throughout Polynesia; the wood-cut will best explain it.

MALE COSTUME, TAHITI.

The men of Tahiti care little about music, but the women appear to be passionately fond of it, and have very correct ears. Many of them have rich contralto voices, and can descend to very low notes, while others do not differ in this respect from the females of our own country occasionally one may be found that can sound exceedingly clear and very high notes. Their voices accord well with each other, and a party of four or five will make excellent harmony.

If they ever had any native music, it has long been forgotten, and no other singing is now heard but hymns and sailors' songs; you observe, however, a peculiar nasal sound, particularly in those who indulge in the latter class of singing.

Social amusements are prohibited by severe penalties, although the people are evidently fond of them: I neither saw nor heard myself of any dancing or theatrical amusements during our stay. Some of the officers, however, persuaded a few females to exhibit a dance, upon the strict assurance that they should not be informed against. Mats were spread upon the floor, on which two of the girls stood up to dance, while the others sat cross-legged around. One of the latter began by uttering a few words of no delicate import, in reply to which all the others made a sort of grunt, with the mouth shut. To this succeeds another set of sounds uttered with the mouth open.

To this all keep time, by drawing up the legs, thrusting out the arms, and making all sorts of contortions. In the meantime the two dancers proceed, twisting their bodies into all kinds of lascivious postures. Little can be said for the gracefulness of these motions, although many have described them as such. The whole finishes by a simultaneous clapping of the hands.

The party that was despatched for the purpose of making an attempt to reach the top of Orohena, consisted of fifteen persons, including four natives as guides, and an American of the name of Lewis Sacket, as interpreter. This man was from the State of New York, and was admirably qualified for his duties.

By the advice of the Rev. Mr. Wilson, the party took the route across the island which follows the Pappino valley. The distance on this line, to Lake Waiherea, is no more than twenty-five miles, while by that which follows the shores, it is fifty miles before the point at which the ascent begins is reached. None of the guides were acquainted with this route, and it was therefore necessary to find a person who was. For this purpose they in the first place proceeded towards the eastward from Matavai, for about five miles, to the mouth of the river Pappino, which they reached about 2 P. M. Here they found a guide, and were informed that the stream was much swollen: they however determined to go forward, and were accompanied by a troop of boys and girls with flowers. Before they had proceeded far they reached a place where it was necessary to ford the stream, which they found difficult on account of the rapidity, although the water was only three feet deep. Other fords of the same description occurred every few rods, until they at last reached one in which the water reached to their necks. This was of course dangerous to those who could not swim, but all crossed in safety. A young native, as if in derision of the difficulty which they appeared to experience, and of their effeminate bringing up, dashed into the flood, and was seen plunging down the rapids in sport, and evidently with great enjoyment, although frequently wholly immersed in the foam. Our gentlemen had now an opportunity of comparing their own awkwardness and want of ease in the new position in which they were placed, with the agility and freedom of motion exhibited by the natives.

The whole afternoon was thus spent in travelling about three miles, at the end of which they reached some huts, in which they passed the night comfortably. These huts they were informed had been erected by natives, who wished to enjoy a carouse of ava, far from the notice of the judges, and free from the intrusion of their spies. Here they were able to indulge in their old habits of debauchery, which not unfrequently ended in riot and bloodshed.

In the morning, after three hours' travelling, during which they frequently crossed the stream, they reached a place where a branch of it came in from the southwest. Some natives assured them that by taking this route they might reach the top of Pitohiti, and stated that this had been done during the last war by some natives, who occupied it as a place of refuge. In support of this statement, they mentioned that these refugees had found the birds so tame that they might be taken by hand. This circumstance, which, from its apparent incredibility, might seem to disprove the statement, is in fact almost positive evidence of its truth, when compared with what we had seen of the birds in uninhabited islands, where their habits would naturally correspond to those secluded in the mountain solitudes of Tahiti.

Here they learned that it would be impracticable to accomplish both objects of their mission within the time for which they were furnished with provisions. It was therefore resolved to divide the party into two detachments, one of which should proceed towards the lake, and the other endeavour to ascend the mountain.

It was in this place that the last battle was fought between the Christian and heathen parties. Paura, their guide, pointed out all the places where any remarkable incidents of the conflict had occurred. He seemed to take particular pleasure in drawing the attention of the party to two places. In one of these a captain had his head beaten to pieces, and the other was a precipice, several hundred feet in height, over which the defeated party had been driven and dashed to pieces.

The detachment for the lake pursued its route, and before dark reached the solitary residence of a native family, called Waiipi, where they were hospitably received, and lodged in a building used as a family chapel. This dwelling is situated in a romantic gorge at the point of a mountain, and its existence appears to have been unknown to the white residents of the coast.

The next day this detachment proceeded up the bed of the torrent, which was even more swollen than before. They were now surrounded with the wild banana or fahie (Musa rubra), having its upright spikes loaded with its beautiful fruit. Besides these, there were many tree-ferns from forty to fifty feet high. Most of the trees were covered with parasitic plants, which grow with great luxuriance. Leaving the bed of the torrent, they soon reached the dividing ridge, which from observations with the sympiesometer, is twenty-seven hundred feet above the sea. The summit of this ridge was only a few paces in width, and was covered with groves of fahies, clinging, and as it were bound by numerous vines, to the rock. In these respects, the surrounding peaks closely resemble it.

The view from the point of the ridge which they had reached, is magnificent. The lake lay almost beneath them, at a depth of about one thousand feet, surrounded on all sides by perpendicular cliffs, and appearing as if inaccessible, while numerous streams rushed in silvery foam down the rocks; and the lake itself seemed diminished in size by the vastness of the precipices which enclose it.

In spite of the steepness of the cliffs, the descent to the lake was accomplished without accident, by scrambling down the bed of a small stream, although they were compelled to stop from time to time, resting upon their staves, or clinging to the shrubs and roots, while the stones they had set in motion rushed onwards, accumulating others in their course, until the united mass equalled an avalanche.

When they reached the edge of the lake, their guides constructed a hut, in which they passed the night. The next day Lieutenant Emmons made a survey of the lake, and sounded its depth from a raft. It was found to be half a mile in length, a third of a mile in breadth, and in shape nearly oval. The depth in the middle was ninety-six feet, whence it gradually decreases to the edge. It had rained the whole of the preceding night, and the lake was observed to rise about five feet in twenty hours. As far as could be discovered, it has no outlet; but the natives assert that if a bread-fruit be thrown into the water, it will make its appearance at a spring, which gushes from the hill-side, about two miles north of Ooaigarra, and near the sea The height of the surface of the lake, measured by the sympiesometer, is about one thousand seven hundred feet above the level of the sea.

This detachment suffered not a little from the continual rains, by which they were kept constantly wet, and from being obliged to sleep in their damp clothes. Their guides also became apprehensive that they would experience much difficulty in their descent to the southern side of the island, in consequence of the Ooaigarra, by whose valley the only route lay, becoming so much swollen as to be dangerous. It was, therefore, resolved to set out without delay; but before their preparations were completed, they were joined by the other detachment.

This detachment having failed in reaching the summit of Orohena had followed the route of the others. The guides who led the mountain detachment were not found as skilful as they had pretended. On the first day, after many unnecessary turnings, they stated that it would be necessary to halt, as the hour was late, and there was a risk of accident from falling, or from being crushed by stones, for which their queen would be held responsible. They then, after much searching, led the detachment to the shelter of a vast projecting boulder, where the ground was dry, and afforded sufficient room to lodge fifty persons. This place was estimated to be two thousand feet above the sea, and commanded a splendid view over two rich valleys, beyond which the sea was visible.

The fahies were in great abundance around them. This plant is not found at levels lower than six hundred feet, and is in its greatest perfection at the height of fifteen hundred feet. It is the vai of Cook, and is thought to be the banana in its wild state. It, however, differs much in the manner of its growth from the cultivated variety, for the fruit grows upon an upright spike rising from a crown of leaves. The fruit has the same shape as the banana, but is twice as large, and is of a deep golden hue, with pulp of dark orange colour. It is destitute of seeds, has a taste resembling that of the common banana, but of a higher flavour, and the natives are very fond of it.

This shelter was reported to be the favourite retreat of wild hogs, but the detachment was not disturbed by them, and passed the night without any other disquiet but that arising from the possibility that the projecting boulder, under which they were lodged, might fall and crush them. Old Paura made fire by rubbing a pointed stick in a groove formed in another, and by its blaze they, after a change of clothing, found themselves very comfortable.

Mr. Dana, who was one of this detachment, was so unwell the next morning as to be compelled to return, taking one of the guides with nim. The other gentlemen, Messrs. Peale, Pickering, Couthouy, and Brackenridge, proceeded forwards; but they soon found that the guides had no intention to lead them to the top of the mountain, during the continuance of the heavy rains, but pursued a course so devious and circuitous as to make it probable that they would lose the chance of seeing either the mountain or the lake. They felt assured that the ascent, under more favourable circumstances, would have been practicable, but were satisfied that it would require more time than they could spare. Notwithstanding their disappointment, they were compelled at last to admit that the guides had acted wisely; for during the continuance of such rains as afterwards fell, the ascent of the mountain would not only have been difficult, but extremely dangerous.

Compelled to abandon the attempt, they followed and overtook Mr. Dana, and then turned up the valley. About noon, after having crossed the stream so often that a fourth of the way might be said to be by water, they reached the residence of the native family Waiipi.

Dr. Pickering is of opinion that this dwelling would be an excellent station for a botanist. It has a rich field around it, and is, besides, within a short distance of the most elevated parts of the island.

They did not stop at this place, but proceeding forwards reached a spot called by the natives Opua, where, in a shelter of the same kind as that which has been described as their previous bivouac, they spent the night. The rain fell in torrents, and about midnight a violent thundergust passed through the valley.

The next morning they went on towards the lake, and on reaching the crest of the intervening ridge, the weather moderated, and gave them an opportunity of enjoying the extensive prospect it commands. Besides the lake at their feet, as seen by the other detachment, they had a view of the cleft peak of Orohena, at the distance of about seven miles, rising from five thousand to six thousand feet above the spot where they stood. Descending the steep sides of the basin of the lake, they joined the other detachment about noon.

It has been stated that the lake had been observed to rise five feet. It was now evident that it had reached as high a level as its waters ever attain. Many plants (such as the Polygonums) which could not live long under such circumstances, were found entirely submerged, and the water had reached the woody plants on the shores, and threatened their speedy destruction.

The proposed line of descent lay on the opposite side of the lake, and the shores are so precipitous as to prevent walking around it. It became necessary, therefore, to cross it upon a raft, which was a slow process. The natives swam over. After crossing the lake, the journey was continued, but it became necessary to stop before night, in order to build a shelter and make a fire. The latter was a difficult operation, where every thing was dripping with water. A hut was soon built of boughs of the banana, and thatched with its leaves. This, although not perfectly water-tight, served their purpose.

Their provisions fortunately held out. The natives after every meal had been careful to gather up all the remnants of food, and to wrap them in banana-leaves, by which all waste was avoided. They had been disappointed in obtaining any game, or any fish from the lake, both of which had been calculated upon. Mr. Peale saw nothing to shoot except a few birds, which were limited to four or five species, among which were swallows, tropic-birds, and small herons. In the lake, nothing was seen, although it had been represented as full of huge eels.

In the morning they rose early and began the descent, which was slow and fatiguing. They found the Ooaigarra very rapid, and were compelled to cross it frequently. In this their clothing was an encumbrance, for it remained wet, and was consequently heavy, while the natives, being naked, became dry in a few minutes after they emerged from the water. This toilsome descent occupied about four hours.

The first inhabited place they reached was Mirapahé, where the Ooaigarra runs into the sea. This is situated on the south side of the island, near the isthmus by which the two peninsulas are united.

Most of the party were now exhausted by fatigue, so that it was with some difficulty that any of them proceeded onwards to Ooariteeva, while a canoe was procured for those who were most overcome. Here they were hospitably received by Teharo, the son of the governor, Taati, who was himself absent, and took up their quarters at the house of the latter.

It was Wednesday evening, (by the account of the island,) and the natives were attending *lecture* in the church. Some of our gentlemen entered the congregation, and excited some sensation among the younger part of the assembly; far less, however, than would be evinced should a Tahitian enter one of our churches. The service was performed by a native, and consisted of praying, reading the Bible, and singing. The old people appeared very devout, and it was remarked that many of them were provided with spectacles, to enable them to follow the reader in their own books.

The reading was performed in a low monotonous tone; and the hymn was sung to an old English tune, with considerable taste, by the female part of the congregation.

The assembly consisted of about one hundred persons, three-fourths

A. T. Agate

Sherman & Smith.

BROOM ROAD, TAHITI

of whom were women, all dressed in the usual loose calico gown, with large straw bonnets, and barefooted.

In spite of the devotion manifested within the church, the conduct of the women after the service was concluded, left room for believing that their former licentiousness was not entirely overcome by the influence of their new religion.

When the service was over, nearly the whole congregation collected, attracted by the sight of so large a number of white strangers. According to their own account, the party, in their tattered and soiled garments, was not in a condition to produce a very favourable impression on these Polynesians. Still, it appeared from an address made by their preacher, that he thought that in spite of their forlorn condition, they possessed many things which the natives were likely to covet, for he exhorted his flock to depart, telling them that some of the strangers' articles might be missing, and "then there would be trouble."

Here was witnessed a gorgeous sunset;—the clouds to the west being lighted up as it were with burnished gold, while to the east, the rain-clouds overhanging the mountains, exhibited a triple rainbow, which at that hour was of course a complete semicircle.

In the morning they set out early, and walked to Papara, over the "Broom Road." This is about twelve feet in width, and is well made, being raised above the level of the ground, and having ditches on each side. Over the streams which cross its direction, are bridges of loose planks.

Previous to reaching Papara, they came to a large mound on a point of land near the sea-shore. This is the same that is described by Cook. The steps of which he speaks are now almost obliterated, being entire only at one of the corners, and there is little appearance of the stones having been squared. It is now about forty feet high; the space on the top is about twenty feet wide; the base is one hundred and forty by fifty feet; and it is almost entirely overgrown with bushes.

An account of the object of this structure was given by the old chief of Panawea. As he is one of the most intelligent chiefs, and his statement differs in some particulars from that given by Cook, I think proper to insert it. This chief stated that it had not been used as a place of sepulture or as a morai, but was a sort of temple or high place on which the image of one of their gods had stood; that formerly many such mounds had existed upon the island, but that whenever war broke out between two districts, each was desirous to obtain possession of the other's tehee or idol, for the loss of it was an acknowledgment that the god was less powerful than that of the other district; that when one district had been subjugated by another, and the two districts united into one, the mound of the conquered party was destroyed. From this account it would appear that these structures had served to mark the boundaries of the several independent districts.

This account is rendered more probable from the fact that a morai of quite a different character exists within a mile or two of this mound. This is situated in the midst of a grove of large bread-fruit, poorou, and aiti trees. It is, like the other morais, an enclosure of quadrangular form, about sixty yards in height, and thirty in breadth. The wall is about ten feet thick and four feet high. The lower part of it is composed of five rows of round stones, of uniform size, each about the size of a man's head, which the natives now call turtle-heads, arranged like cannon-balls in an arsenal. On these lies a course of square stones. The corners of the wall are strengthened by square blocks. There was also a flat stone, placed upright, at each corner, and another in the middle of the enclosure,—the last of which is called a preaching stone.

The chief of Panawea likewise stated, that there used to be in each of these morais a sort of platform, supported by stakes, on which were laid the bodies of human victims, and the carcasses of hogs, on which the tehee or god was supposed to feed: that the dead brought to the place for interment were not buried immediately, but were placed in a hut until all the flesh was consumed, after which the bones were carefully cleaned, and, with the exception of the skull, wrapped in many folds of tapa, and deposited in the enclosure; that the skull was taken home by the nearest relative, and kept as a talisman. He added, that only the bones of chiefs were honoured in this way; while those of lower rank were deposited on the outside of the enclosure.

Other accounts, however, state that these morais were never used as ordinary places of burial, but that the bodies of enemies slain in battle were brought to them, and consumed in a fire made on a thick part of the wall.

The party saw an old man, who had his deposit of bread-fruit in one of these morais. The bread-fruit in its crude state will keep only for a short time; but if buried in pits, it ferments and forms a substance which may be long preserved, called mahi. This substance has a taste like that of bee-bread, and is used at sea, in the voyages the natives make to the Paumotu Group, and in their excursions in search of pearls.

The next place the party reached was Papara, at which Mr. Orsmond of the Mission resides. Taati, who has been named as one of the great chiefs, has his usual abode here, but was absent on a visit to the ships. Mr. Orsmond was kind enough to offer to such of the party as felt too much fatigued to proceed farther on foot, a passage to Papieti in his boat.

The coral reef is interrupted in front of the small bay which makes up to Papara, and the surf dashes with some violence against the shore; but the passage, except for a short distance, was made within the reef, and in smooth water.

As Taati was absent, one of his sons, (and he is reported to have a dozen,) did the honours of the house, and provided them with breakfast. This consisted of pig, taro, and bread-fruit, the standing dishes of Tahiti. The table was covered with a cloth made of Russia sheeting, and furnished with plates, cups and saucers, knives, forks, and spoons.

In this place, also, a large portion of the population, and particularly of the women, assembled to have a look at the strangers. It was remarked that there was more beauty among the young girls than they had before seen upon the island. Many of them had their heads gaily and tastefully ornamented with flowers.

On leaving Papara they were accompanied by a bevy of these gay damsels, laughing and enjoying themselves with each other, and with the guides. The latter had also stuck the scarlet Rosa Sinensis through the lobes of the ears, and decorated their heads with wreaths after the manner of the girls.

At Papara, the coast begins to wind to the northwest, and gradually inclines more to the northward. Their route again lay along the Broom Road, which extends completely around the island. It is in places almost arched over by trees, that grow on each side of it, and form a delightful shade. It may be termed a national work, and but for its want of width, would bear comparison with many of the turnpikes in the United States. The streams which intersect it have all been once crossed by good bridges of plank, but these are now going to decay.

At short intervals are groves of cocoa-nut trees, planted along the road; these are called the queen's, and travellers are at liberty to help themselves to the fruit. The private groves whose fruit it is wished to protect, are tabooed. This is indicated by tying a girdle of leaves around them, and this simple signal is still respected, although the superstition on which its sanctity was founded has ceased to prevail. The natives travel a great deal on this road, and some use it in going to visit places on the opposite side of the island, in preference to passing over the mountains.

At sunset they reached Otapuna, which is a large village, pleasantly situated on a low point of land, covered with orange, bread-fruit, and cocoa-nut trees. In this place Utami, one of the principal chiefs, who is governor of this district, resides. With him our travellers took up their quarters. His dwelling differs from other native houses, in being divided into rooms, and having floors of plank; indeed every thing about his establishment indicated more refinement than they had yet met with. His table was set out after the European fashion, and he gave them tea made of a native herb.

Utami is a very intelligent person, and had been one of Pomare's great captains. He related many anecdotes of the struggle in 1815, which by his account appeared to have been a sanguinary one. He stated, that on the retreat of the heathen party, all their prisoners were put to death, by beating out their brains with clubs; that numbers of them had been forced over precipices; and that many had fled even to the summits of the mountains, where they remained until a general pardon and amnesty had been proclaimed by Pomare. Even this was not accepted by all, for many remained in the wild and dreary regions of the mountains, refusing to embrace the new religion; and of these some had continued to exist until within the last five or six years. Utami was of opinion that this remnant was now extinct.

Among other things, he informed them that the wild race which had previously inhabited the island, were accustomed to roam all over it; but that since they had been subdued or exterminated, none but the cutters of sandal-wood had been accustomed to enter the mountainous regions. These had been in the habit of ascending the most difficult peaks by the aid of ropes, and of steps cut in the rock, but he was of opinion that no one now remained who knew the way.

Utami was a fine-looking white-headed man, of giant proportions, and when speaking of his former deeds in arms, showed a great deal of fire and animation. Before he retired to rest he read a chapter in the Bible, and made a prayer, with much apparent devotion. He seems to have a great disposition to adopt European improvements, and was engaged in building a house, which when finished will be the best in the island. The floors will be all laid with plank; it will have partitions of boards, and large windows with Venetian blinds.

This village is the next in size and importance to Papieti.

When this party first descended to the beach and reached Mirapabé, they found a vessel on the stocks there. This enabled them to obtain some information in relation to the naval architecture of Tahiti. The vessel was a small schooner, and the building of it was superintended by a Yankee. The timber employed was that called by the natives mape, (Inocarpus edulis,) which is said to be of excellent quality. The supply of this wood is, however, limited. The poorou (Hibiscus tiliaceus,) is also employed in ship-building, but it can only be procured of small size, and is therefore unfit for the structure of many parts of a vessel.

Several vessels of about one hundred and thirty tons burden have already been built upon the island. These have been employed in the trade to New South Wales, whither they carry sugar, cocoa-nut oil, and arrow-root, and whence they bring back in return hardware, cloths, calicoes, &c. In the ports of New South Wales they pay the same duties and charges as British bottoms.

The commercial resources of these islands are very limited; most of the vessels that visit Tahiti are those belonging to our whaling fleet: these average less than a hundred annually. From them the natives are enabled to dispose of some of the supplies they raise, and in return obtain such articles as will promote their comfort and add to their pleasure. The whale-ships, for the most part, have articles of trade which they barter with the natives, so that little money is required to carry on their business. The natives, particularly the chiefs, are however well acquainted with the value of money. An estimate has been made that each of these vessels introduces goods into the islands to the amount of $500 each, making a total amount of $50,000; but I very much question whether it can reach this extent; and if this amount be sold, it must include the profits: half the sum, I should think, was a large estimate.

The few other vessels that visit the islands bring little cargo; if two arrive at the same time, they destroy each other's ventures by glutting the markets.

The pearl-shell fishery of the Paumotu Group centres here. I was told it was principally in the hands of the French consul. For a few years before our arrival, viz., from 1832 to '38, it had been very productive. The amount obtained was about nine hundred tons, which was estimated to be valued at $45,000 to $50,000; the greater part of this was sent to France. Of the agricultural products they have little to dispose of as yet; neither is the island susceptible of any very extended operations, to induce vessels to visit it exclusively for its trade or productions. The three chief articles of production are sugar, cocoa-nut oil, and arrow-root. The following statement was furnished me of the quantities produced.

SUGAR.

Tahiti,	105 tons, estimated value			$8,000	
Eimeo,	32	"	"	"	2,000	
Raiatea,	15	"	"	"	1,300	
						11,300

COCOA-NUT OIL.

Tahiti,	55	"	"	"	3,500	
Eimeo,	20	"	"	"	1,500	
Huaheine,	60	"	"	"	4,500	
Borabora and Raiatea,	50	"	"	"	3,800	
						13,300
						24,600

Of arrow-root there are about fifty tons produced, valued at nearly $4,000, which revenue, included with the above, will make the sum of nearly $28,000 as the value derived from the agricultural productions of all these islands. If they were under proper cultivation, this amount would be greatly increased; and from the estimate of a competent person, it was believed that the productions might yield, if properly attended to, a revenue of $300,000, as coffee, cotton, and indigo might be added to the above articles; but this is undoubtedly an estimation one-half too great, and would require an amount of labour that the present native population are inadequate to perform, and which their climate, wants, and desires will never probably excite them to, or render necessary.

TRADING CANOE.

CHAPTER II.

CONTENTS.

SAILING OF THE PORPOISE—APPLICATION FROM "JIM" THE PILOT—SURVEY OF HARBOURS—DOLPHIN BANK—THE VINCENNES PROCEEDS TO PAPIETI—VIEW OF PAPIETI—INTERVIEW WITH THE CHIEFS—GENERAL FREYRE—HITOTI—SECOND ATTEMPT TO ASCEND OROHENA—GEOLOGICAL STRUCTURE OF TAHITI—VILLAGE AND BAY OF PAPIETI—COMMERCE OF PAPIETI—POPULATION OF TAHITI—DISEASES—CRIMINAL TRIAL—CULTIVATION—USEFUL PLANTS—ASCENT OF MOUNT AORAI—ABSENCE OF FOSSILS ON TAHITIAN MOUNTAINS—RETURN FROM MOUNT AORAI—MANUFACTURES OF TAHITI—REMARKS ON THE CONDUCT OF FOREIGN VISITERS—THEATRICALS BY THE CREW OF THE PEACOCK—VINCENNES SAILS FOR EIMEO—VIEW OF EIMEO—TALOO HARBOUR—LANDING AT EIMEO—RESIDENCE OF MR. SIMPSON—SCHOOL FOR THE CHILDREN OF MISSIONARIES—CHARACTER OF THE NATIVES—SCENERY AND STRUCTURE OF EIMEO—SUPPLIES TO BE PROCURED THERE—VILLAGES—COTTON FACTORY—SUGAR PLANTATIONS—OTHER PRODUCTS—VON SCHANTZ'S CHART OF TALOO HARBOUR—CONCLUSION.

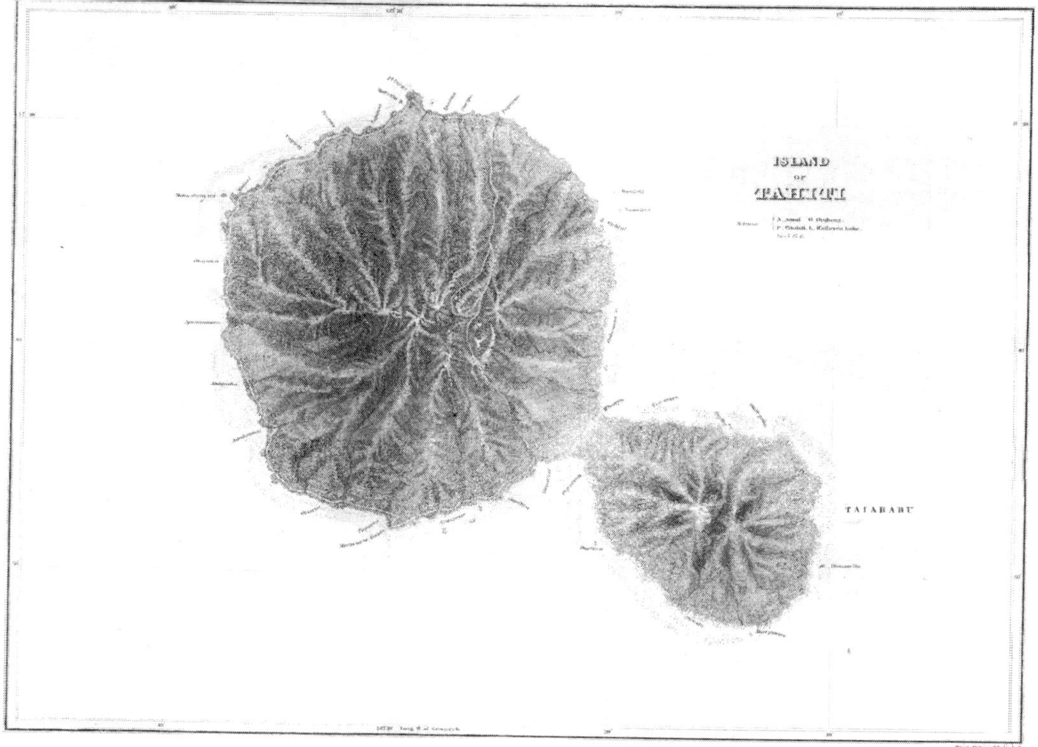

ISLAND
OF
TAHITI
TAIARABU

CHAPTER II.

TAHITI AND EIMEO.

1839.

The Porpoise, having been refitted, was sent to sea on the 20th September, for the purpose of again visiting the west end of Nairsa or Dean's Island, with Krusenstern's and Lazareff. She was also ordered to pass over the supposed locality of Recreation Island, and then to meet the Vincennes at Rose Island, the easternmost of the Samoan or Navigator's Group.*

A stormy evening having occurred previous to our leaving Matavai Bay, "Jim," the pilot, desired to see me; on his coming into the cabin, to my great amusement, he urged me to allow him to go to Papieti, where he was sure he would be wanted; and when I asked for what purpose, he told me that the "thunder and lightning would bring in ships of war." He was displeased when I laughed and said, that as he was engaged on board my ship, I would wait until I saw the ships before I could give him permission. He then reminded me of the night before we arrived, when there was plenty of thunder and lightning, and that he had told me as soon as he came on board that he expected us. He went on to repeat that he was sure that they would want him early in the morning at Papieti, but I persisted in my refusal; and in the morning he appeared much disconcerted to find that there was nothing in sight out of which he could make a ship of war.

The surveys of the four harbours of Matavai, Papaoa, Toanoa, and Papieti, had been, as stated in the preceding chapter, commenced soon after our arrival; they had made considerable progress before the

* A copy of the orders will be found in Appendix III.

departure of the Vincennes, and were subsequently completed by the officers of the Peacock.*

I was desirous, in addition, of ascertaining with precision the least depth of water on the Dolphin Shoal, and had employed boats in making accurate soundings. The results, however, were so unsatisfactory, on account of the ever-varying level of the tide, that I could not hope that they would serve for a reference by which to ascertain the future increase of this bank, by the growth of the coral. I therefore planted a large stone, with a horizontal surface, and having found the shoalest place on the bank, carefully measured the difference in the height of the surface of the stone and the rock of the bank, by an excellent spirit-level. I then caused a mark to be traced upon the surface of the stone pointing towards the place on the bank where the levelling-staff had been placed. The difference of level was twenty-five feet seven inches. Old Taua, the chief of the district, promised me that he would taboo the stone, and that it should not be touched or disturbed on any account; and, to preserve it from any accidental disturbance, it was set four feet into the ground.

The Vincennes moved to the harbour of Papieti on the 22d September. At the same time, orders were given to the Peacock and Flying-Fish to take on board their articles from Point Venus, and to follow as soon as they had done so. The tender required some repairs, which could be done with more safety at Papieti. Both vessels joined us in that harbour on the 24th.

In proceeding to Papieti, we left Matavai Bay in the morning, and within a few hours had anchored in the harbour of the former place. No soundings are to be had beyond the line of reefs, and consequently there is no anchorage; the outer wall of the reef surrounding the island is in fact perpendicular, with the exception of some projecting patches in Matavai Bay, and to the eastward of Point Venus. On the latter the French frigate Artemise struck, in 1836.

At the season of the year when we made this short passage, there is some danger to be apprehended in entering the harbour of Papieti, and much caution is therefore necessary. The trades at this season are irregular, and the winds which prevail are light; they also are most apt to fail at the critical moment when the vessel is at the entrance of the narrow passage through the reefs, in which case the current, which rushes strongly out and sometimes across the passage, may cause a vessel to drift upon the western reef. The proper mode of guarding against this, is to keep the vessel as close as possible to the eastern reef.

* The chart of these harbours will be found in the Hydrographical Atlas.

The scenery at Papieti is remarkable; the background is filled up with a number of pinnacle-shaped mountains, jutting up in a great variety of forms; beneath, and directly in front of them, lies the semi-circular harbour, surrounded by the white cottages and churches of the village, embosomed in luxuriant foliage; these dwellings have a peaceful and home-like look, to the eye of an American. In front, the little coral island of Moto-utu forms an embellished foreground, and serves to break the regularity of the line of the harbour, while by concealing its extent, it gives it an air of greater magnitude than it in reality possesses. To my eye, this view combined within itself a perfect picture of Polynesian scenery, enhanced in beauty by the signs of civilization, among which was the national flag of Tahiti, waving from a fortress on Moto-utu.

The purpose of my visit to Papieti had originally been to go through the ceremony of receiving the great chiefs on board, when, according to custom, presents are made them; but before this was done, I determined that the business, which I had laid before the council, as stated in the preceding chapter, should be adjusted. This was done satisfactorily on the 22d, when they assented to all that had been asked of them. I am convinced that their conduct in this matter was dictated throughout by a sense of what is right, and am satisfied that if grievances do exist, it is only necessary to state them, when, if redress is within their power, it will be granted.

Agreeably to my invitation, Uata, who appeared as the representative of the queen, the two princesses, Ninito and Taii, and all the head chiefs, visited the ship, accompanied by the foreign consuls. The ship was dressed for the occasion with flags, and they were received with every mark of respect. Luncheon was prepared for them; and when they were all seated at it, it struck me that I had never seen such a collection of corpulent persons. Previous to eating, one of the oldest chiefs said grace. Their appetites were good; none of the food appeared to come amiss. They seemed heartily to enjoy themselves, and conducted themselves with a propriety that surprised us all. They were cautious in partaking of the wine which was set before them, and seemed evidently upon their good behaviour. This was the case with the high chiefs, who, to the number of about fifteen, had been invited; but, besides these, about an equal number of others contrived to get on board without invitation; the latter thrust themselves forward with eagerness to occupy places at the table, but were compelled to give place to those of higher rank. A second table was, however, prepared for them, at which they took their seats, and did ample justice to what was set before them.

The variety of costume which was exhibited at this banquet was amusing. The princesses were dressed in white frocks, shoes and stockings, and chip bonnets, but looked awkwardly in them, and appeared more like boys in girls' clothes than women. Some of the men wore full suits,—coats, vests, and pantaloons,—of a variety of colours; others had sailors' round jackets; others again had only shirts and pantaloons, all too small, both in breadth and length. Some had black felt hats, of all possible fashions, and others wore them of straw; some had shoes on their feet, others had none.

Paofai's son attracted attention by his ridiculous appearance: he wore a red check shirt, light white pantaloons, that reached only half way down his legs, coarse shoes without stockings, and a short-skirted drummer's coat of blue, plentifully faced with scarlet. The latter was so small for him, that no force would make it button upon him. To finish all, he had a high-crowned conical felt hat stuck upon the top of his head.

After luncheon, they repaired to the deck, to receive the presents prepared for them. These I had been advised, in order to avoid unpleasant scenes, to pack in bundles, assorted to the rank of the parties. In spite of this precaution, much consultation took place among them, and a desire to exchange with one another was manifested. This was particularly the case with our old acquaintance Taua, and his friend Otore, the ex-minister and former favourite of the queen. The presents for the queen and royal family were committed to the charge of Uata, who, as has been stated, appeared as the representative of her majesty.

Otore, who has been just named, is only a petty chief, but had been the queen's favourite and minister, until he was dismissed in consequence of his frequent indulgence in intoxication. He is considered as the greatest orator on the island. He and Taua are boon companions, and were continually on board the vessels, where they so timed their visits that the hour of breakfast was sure to find them either actually seated at table or awaiting an invitation. Although at first welcome, the habitual intrusion of these and others upon the messes, finally became an annoyance, and on board the Peacock they had at last recourse to "clearing the ship of strangers" during meals. Taua did not mind this; and when we left Matavai, he was so kind as to remove to Papieti, in order to be near his friends.

Among other visiters on this occasion, I had the honour of the company of General Freyre, formerly President of Chili, who has chosen Tahiti as his residence. It gave me great pleasure to become acquainted with him, particularly as I had it in my power to give him recent news from his own country and Peru, which he was desirous of hearing. He spoke much of the deprivation he suffered by a separation from his family, and from the want of society, but uttered not a word of complaint against his enemies.

He lives in a small cottage on the bank of the harbour at Papieti, where he is highly respected; his manner and whole deportment are gentlemanly; he is tall and robust, with a florid complexion, and appears about fifty-five years of age. In the chapter on Chili, his public acts have been sufficiently enlarged upon; and although his political course may have been much condemned, I can bear testimony to the high estimation in which his private character is held in his native country.

FISHERY – TAHITI.

There are many pretty walks about Tahiti; the small streams, flowing through luxuriant woods, add much to its beauty; these run bubbling along to the sea, passing many cool and pleasant places: their entrances are usually closed up by the natives, for the purpose of taking fish, a sort of dam being constructed, over which the waters flow, and the natives, standing on the outside up to their waists in water, are often seen taking the fish in baskets. The sketch by Mr. Agate is characteristic.

On arriving at Tahiti, or indeed at any of the islands, respect is naturally due to the chiefs; this, I am assured, was felt by us all; but long before sailing we became disgusted with seeing these large and noble-looking men passing from ship to ship, even including Paofai himself, soliciting foul linen to wash, and performing other services that were not in keeping with their rank. There is one, however, whom I must do justice to,—Hitoti. He maintained the character given him by Captain Beechey. I was much pleased with his whole deportment on his visit to me, and also when I saw him at his own house; he paid but two visits to the ship, and those within a day or two of our departure. That he did not visit the vessels before, was in order, as was supposed, to avoid the suspicion of trespassing on our liberality; he refused to accept any presents, and would only drink wine when requested, performing all the little courtesies of the table with grace and politeness.

On his visit to the Peacock, Lieutenant Emmons and Mr. Hale being the only gentlemen on board, received him with the attentions due his rank; when taking leave, he requested to know their names, which were given to him in English orthography; he at once took out his pencil, and with great readiness wrote them in the Tahitian dialect, as "Emaani" and "Helavi!"

Dr. Pickering and Mr. Couthouy being desirous of making another attempt to reach the top of Orohena, I willingly gave them the longest leave possible, to effect their object. They determined on attempting the ascent of the ridge leading directly up from Matavai Bay, as the one that had appeared to them most practicable. Lewis Sacket, who has already been spoken of, was again their interpreter; for guide, Mr. Wilson recommended an old man by the name of Vahaore, who was said to be the only person now living who had visited the top of Orohena; in this selection they were fortunate: Vahaore had been in his youth a great warrior, and his looks did not belie it; he was of the middle size, thin and sinewy, and with a fine eye; although past sixty years of age, he had scarcely a gray hair; his gait was firm and his carriage erect; he was constant in his attention to his duties, and rarely spoke unless relative to his business.

After the guides had been engaged, our gentlemen passed the night in a native house close by, ready to start at an early hour. At daylight, Vahaore and his son were provided with ropes; the reason the old man gave for taking his son was that he might be able to learn the way. They now set out, and by nine o'clock had reached a higher point than at any time on their former journey: this was about three thousand five hundred feet, and was attained after having walked six miles; when they had reached the altitude of fifteen hundred feet they no longer found any paths; on arriving at this point, they halted for some time to make collections of land-shells, and some very interesting specimens were obtained of Helices, Patulas, Cyclostomas, Curocollas, and Pupas; after this they continued ascending, the ridge gradually becoming narrower, until they reached a spot on the ridge where there was not room for one person to pass by another, and where they could look down a precipice on each side to depths of two thousand feet.

Plants that were below of small size here grew into large woody shrubs; among them a species of Epacris was found growing luxuriantly along the crest of the ridges, and magnificent arborescent ferns on the mountain sides, some of them forty feet in height; another species was seen whose fronds were more than twenty feet in length. Their path was much impeded by the tangled ferns and wiry grass (Gleichenia), which it was impossible to get through without the aid of a knife or a hatchet. They had now reached four thousand five hundred feet, the highest point yet attained, according to the guide, by white men; two o'clock had arrived, and as there was no place where they could encamp, or any chance of reaching a point suitable for passing the night in, by the advice of Vahaore they allowed him to look for one. The mountain top was still estimated to be six miles distant; they had little doubt that it could be ascended by following the ridge, and it was thought that they could accomplish the task if time permitted. The day was fine, and they enjoyed a view of the whole mountain, which appeared as if it were the centre, from which the different ridges of the island radiate in ten or twelve directions towards the coast, having deep and narrow valleys between them, through which the mountain torrents rush; these valleys spread out as they approach the coast, and the ridges become more rounded and accessible.

After reconnoitring the ground for some time, Vahaore recollected a place where they might pass the night, which he thought was not far distant. He therefore immediately began to break a road, which he continued for about a quarter of a mile along the ridge. He then reached a place where the descent might be made, which, however, to all appearances, presented as few facilities for the purpose as any they had before looked at. They, however, tried it, and after a hard scramble reached, about sunset, the place he sought. The descent was estimated to be about two thousand feet, and was performed partly by leaping from tree to tree, and partly by lowering one another by ropes over precipitous ledges from ten to twelve feet in height. In the words of Sacket, "No man in his senses ever went down such a place before, and none but a fool would attempt to do so again." At the foot of the descent lay the first valley, and they found themselves among groves of the wild banana (fahies).

A temporary shelter was soon constructed on the banks of a mountain stream, which ran headlong by. In this, Vahaore amused himself by catching eels, a sport in which he was expert, and which he performed in the following manner: having found a small basin at the foot of the cliff, in which an eel was concealed, he placed a large flat stone in the middle of it, and began to bale out the water with his hands; he next disturbed the fish, which sought shelter beneath the stone, when, by cautiously introducing his hands, he contrived to grasp it, and by a sudden jerk threw it thirty or forty feet into the woods, where he easily secured it. In this manner, two eels were taken, nearly four feet in length, and as thick as a man's arm. The eels were not skinned, but carefully cleaned and washed; they were then wrapped in leaves, and cooked in the usual Polynesian manner. Thus prepared, they proved a great delicacy. These eels, although much esteemed by the natives, appear to be almost unknown to the white residents.

The idea of ascending the peak was now abandoned, and in the morning they set out on their return to the coast. In their route, they crossed several spurs of the main ridge; about noon they again halted, and employed themselves in making collections, while Vahaore again went a fishing. He soon returned with three eels, the largest of which was upwards of three feet in length. These eels are of a uniform dark olive colour on the back, which passes on the belly to a dirty white.

Early in the afternoon they reached the lower valley, with the stream called Pappiamatia, about twenty yards wide, and from two to three feet deep, running down it. About five miles from the coast they passed a range of basaltic columns, one-fourth of a mile in length; the cliff presenting a perpendicular wall, rising up at the east side of the stream, formed columns, one hundred and fifty feet high; the number of sides varied from four to seven, and their diameter from nine inches to two feet. The ridge at whose base this was found, was upwards of two thousand feet high. The party reached Matavai Bay at sunset.

The ridge that our gentlemen followed, is considered by them as the most feasible route to the summit of Orohena; the greatest difficulty with which the attempt is attended arises from the dislike of the natives to visit places where they have not been before, and their anxiety to keep themselves within the region of the fahies.

Pitohiti might also be reached by the same route; there is little doubt that the latter has also been ascended by following the western branch of the Pappino valley, a route which was also considered feasible by our party; a third route might also be found by following the main branch of the Pappino, which, as will be seen on the map, rises behind Pitohiti.

This excursion furnished more full information in relation to the geological structure of the island than had before been obtained. This is exclusively volcanic, and the rocks are either compact basalts, or conglomerates of basalt and tufa, although no active volcano exists, nor any well-defined crater, unless Lake Waiherea can be considered as one. Through these rocks olivine and pyroxene are copiously disseminated; cellular lava was found in some places, but neither pummice nor obsidian; quartz and mica were not observed, nor any carbonate of lime, except in the form of coral rock.

There is no conformity between the rocks of the centre of the island and those which in most places extend inwards for a few miles from the coast. The former are usually compact, of columnar structure, and exhibit no appearance of horizontal stratification; the latter lie in horizontal layers, composed of scoriaceous and vesicular lava. In both of these structures, singular twistings and contortions were observed. Many dikes were seen to occur, not only in the mountains, but near the sea-coast; these were from three to six feet in width.

All the rocks of the island appear to be undergoing rapid decomposition. Even in places where the rock seemed to have retained its original form of sharp edges and pointed pinnacles, it was found so soft, to the depth of a foot or more, as to crumble in the hand. The earth thus formed varies in colour from that of Indian red to a light ochrey tint; in consequence, many of the hills are of a red hue, and one immediately behind Papieti, takes its name (Red Hill) from this appearance.

This decomposed earthy matter, mixed with the abundant decayed vegetation of a tropical climate, forms, as may be readily imagined, a soil of the greatest fertility, adapted to every kind of cultivation. On the higher grounds, the soil thus constituted has the character of a clay, and is in wet weather slippery and unctuous; in lower positions it is mixed with lime derived from coral and shells, which often tends to augment its fertility.

Iron abounds throughout; on the mountains to such an extent that compasses were found of little use from the local attraction by which they were affected; and on the shore, the sand was composed in part of iron, which could be separated by the magnet.

Water gushes out near the coast in copious springs, but none of them were found hot, nor were any warm springs reported to exist.

Papieti, in whose harbour we were now lying, is one of the largest villages on the island; being the ordinary residence of the queen, and the abode of the foreign consuls. The foreign residents are also, for the most part, collected here. Among all its dwellings, the royal residence, and the house of Mr. Pritchard, are the only ones which possess the luxury of glazed windows. The houses of the foreigners are scattered along the beach, or built immediately behind it.

The bay of Papieti is the safest, and its port affords the greatest facilities for the repair and supply of vessels, of any belonging to the island. For the first purpose a wharf and warehouse have been constructed, which are let to those who wish to use them. We occupied them for ten days, for which we paid thirty dollars. The tender was hove out at the wharf, and her equipment secured in the warehouse. A limited supply of ships' stores and chandlery is kept for sale, and may also be purchased from the vessels which frequent the port.

The greater part of the commercial business of Tahiti is transacted here, whither the articles for export from other parts of the island are brought to be re-shipped. The number of vessels which visit this port annually, is about sixty, of which the largest portion are whalers; the remainder are transient merchantmen, or regular traders from New South Wales. The latter bring cotton fabrics, which they exchange for sugar, molasses, arrow-root, and cocoa-nut oil. The value of the exports in this direction is supposed to be about $35,000.

The amount of American manufactured goods imported into the island is estimated at an equal sum; they find their way here in transient ships from the coast of South America, and the supplies furnished our whale-ships are generally paid in American goods.

It is almost impossible, in the absence of all statistics, to arrive at any correct statement of the amount of foreign manufactures annually consumed here; but the quantity is evidently on the increase.

By a regulation of the colonial government of New South Wales, Tahitian vessels are allowed to enter their ports on the same footing with the English. There are several vessels engaged in the trade, and others building.

The position of this island, in the vicinity of the cruising-ground of our whale-ships, and the resources it possesses for supplying shipping, make it a desirable point of rendezvous.

The following statement will show the number and value of American vessels visiting this island in the years 1836, '37, and '38.

	WHALE-SHIPS.	TONS.	VALUE OF CARGO.
1836	52	18,090	$1,307,500
1837	57	20,500	1,817,000
1838	42	15,000	1,268,000
	9 merchant-ships,	1,700	75,000
		55,290	$4,467,500

A census recently taken, gives for the population of Tahiti nine thousand, and for that of Eimeo one thousand. When this is compared with the estimates of the navigators who first visited these islands, an enormous decrease would appear to have taken place. The first estimates were, however, based on erroneous data, and were unquestionably far too high; yet there is no doubt that the population has fallen off considerably in the interval. The decrease may be ascribed in part to the remains of the old custom of infanticide, in part to new diseases introduced from abroad, and the evils entailed upon them by foreigners, and in part to the transition now going on from a savage to a civilized life.

Whatever may have been the case, during the first years after it was visited by Europeans, the population for the last thirty years has been nearly stationary; the births and deaths are now almost exactly in equal numbers. One of the oldest of the missionaries informed me, that although he saw much change in the character and habits of the people, he could perceive none in their apparent numbers.

Tahiti does not appear to be afflicted by many diseases. Some have been introduced by foreign ships, and among others, the venereal; from which the natives suffer much, being in possession of no method of arresting its ravages, and ignorant of the proper mode of treating it. In connexion with this subject, the want of a physician as a part of the missionary establishment, struck me as an instance of neglect in its managers; and I was surprised to hear that the London Society did not employ any medical men. From this cause, not only are the natives deprived of the benefits which might so easily have been conferred upon them, but the missionaries themselves are compelled to pay, out of their private purses, for medical aid, when it can be procured. They are even at times wholly without a physician. This happened to be the case at the time of our arrival, when a medical practitioner who had formerly resided on the island, had just taken his departure.

The effects of intoxication from ardent spirits and ava are said to have swept off many of the inhabitants. Secondary syphilis is in some cases severe, but their usual vegetable diet and simple mode of living, together with frequent ablutions, tend to mitigate this disease. Its continued prevalence, as well as the severity of some of the cases, are ascribable to the inordinate use of mercury, administered by a physician who was accustomed to distribute it in inordinate quantities among the affected, who were of course ignorant of its nature and consequences.

While lying at Papieti, we had an opportunity of seeing the manner in which justice is administered in criminal cases. The court was held in the council-house, an oblong building, in the native style; the alleged crime was assault with intention of rape. The judges were seated on mats, having Paofai, their chief, a little in front of the rest; and the audience sat or stood around. The culprit was a petty chief, called Ta-ma-hau, a man of huge size, and apparently somewhat of a bully; he stood during the trial leaning against one end of the house, with an air of cool indifference. His accuser was a damsel not remarkable for personal beauty; she sat near the door among a number of other women. The witnesses were patiently heard, and the matter argued, after which the six judges severally gave their opinions and made remarks on the evidence, to which Paofai listened in an attentive and dignified manner, expressing, as occasion demanded, his assent or dissent. He then pronounced the verdict of the court, by which the prisoner was acquitted, but did not dismiss him without a brief and merited admonition. It appeared, that although not guilty of the crime alleged, he had while intoxicated addressed indecent language to his accuser.

Cultivation has undergone a great change within a few years, from the introduction of the guava, which has overrun the lower plain; the pasturage has not only suffered, but to its destructive effects are attributed many evils. Ten years prior to our visit, about which time the guava was introduced by the missionaries, the plain, from the sea to the base of the hills, was covered with verdure; and now it is overrun with an almost impenetrable thicket, before which all other vegetation disappears. I am inclined to think, that although this tree is now looked upon by the natives as a great curse, it will in time be beneficial to them, and cause them to become industrious, when they are obliged to get rid of it to make room for their sugar-cane, cotton, and indigo plantations; which products succeed remarkably well, can be raised at small cost, and will before many years be in great demand.

The cocoa-nut trees were also reported to have been decreasing, but our inquiries did not confirm this statement.

The manner of ascending the trees by the natives, has been frequently described, but can scarcely be imagined until witnessed; the feat is performed by leaping without any cessation, even in climbing the highest tree; the body of the tree being rough or composed of rings, affords some hold for the thong which spans the tree between the feet; at every jump, the body is thrown entirely free from the tree.

The bread-fruit tree is also said to have decreased, and this is no doubt the case; the seeds are said to be often abortive at Tahiti, for which reason the cultivation in this way has been neglected of late, and the plants raised in other modes have become less productive in consequence; its timber is used for many purposes; the fruit was not in season while we were at Tahiti.

Wild sugar-cane was found in the interior, commonly growing in tufts, but so small in size that it was with difficulty recognised; the cultivated kind is derived from this, and is also of small size.

The fruits we met with were oranges, lemons, limes, shaddocks, pine-apples, papayas, bananas, figs, vi-apple, fahies, cocoa-nut, and bread-fruit; the six first mentioned have been introduced since Cook's time.

The vegetables are sweet-potatoes (Convolvulus), yams of small size, taro (Caladium esculentum), the ape (Caladium macrorhizon), turnips, onions, and leeks; but there were no common potatoes cultivated. I gave Mr. Wilson some of the yellow Peruvian potato (Papas amarillas), but he informed me that all their attempts to raise potatoes in the low ground had failed.

The tacca, from which arrow-root is manufactured, grows in quantities, but we did not see it cultivated.

In the botanical researches it was remarkable that not a single stem of paper mulberry (Broussonetia) was found, although former visiters speak of it as the tree from which their cloth was made.

There are a vast variety of ornamental shrubs, and many aromatic plants, which the natives use to perfume their cocoa-nut oil.

The tutui tree (Aleurites triloba), the nut of which is used in tattooing, is very common all over the island.

Tobacco is grown in small quantities.

Mr. Henry informed me that grapes succeeded well on the south-east side of the island.

The price of labour is from two to four dollars a month, but for occasional labour fifty cents a day is usually paid.

Wild hogs are said to be numerous in the mountain region; none of our parties, however, met any. Horses are possessed by many persons on the island, and goats were seen. Dogs and cats were abundant. The island is well supplied with cattle; they are suffered to run wild, and frequent the neighbourhood of the hills, whither they are obliged to go for pasturage, which is now very scarce on the island, on account of the thick growth of the guava.

After the departure of the Vincennes, a party from the Peacock, consisting of Mr. Dana and some others, obtained leave of absence from Captain Hudson for five days, with the design of ascending Mount Aorai. They commenced the ascent immediately in the rear of Papieti, and by noon on the second day had reached an elevation of five thousand feet, where they stood upon a platform about twelve feet square; thence they looked down eastward two thousand feet into the Matavai Valley; to the westward they had a gorge about a thousand feet deep running into Toanoa Valley; to the south, the platform on which they stood was united by a narrow ridge with Mount Aorai, which was apparently only a short distance before them. In this place they were compelled to pass the night, by a fog which enveloped them, through which the guides were unwilling to lead them, refusing to proceed further along the dangerous path until the clouds should clear away.

The next morning was clear, and they pursued their ascending route along the edge of a ridge not more than two or three feet in width, having on each side an abyss two thousand feet deep. Seen from this ridge, looking south, Mount Aorai seemed a conical peak, but as it was approached it proved to be a mountain wall, whose edge was turned towards them. The only ascent was by a similar narrow path between precipices, and surpassed in steepness those they had already passed. The width of the crest seldom exceeded two feet, and in some cases they sat upon it as if on horseback, or were compelled to creep along it upon their hands and knees, clinging to the bushes. At last they reached the summit, where they found barely room to turn around. The ridge continued for only a short distance beyond them, being then cut across by the Punaania Valley.

From the summit of Aorai they had a magnificent view; to the south, it was speedily bounded by the peaks of Orohena and Pitohiti, whose steep sides rose from the valley beneath them; to the east, they had the rapid succession of ridge and gorge which characterizes Tahitian scenery; to the west, over a similar series of jagged ridges, Eimeo and Tetuaroa stood out from the horizon of the sea in bold relief; to the north, they looked down upon the plain, studded with groves of cocoa-nut and orange, and upon the harbour with its shipping, and the encircling reefs of coral.

A short distance below the summit of Mount Aorai, a mass of turrets and pinnacles, which from its singular outline is called the crown, runs along the top of a narrow ledge.

Except the plain of the coast, no level land is in sight but the valley of Punaania; this is divided from that of Matavai by a ridge of the usual edge-like form, running upwards towards Orohena.

Very few of the natives who are now alive have been on the summit of Aorai; their paths in this direction, as in other places, do not lead beyond the limit of the groves of wild banana (fahie). Beyond the height at which these cease to grow, the ground is chiefly covered with a wiry grass (Gleichenia), which springs up in many places to the height of ten feet, and is every where almost impenetrable. When this was not too high, they broke it down by casting their bodies at full length upon it; and when of larger growth, they had recourse to cutting away or breaking its stiff and crowded stems, until they had formed a way beneath it, whence the light was almost excluded.

The want of water, which after a few days of dry weather is seldom found even in the elevated valleys, was an additional discomfort. It is to be recommended to future travellers in the mountains of Tahiti to make provision against this inconvenience. The party was so much distressed from this cause as to enjoy the dew upon the leaves as a luxury.

Mr. Dana reported that the visit to Aorai conclusively settled one questionable point in the geology of the island. He found upon its summit neither corals nor "screw-shells," which vague rumours have long located on the top of the Tahitian mountains. Every one who has visited this island has probably heard that such formations exist in these lofty positions; but the report rests wholly on native authority. Moera, the guide who accompanied the party, and who resides near One-Tree Hill, insisted that he had seen both, and promised to show them. On reaching the summit, he began digging, and the rest of the party aided him. He soon brought up what he called coral, but which proved to be a grayish trachytic rock; and, although he continued to dig for some time longer, he could find nothing which he could venture to exhibit as screw-shells.

In their descent from Mount Aorai they followed the western side of the valley of Papoa, along a narrow ledge, similar to that by which they had ascended. After proceeding for two hours they reached a small plain, which speedily narrowed to a mere edge of naked rock, with a steep inclination; this they were compelled to traverse on their hands and knees, taking the greatest care to avoid detaching the rock, which in many places overhung a precipice; next followed a perpendicular descent of about twenty-five feet, down which they let themselves by ropes; this difficulty overcome, the rest of the route presented no dangerous features, and was performed in safety.

The manufactures of Tahiti are of little amount. Among them is that of arrow-root from the Tacca pinnatifida, which employs a portion of the population. Cocoa-nut oil is also made, and preserved for use in pieces of bamboo, cut off at the joints, when the natural diaphragms form a bottom, and the piece is thus a convenient bucket.

This oil is often scented with aromatic herbs, to be employed by the natives in anointing the hair and body; it is also used for burning in lamps, and is exported in considerable quantities. The lamps, which are always kept burning in their houses at night, are made of the shell of a cocoa-nut. The wick is formed of wild cotton, and is kept upright in the centre of the bowl by two elastic strips of cocoa-nut leaf crossing each other at right angles.

Sugar is beginning to attract attention, and some attempts have also been made in the culture and preparation of indigo.

Making straw or chip hats is a favourite occupation among the women, whose former employment of making tapa has, as was stated in the preceding chapter, been much diminished by the introduction of European fashions.

I have also before referred to the abortive attempt of the missionaries to introduce machinery for the manufacture of cotton, which will be again mentioned in speaking of the island of Eimeo, where the experiment was made.

Before closing my remarks on Tahiti, I consider it my duty to say a few words in relation to the transgression of the local laws by many of the vessels which visit it, and some of which, I regret to be compelled to confess, bear the flag of the United States. I have particular reference to the license always allowed to the crews, and in which the masters and officers often themselves indulge, in making brothels of their ships. They also do not scruple to retail ardent spirits to the natives, although they well know that it is contrary to a law of the island, most strictly enforced on shore. Such conduct not unfrequently gives rise to difficulties very prejudicial to the interests of the owners; but it is still more disgraceful when considered in its destructive effect upon the people whose hospitality they are enjoying, and as a practice that they would not dare to indulge in, when in the ports of any civilized nation.

The influence of the example of these visiters upon the natives is demoralizing in the extreme, is calculated to retard their advancement in civilization, and throws countless difficulties and obstructions in the way of the laudable exertions of the missionaries.

Little idea can be formed by those who have not witnessed it, of the extent to which the practice of vending spirits is carried, not only at Tahiti, but throughout the Polynesian islands. I am satisfied, that if the owners of the vessels which indulge in it were aware of the traffic, and had a just sense of their own interest, they would interdict the sale of this pernicious article, and prohibit the carriage of it in their ships.

Captain Hudson, who was much troubled with the illicit supply of spirits to his men, and was aware of the fact that the practice of vending it was contrary to law, endeavoured to discover the parties engaged in this traffic. He did this not only for the sake of his own crew, who, when questioned, stated that their intoxication was produced by gin, bought at the rate of three dollars a bottle, but to aid the natives in their exertions to prevent the infraction of their laws by the white residents. In pursuance of these objects, he called a meeting of the chiefs, and stated his complaint. They forthwith ordered search to be made for the offenders by the police, by which some of them were discovered and immediately fined. At the examination, however, the chiefs stated to Captain Hudson, with what truth I do not pretend to say, that seventy cases of *gin* had been landed by our own consul, from whom they believed that the retailers had obtained it, while the main stock being upon his premises, under the United States flag, was protected from search.

The repairs of the Flying-Fish were not completed before the 10th October, up to which time the Peacock was detained, not only in order that they might sail in company, but because her officers were still engaged in the survey of the harbours. In the interval of leisure which was thus afforded them, the crew of the Peacock asked and obtained permission to get up a theatrical entertainment, for the amusement of the natives and themselves. The council-house was placed at their disposal for the purpose by the native authorities. The play chosen was Schiller's "Robbers," the parts of which had been rehearsed at sea, in the afternoons—a task which had been the source of much amusement. An opportunity was now presented of getting it up well: the dresses having been prepared, the day was appointed, and when it arrived the piece was performed; the acting was thought by the officers very tolerable, and finally gave great delight to the natives. The latter, however, were somewhat disappointed in the early parts of the performance, for they had expected an exhibition of juggling, such as had been given for their entertainment on board of a French frigate. While under this feeling, they were heard to say there was too much "parau" (talk). After they began to enter into the spirit of the performance, the murders took their fancy; and they were diverted with the male representatives of the female characters.

A number of comic songs, which formed the relief of the more serious play, were exceedingly applauded; among others they laughed heartily at "Jim Crow" sung in character, and could not be persuaded that it was a fictitious character.

On the 25th September, the Vincennes sailed from the port of Papieti for the island of Eimeo. The distance between its reef and that of Tahiti, measured by the patent log, is ten miles.

Eimeo is a beautiful object in the view from Tahiti, and its beauty is enhanced on a nearer approach; its hills and mountains may, without any great stretch of the imagination, be converted into battlements, spires, and towers, rising one above the other; their gray sides are clothed here and there with verdure, which at a distance resembles ivy of the richest hue.

Drawn by A. T. Agate. EIMEO. Engraved by Smillie & Hinshelwood.

Taloo harbour is an inlet about three miles in depth, situated in a glen enclosed by precipitous sides rising in places to the height of two thousand feet; at its head is an extensive flat of rich alluvial soil, now employed in the culture of sugar, and studded with trees, shrubs, and other interesting objects. The ship lay at anchor close beneath a high mountain on the left, in contrast with which her dimensions seemed those of a cock-boat.

I had been furnished with letters to the Rev. Mr. Simpson, who is stationed as missionary at Eimeo; when we landed, he met us upon the beach, and gave us a most cordial reception; we were soon surrounded by nearly all the natives in the place, male and female, old and young, who followed us with expressions of wonder; their conduct reminded me of the manner in which an Indian chief is run after in the streets of our American cities. In spite of their excitement they were all extremely civil, and said they only wished to look at us, although some were disposed to feel us.

Mr. Simpson led the way to his house, passing by a thick and well-built stone wall, the only one which I had seen used as an enclosure in these islands; on my inquiring if it was the work of native labour, I was informed that it had been erected by an Irishman, who is now the overseer of Mr. Simpson's sugar plantation. This wall encloses a large lawn, with a number of fine bread-fruit trees; on each side of the walk was a row of low acacias, which were at the time in full bloom, with flowers of many colours,—yellow, orange, red, and variegated; at the end of the walk was a low thatched white cottage.

Mr. and Mrs. Simpson have the care of a school for the children of missionaries and respectable white parents: these are kept entirely separate from the children of the natives; the reason assigned for this exclusiveness is, that the danger of the former receiving improper ideas is such as to preclude their association with the latter. This may be good policy as far as the white children are concerned, although I doubt its having a good effect on their minds if they are destined to spend their lives among the islands. The habit they will thus acquire of looking upon the natives as their inferiors, cannot fail to have an injurious influence on both. The exclusiveness is carried so far, that the children of whites by native women, although they are united in the relation of husband and wife, are not admitted into these schools, because, as they say, they do not wish their children to be contaminated by intercourse with such a mixture of blood. In pursuance of the same policy they have, as it is said, procured the enactment of a law prohibiting marriage between whites and the natives.

This, I must say, appeared to me the worst feature I had seen in the missionary establishment. It is placed here for the avowed purpose of reclaiming the natives from idolatry, and the vices which are its concomitants. In doing this, their most successful efforts have been in the conversion and moral improvement of the young; yet they bring up their own children to look down upon them as beings of an inferior order. In becoming acquainted with this feature, I no longer wondered at the character, which I was compelled by a regard for truth to give, of the children of missionary parents in Tahiti.

The missionaries are now aware that their proper plan is to devote their time and attention to the young; and in pursuance of this object, Mr. and Mrs. Howe have lately arrived from England, for the purpose of establishing an infant school.

It is to be regretted that the schools of manual labour have, for what reason I could not learn, been discontinued. Some of the natives who had been instructed in them evinced a knowledge of the trade of the carpenter, and furnished the ships with very good boards sawn by themselves.

The natives of Eimeo have an advantage over those of Tahiti in being free from the influence of evil example; many of them are industrious, and possess a proper feeling of the benefits they have derived from the missionaries, of whom they speak, whenever questioned, as friends.

Three of our crew having become enamoured of these islands, deserted while the Vincennes lay at Eimeo. They left the ship about ten o'clock at night, soon after which their absence was discovered, and parties sent out in every direction to intersect the roads and drive them to the hills. This was effected the following morning, and a large party of natives was employed to hunt them up. This task they speedily performed, and at last drove the deserters to one of the highest ridges, in full view of the ship. Here the runaways appeared at first disposed to make fight with stones; but when they saw the odds against them, and witnessed the alertness of the natives in leaping from cliff to cliff, they thought it best to give themselves up; which they did to three natives, naked except the maro, and armed respectively with a rusty sword, an old cutlass, and a piece of iron hoop. These bound their hands, and led them down to the shore, whence they were brought on board, where the three natives received the reward offered for their apprehension. The chase and capture was an amusing sight to those who watched the proceedings from the ship.

Eimeo has, if possible, a more broken surface than Tahiti, and is more thrown up into separate peaks; its scenery is wild even in comparison with that of Tahiti, and particularly upon the shores, where the mountains rise precipitously from the water, to the height of twenty-five hundred feet. The reef which surrounds the island is similar to that of Tahiti, and as we have seen to be the case there, no soundings are found on the outside of it. Black cellular lava abounds, and holes are found in its shattered ridges, among which is the noted one through which the god Oroo is said to have thrown his spear.

While we remained at Eimeo, I visited Papoa or Cook's Harbour, which lies to the east of that of Taloo. There is a marked resemblance between the two ports, except that the shores of Papoa are not quite as precipitous as those of Taloo, and the entrance of the former not as practicable.

Wood and water may be had at both harbours in abundance, but in other respects the island is not well adapted as a place for the supply of ships. No more than a single ship would probably be able to find refreshments at a time. It is, therefore, seldom visited, and its surplus produce is carried to Tahiti for sale. Notwithstanding, the articles of traffic are quite as dear as at Tahiti.

The inhabitants of Eimeo reside upon the shores, and there are several large villages on the southern side of the island; among these is Afareaitu, at which the Rev. Mr. Hale, whose recent arrival has been spoken of, is about to take up his residence.

It was in this island that the establishment of a factory for spinning cotton, and weaving cloth and carpets, was attempted by Messrs. Armitage and Blossom, who were sent out for the purpose by the London Missionary Society. Its failure and cessation after a fair trial have already been mentioned.

It has been seen that the alluvial plain at the head of the harbour of Taloo, is partly occupied by plantations of sugar. The cane is of superior quality, and the climate well adapted to its production; the plant is indeed indigenous, and it is well known that the variety of it found at Tahiti has been introduced advantageously into the West Indies. At Eimeo the crop is liable to injury from the ground-rat, and there are difficulties attending the management of the crop, which cause the cultivators to speak despondingly. About one hundred tons, however, are made annually.

Coffee, cotton, and all other tropical plants, succeed well at Eimeo, and the quantity of tapa manufactured is greater in proportion than at Tahiti.

I took the opportunity of my anchorage in the harbour of Eimeo, to verify the chart made by Captain Von Schantz, of the Russian ship America, and found it accurate. I have added some soundings, and laid down the topography of the shores, and the outline of the reefs, more minutely than he had attempted.

On leaving Eimeo, I bade adieu to the Tahitian islands; but I cannot close the portion of the Narrative which is devoted to them, without again expressing the pleasure I and all my officers derived from our intercourse with the missionaries, and our obligations for the kindness received from them and other residents. Among those to whom we are indebted, I cannot refrain from naming George Pritchard, Esq., H. B. M. Consul, of whose strenuous exertions to advance the welfare of the people, and sustain the government in its efforts to promote their best interests, I became by observation fully aware. It is to be regretted that his very activity in thus labouring in many ways for the good of the community in which he resides, should be the probable cause of unkind and unfounded imputations, from those actuated, if not by motives positively bad, at least by a less enlightened or less ardent zeal.

BEATING TAPA.

CHAPTER III.

CONTENTS.

DEPARTURE OF THE VINCENNES FROM TAHITI—BELLINGHAUSEN'S ISLAND—ROSE ISLAND—MANUA—ITS DESCRIPTION—THREATENED WAR—ITS CAUSE—CANOES OF MANUA—APPEARANCE AND CONDUCT OF THE NATIVES—VILLAGE—DRESS OF THE NATIVES—PRODUCTIONS OF MANUA—OLOOSINGA—ITS DESCRIPTION—HOUSE OF THE KING—HIS ENTERTAINMENT—RETURN TO THE SHIP—CORAL REEF OF OLOOSINGA—OFOO—TEMPERATURE DURING THE PASSAGE—PLAN OF THE OPERATIONS OF THE SQUADRON—APPEARANCE OF TUTUILA—HARBOUR OF PAGO-PAGO—ANCHORAGE—TOA—OBSERVATORY—GEOLOGICAL STRUCTURE OF TUTUILA—APPEARANCE AND CHARACTER OF ITS INHABITANTS—MASSACRE OF M. DE LANGLE—VILLAGE OF PAGO-PAGO—GOVERNMENT—WAR-MATS—VISIT TO TOA—HIS FEAST—PRICE OF PROVISIONS—WAR-SONG—BATHING—MODE OF LIVING—EMPLOYMENTS AND AMUSEMENTS—MR MURRAY, THE MISSIONARY—CUSTOMS OF THE NATIVES—PUBLIC WORSHIP—MISSIONARY—JUNCTION OF THE PEACOCK AND FLYING-FISH—SURVEYS—CLIMATE—VISIT TO HEATHEN VILLAGES—SAVAGE FEAST—TEMPERATURE AT THE TOP OF MATAFOA—SUPPOSED MURDER—REMARKABLE PHENOMENA OF THE TIDES.

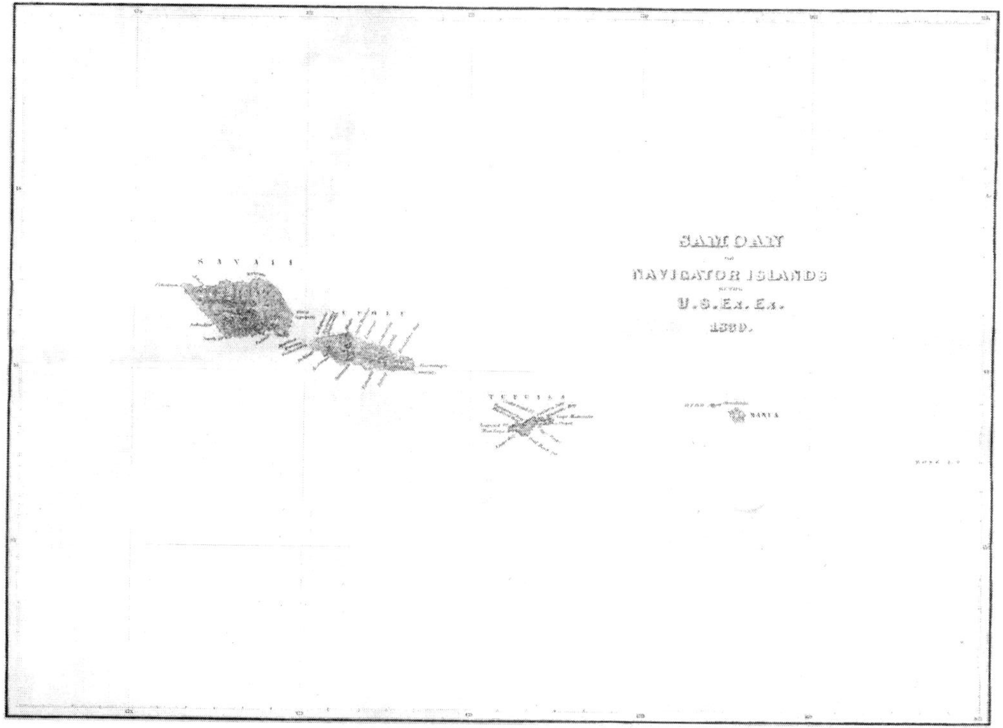
NAVIGATOR ISLANDS
U.S. Ex. Ex.
SAVAII
TUTUILA
MANUA

CHAPTER III.

TUTUILA.

1839.

On the 29th of September, at daylight, having the wind from the northward and eastward, we got under way, and made sail to the westward, passing the Society Island Group: viz., Sir Charles Saunders' Isle, Huaheine, Tahaa, Borabora, Maufili, and Moutoiti. All of these, with the exception of the last, are high lands.

On the 30th, we made Bellinghausen's Island, which is a low coral island, similar to those which have been already described. It was uninhabited, and is of a triangular form, with the usual vegetation, with the exception of cocoa-nut palms. We landed upon it, and made the magnetic experiments.

Birds were in great plenty, and as tame as we had found them at other uninhabited islands we had visited. No lizards or rats were observed, nor was the common fly seen. The lagoon had no passage into it at low water, but the tide flowed into it over the reef.

During the time of our stay on the island, the tide rose and fell upwards of two feet, and it was high water at 8 A. M. Many specimens of fish were obtained here, of which the department of Natural History will treat.

In the afternoon, we again made sail to the westward, for Rose Island, and on the 6th of October, we passed near the locality of the Royal George Shoal, but saw nothing of it.

On the 7th, which was the day appointed for our rendezvous off Rose Island, we came in sight of it, and at the same time descried the Porpoise. That vessel had passed by Nairsa or Dean's Island, and connected the survey of it with that of Krusenstern's and Lazareff. Both of these were found to have entrances into their lagoons; they are uninhabited, though occasionally visited by the natives of Nairsa Island. The position of Recreation Island was passed over, but no signs of land discovered.

Rose Island, the most eastern of the Samoan Group, was discovered by Freycinet, who gave it its name. It appears, at first, like a round knoll of land, but on a nearer approach, this is found to arise from a large clump of Pisonia trees, similar to those found growing in the low archipelago. It is a low annular coral island, of small dimensions, inundated at high water, with the exception of two small banks, one of which is entirely covered by the clump of trees. The other is formed of dead coral, without any vegetation. The tide was found here to rise about four and a half feet, the flood setting to the eastward. The breakers on its weather or southeast side are heavy; and there is an entrance into the lagoon, having four fathoms depth of water through it. The lagoon has from six to twelve fathoms in it. A remarkable coral formation, like a submerged tree, thirty feet in diameter over its top, was found in the centre of the lagoon, rising to the level of low water, and having all around it a depth of six fathoms. The currents set regularly out and in to the lagoon, according to the state of the tide. In stormy weather the sea must make a complete breach over the reef.

Some boulders of vesicular lava were seen on the coral reef; they were from twenty to two hundred pounds weight, and were found among blocks of coral conglomerate.

Birds were seen flying over the island, and on landing we found them in great numbers and very tame. The frigate-birds, and boobies (sula), whose nests had before been observed on low bushes, were here found on the tops of trees fifty feet high. The noddies laid their eggs on the parts of the island destitute of vegetation. Tern were in great numbers. their breeding-place was in a thicket on the weather side of the island, or that which was exposed to the wind and sea, and was remarkable from the regularity with which the eggs were placed, about three feet apart, without any nest, and, with but few exceptions, out of many thousands, each egg lay separately. The colour of the eggs is a dirty white, mottled with brown. The noise made by these birds when disturbed was almost deafening; but on making a loud sound, such as the firing of a gun, their cries would cease for a moment or two, producing a singular stillness.

Several small turtles, similar to those seen at Honden Island, were observed here. One of them was taken, but its flesh proved coarse, and was drier than that of the green turtle: they feed upon a species of fucus that grows upon the reefs. Here we made observations for intensity and dip.

On the 7th, we left Rose Island and stood to the westward, making at sunrise the island of Manua, which is two thousand five hundred feet above the level of the sea. It has the form of a regular dome, rising in most places precipitously from the water to the height of three or four hundred feet, after which its ascent appears more gentle and even. It is sixteen miles in circumference, is well covered with a luxuriant vegetation, and has many cocoa-nut groves on its north-west side.

On approaching it, Oloosinga was in sight, and shortly after Ofoo. These two islands lie to the northwestward, at the distance of about four miles.

The boats were lowered, and sent to trace the shores of the island of Manua, for the purpose of surveying it; whilst the Vincennes and the Porpoise passed on each side.

This island is inhabited. The principal settlement is on the north-west side, and there is anchorage for a small vessel near the shore, where there is a cove to land in, with but little surf during the fine season, or from April to November. It has a shore-reef of coral, and the soundings extend off some distance, eight fathoms being found four hundred yards from the shore.

Some large blocks of vesicular lava were seen on its northeast point, but the general structure was a conglomerate of a drab colour, in horizontal strata; yet the beach was of light-coloured sand, formed by a mixture of coral and shells.

Our arrival off Manua was opportune. According to the statement of one of the brothers of the king, who spoke a little English, hostilities had been threatened between the "missionary" party, and the "devil's men." A native missionary, resident in the island, had already prevented a battle, by telling them that if they wished to fight with each other they must first kill him. Through his influence and exemplary conduct, peace had hitherto been preserved. It was stated that several "very bad" white men were on the island, and that they made "plenty of fight;" but that on seeing "mannawa" (man-of-war), they had gone into the "bush."

Eight of these men had deserted from an English whaler, whose boat they had stolen. Three of them came alongside of us next day, clad after the manner of the natives, and were very anxious to be taken off the island.

The canoes of these islanders were the best we had seen. They are built of a log, having upon it pieces fastened together, to raise them sufficiently high. They are thirty or forty feet long, and are partly covered in at both ends. Some of them are capable of containing twenty or twenty-five men, and are very swift. The chief usually sits cross-legged on the forward platform or deck. They have an out-rigger, which is not so far removed from the canoe, and renders them more liable to be upset.

Several of the natives came on board. They were a finely-formed race, and appeared lively and well-disposed, though in a much wilder state than those of the Society Islands.

Our party, on landing, were immediately surrounded with natives willing to trade, and calling out for "bacca" (tobacco), which is in great request among them. Fish-hooks were also much sought for. A fowl, a bunch of bamboos, and a dozen of cocoa-nuts were procured for a small one.

They seemed willing to exchange any thing they had, viz., baskets, mats, spears, clubs, &c., to obtain these articles. They were not found altogether honest, though this did not consist in stealing, but in selling their articles twice over; for after we had made a purchase from one, another would claim the article as belonging to himself, and insist on also receiving a price for it.

Near the village are thick stone walls, intended to all appearances for defence. The houses are elliptical, supported on stout posts, about four feet high, from which the roof or thatching rises to the height of twelve or fifteen feet; they are generally erected on a raised terrace of stone, two feet above the ground. The floors are covered with coarse matting.

The king or chief of these islands resides at Oloosinga, in consequence of its being more easily defended.

The dress of the natives consists only of the maro, made of the leaves of the Dracæna, which has a graceful appearance. The leaves are slit, and form a kind of short petticoat.

The tattooing is of the same kind, as will be described in the general account of the Samoan Islands.

These islands furnish pigs, fowls, sweet-potatoes, fruit, and some taro. The vegetation was thought to be more luxuriant than at Tahiti, and the climate moister.

Many running streams were observed coursing down the sides of the island. When off the eastern end, we were much surprised to see the natives plunge off the rocks into the heavy surf to reach our boats.

After our party reached the ship, we made sail for Oloosinga, where I went on shore to see the king or chief, who was old and decrepit. His name is Lalelah. His brother, and presumptive successor, was with him, and met me as I landed from the boat. His mode of salutation was by taking my hand and rubbing the back of it against his nose.

The old man, I was told by the interpreter, could speak a little English, but I could not understand him. This he attributed to his age, and would not admit that it was owing to his ignorance of the language. They led the way to his hut, situated under a mural precipice twelve hundred feet in height.

The island of Oloosinga is a narrow ledge of rocks, rising nearly perpendicular on both sides, and is three miles in length. So precipitous is it at its ends, that it is impossible to pass around it on the rocks. The strip of land is about five hundred yards in width, on which bread-fruits and cocoa-nuts grow in great profusion and sufficient abundance for all the wants of the natives. They told me that this island had been chosen as a place of safety, since the other became unsettled in consequence of the wars of the Christian and Devil's parties; and that the island of Manua had formerly been the residence of the king, but that he found himself unsafe there, and had taken up his abode at Oloosinga, on its northwestern side.

His house was elliptical in form, and thirty feet long, erected on a well-flagged terrace of stone, about four feet above the ground. It was well shaded with cocoa-nut and bread-fruit trees, and was supported around by ten stout posts, with three others in the centre reaching the top. The roof came down within three and a half feet of the ground, and projected as eaves about eighteen inches or two feet. In the centre the hut was fifteen feet high and well thatched.

The whole floor was ordered to be spread with fine mats, which were carefully unrolled, and laid over the coarser ones on the floor. The king then seated himself in the centre, and desired me to take a seat between himself and brother. Shortly afterwards two large wooden trays were brought in, filled with cooked bread-fruit and covered over with leaves. One of these was placed before me, when the king made a long speech, giving me welcome and offering food to eat. I was then desired to hand some to the king and his brother, and to others who were pointed out to me. This I did, but unfortunately continued my task, and handed it to one of the Kanakas, or common people, who were sitting close around us; much displeasure was evinced, accompanied with angry looks. I now looked around for my men, but they were out of sight, on their return to the boat. In order to make the best of my situation, I asked what was meant, and feigned to be quite ignorant of having given any offence. After a minute they were apparently appeased, and pleasant looks were restored.

They handed round a shell containing cocoanut-oil to dip the bread-fruit in, and another containing salt water. After we had eaten, they began a careful examination of my clothes, and appeared much pleased with the buttons. My pocket-handkerchief was taken out of my pocket, and spread on the mat to be examined by the king. His brother took off my hat and put it on the top of his large bushy head. They then had ava made, of which I could not partake, after seeing the process of making it. It is first chewed by the women and thrown into a large bowl; water is added to it, and it is then strained through leaves. This was partaken of by them all, while they gave me a fresh cocoa-nut.

They were becoming more familiar every moment, and it was getting late, so I thought it time to make a move. I therefore rose up, and was followed by the natives, in number upwards of a hundred, including the king and his brother, to the boat. I looked carefully around for arms, but saw none among them. My boat was aground: the king, his brother, and several others, got into it, saying they must have some presents. They seemed disposed to resist, and showed a determination to contest our getting off. I on the other hand was determined to get rid of them, and peaceably if I could; I therefore ordered the boat's crew to arm themselves, and drive every one of the natives from the boat, at the same time intimating to the king to use his authority, which I found, however, existed only in name. We thus succeeded in getting clear of the crowd, until we had no more than eight left; to each of these I presented a small fish-hook, and ordered them to get into the water, which was about a foot deep, and go; this they did, one by one. At last came the king and his brother's turn, to whom I presented, with great ceremony, first a small and then a large fish-hook; after which they left me, apparently in great good humour. I was heartily glad to be rid of such rapacious troublesome fellows so easily and without a fight. We then pushed our boat off. When just beyond the reef, in taking up our anchor, the boat had the appearance of returning again on shore. On seeing this, a great shout was set up by the natives, and one of them immediately advanced with my powder-flask. He said it had been taken by a boy out of the boat, and had been dropped into the water, to be picked up after we had shoved off. I gave the man a small present for his apparent honesty; but I am inclined to believe it was the fear of detection, and the belief that we had missed the article, and were returning for it, that induced them to give it up so willingly. It was some time before he could be made to understand what the reward was for, but when he found it was for his honesty, he laughed heartily.

This having excited our suspicions, the boat's crew informed me that a canoe that was paddling off had been alongside of the gig, and that they felt satisfied that the natives had taken something from us. It being in our course towards the ship, we gave chase, and being favoured by the wind, soon overtook the canoe, to the great fright of the two natives, who were paddling with all their might, and whose eyes were full of tears when overtaken. They had nothing at all in their canoe, and after examination it proved we had lost nothing. To console them for this alarm, I gave them a few trifles, and they became easy and cheerful.

The coral reef around this island was different from any I had hitherto seen. It consisted of two regular shelves, the outer one from fifty to sixty feet wide, and the inner in places measuring one hundred and forty feet. A distinct mark of high water was measured along the beach, and found to be twenty feet above the ordinary sea-tide, which has from four to five feet rise.

The rock at Manua was volcanic conglomerate, with large blocks of vesicular lava lying loose on the coral beach.

Before sunset the boats returned to the ship, having completed the survey of both islands.

Ofoo lies to the westward of Oloosinga. There is a passage for boats of about a fourth of a mile in width between them, and anchorage on the western side. Ofoo resembles Oloosinga; and, from the accounts we received, it has but few inhabitants: those of Oloosinga having made war upon them, and killed the "natives" off. There is a small and comparatively low islet off its western end, near which there is an anchorage. After sunset we bore away for Tutuila, which can be seen in fine weather from these islands.

The temperature in the passage from Tahiti to the Samoan Islands had increased from 77·6° to 81·11° in the air; and that of the water from 79·6° to 81·6°.

As it was my intention to make a thorough examination of this group, I resolved, in order to accomplish it in the least possible time, to divide the squadron, so as to put all the remaining islands under examination at the same time. The island of Tutuila being the most central, and, from the information I had obtained, the best position for my astronomical observations, I selected it for the Vincennes. That of Upolu was reserved for the Peacock and Flying-Fish when they should arrive; and in case of their being detained longer than I anti cipated, I should be ready to take up the survey of the latter, or assist in completing it. The Porpoise was ordered to examine the island of Savaii; and one of the naturalists, Dr. Pickering, was directed to join her, for the purpose of exploring the interior of the island during her operations in its vicinity. Lieutenant-Commandant Ringgold was therefore directed to land him for the purpose, and take him on board when the survey should be concluded.*

On the 10th of October, we had light winds, in consequence of which we did not reach Tutuila that day. At daylight on the 11th we were near its eastern end, and off the island of Anuu.

About eight miles to windward of the harbour of Pago-pago, we were boarded by several canoes, in which were some natives, with a white man, by name William Gray, whom I retained as interpreter during our stay here, and found of much use.

The island of Tutuila is high, broken, and of volcanic appearance. It is seventeen miles long, and its greatest width is five miles. The harbour of Pago-pago penetrates into the centre, and almost divides the island into two parts. It is less varied in surface than the Society Islands; and its highest peak, that of Matafoa, was found to be two thousand three hundred and twenty-seven feet above the sea. The spurs and ridges that form the high land are like those of Tahiti: precipitous, sharp-edged, and frequently rise in mural walls from the water to a height of three or four hundred feet, showing the bare basaltic rock. Above this height, the surface is covered with a luxuriant vegetation to the very top of the mountains; the cocoa-nut tree and tree-fern give the principal character to this beautiful scenery. Dead coral is seen along the shores, above high-water mark.

The harbour of Pago-pago is one of the most singular in all the Polynesian isles. It is the last point at which one would look for a place of shelter: the coast near it is peculiarly rugged, and has no appearance of indentations, and the entrance being narrow, is not easily observed. Its shape has been compared to a variety of articles: that which it most nearly resembles is a retort. It is surrounded on all sides by inaccessible mural precipices, from eight hundred to one thousand feet in height. The lower parts of these rocks are bare, but they are clothed above with luxuriant vegetation. So impassable did the rocky barrier appear in all but two places, that the harbour was likened to the valley of Rasselas changed into a lake. The two breaks in the precipice are at the head of the harbour and at the Pilot's Cove. The harbour is of easy access, and its entrance, which is about a third of a mile in width, is well marked by the Tower Rock and Devil's Point.

* For orders, see Appendix V.

HARBOUR OF PAGO-PAGO.
TUTUILA.

About three miles to the southward, off the mouth of the harbour, there is a coral bank half a mile long, on which the sea breaks in stormy weather: the least depth of water found on it was four and a half fathoms; the depth increases to the eastward, towards the island of Anuu.

As we arrived off the harbour the wind grew light, and finally came out ahead, thus compelling us to beat in to our anchorage, under the direction of Edmund Foxall, a white pilot. He usually comes off to vessels when within two or three miles of the harbour, on a signal being made. We made many tacks before we reached our anchorage, which was in deep water, twenty-nine fathoms. About half a mile from the entrance of the harbour, it bends at right angles. In this position, surrounded by cliffs, the firing of a gun produces a remarkable reverberation, resembling loud peals of thunder.

We were surrounded, as soon as we entered, by a large number of canoes, filled with natives, who all seemed delighted with the ship and the number of men on board. When we had moored, one of the principal chiefs, whose name was Toa, was admitted on board; he was an athletic, muscular man, of large frame, about forty years of age, with a pleasant expression of countenance; he manifested great pleasure in welcoming us. He began by telling me, through the interpreter, that he was a missionary; that he had formerly been a great thief, and a doer of many bad acts, but being now a missionary, he was reformed and stole no more. He told this with such an open expression of countenance and so much simplicity, that I could scarcely forbear smiling. After I had finished asking him questions, he continued eyeing me from head to foot, as if determining my dimensions. I told the interpreter to ask him why he looked at me so intently. He replied, that he had a coat on shore that was too tight for him about the arms and chest, and he believed it would fit me: if so, he should be glad to exchange it for the jacket I had on. Not being inclined to this exchange, I ordered a small hatchet to be given him. This gratified him much, and he instantly went over the ship's side to show it to his friends. This same Toa is chief of the village of Fungasar, about three miles distant from the harbour, on the north side of the island. He learns to read and write, being taught by some of the small children, and attends school regularly. He became of great use to us, and was a constant visiter. During one of his visits on board, he espied some red umbrellas among the presents, and from that time was continually endeavouring to obtain one for his wife, and brought many articles in the hope of inducing us to part with it in exchange for them.

The day after our arrival a place was chosen for our observatory, and the tents and instruments were landed. Understanding that I wanted to see the sun and stars, I was told by Mr. Murray, the white residents, and natives, that I should have little weather for observations for the next fortnight, which proved literally true with the exception of the last two days.

The geological character of this island is similar to that of Manua; it has only a shore-reef of coral, and soundings extend some distance from it. It has many desirable ports or bays on its north side, where vessels may obtain wood, water, and supplies. The best and safest port, however, is that of Pago-pago, on its south side, which affords a safe harbour for vessels to overhaul, and where supplies may be obtained in abundance.

Tutuila is thickly settled round its shores, and particularly at its southwestern end: this is lower and more easily cultivated than the eastern, which is high and rugged. The only communication is by the sea-shore, the hills being too precipitous and difficult of ascent to pass over.

The men of Tutuila are a remarkably tall fine-looking set, with intelligent and pleasing countenances. In comparison with the Tahitians, they would be called sedate.

The women are far from being good-looking, with the exception of some of the younger ones. They are remarkably domestic and virtuous, exhibiting a strange contrast to those of Tahiti. Here there is no indiscriminate intercourse, the marriage tie is respected, and parents are extremely fond of their offspring. The inhabitants are disposed to be hospitable to strangers, although they expect remuneration for it. Travelling is generally believed to be safe throughout the island of Tutuila, and the natives, as far as our experience goes, are not the blood-thirsty race they have been reported to be. The unfavourable estimate of their character has, I presume, been derived from those who first knew them, and particularly from their attack upon the expedition of La Perouse. Of this conflict I obtained the following particulars from the Rev. Mr. Murray, who had them from an old man, who was a witness of the affray. The latter is the only individual now alive in the settlement who was present when it occurred, and his testimony was corroborated by others who had heard of it from those who witnessed the scene.

On the morning of the massacre, the vessels stood in towards the land. About noon the boats went ashore, as recorded by La Perouse, and while on shore, a number of canoes, belonging to the island of Upolu (to which Tutuila was at the time subject), went from the shore, and proceeded directly to the vessels. When these canoes were alongside, a young man in one of them laid his hand on an iron bolt in some part of the ships, with the intention, it is supposed, of stealing it. He was fired upon by the French. The ball passed through his shoulders, and mortally wounded him. The natives, on seeing the effect of the shot on one of their number, were greatly enraged, and immediately left the vessels, and hastened to the shore, where they found the boats that had gone to get water. On reaching them, they began the attack, which resulted in the massacre of M. De Langle, and of those who were with him on shore. When the natives began this attack, the great body of the French were absent from their boats; some were in the bushes gathering plants, and others talking to the females. On the commencement of the disturbance, they all rushed towards their boats, and the confusion became general. The minute circumstances of the affray, farther than the above, cannot now be ascertained from the natives. They are, however, very clear in reference to the cause, and to those who were the actors in it, viz., the natives of Upolu. The Tutuilians maintain that they endeavoured to save the lives of the French; and, on the following day, as soon as they dared to venture from the mountains, whither they had fled during the massacre, they collected the bodies, which they found in a state of nudity, dressed them in native cloth, and buried them in the beach, as they were accustomed to bury their own chiefs. The actors in the massacre proceeded at once to Upolu, which will account for their having been afterwards seen there, and recognised by the French. Our inquiries relative to the spot where they had buried the bodies, were not satisfactorily answered. How the carpenter's son escaped is not known. He is said to be still living at a village on the eastern part of the island. There appears to be mention made of a boy among the missing, in La Perouse's account. Levasii, a chief of the district of Faleletai, was at the massacre of the party of La Perouse. He was then a boy of thirteen years of age. He remembered the occurrence, and that three of the Papalangi were killed.

The perpetrators of the deed were some young chiefs from the district, who were on a "malanga" to Tutuila. At that time Aana district had the rule, or was the "Malo" party, and domineered over the inhabitants of the other islands and districts.

The village of Pago-pago contains about thirty dwellings, and a council-house, which is in use as a church, until the large one they are engaged in building shall be finished. Every village has a council-house for the entertainment of visiters, and the accommodation of meetings.

This island is under several chiefs, each of whom rules over a town, district, or bay. The present chief of Pago-pago is Mowna, the adopted son of the last chief, Pomale, who died not long since, leaving an only son, also called Pomale, who from his great modesty lost his inheritance. Mowna was more crafty than Pomale, and understood well his rival's character. After the death of the old chief, these two young men, about the same age, became candidates for the succession. Mowna, through his intrigues, succeeded in getting the whole family together to decide between them. Both Mowna and Pomale were present, the former appearing dejected, silent, and willing to leave the decision to the meeting; whilst Pomale, when asked who should be chief, said with his usual modesty that he was in favour of Mowna, who was accordingly made chief. Mowna, however, is now so in name only, for Pomale rules in fact. This arises from his good character, and the influence he derives from the missionaries, of whom he is one of the most active and pious supporters, and withal a great preacher. So great is the confidence Mr. Murray has in Pomale, that he is frequently left to take charge of the congregation, during the absence of Mr. Murray in another part of the island.

The greatest restraint on the conduct of the chiefs, appears to be the fear of losing the good name of their ancestors, and of not handing it down to posterity pure and unspotted. This feeling seems to govern their conduct, and from the information I received, may be made use of as an appeal to them, to avoid doing evil, and to do right.

The missionary, the Rev. Mr. Murray, deserves the greatest credit for this state of things. He has unbounded influence over the natives, and deserves it. The ten commandments are the common law of the island, wherever Christianity has taken root, and any infringement of them is surely punished;—the guilty persons being put out of the church, and denied the privilege of attending worship. They are looked upon as having fallen, and are consequently avoided. This fear of public opinion, I was informed, was found to be sufficient to deter them from the commission of crimes and immoral practices.

The tapa or rugs worn by distinguished chiefs, were preserved, and were formerly much venerated by them. Since the introduction of Christianity, however, such has been its influence that they will now readily part with any thing of the kind. Pomale was induced to let us have those in his possession, and also exchanged the "war spirit" mat for a small present for his wife.

On the 17th, our friend Toa gave us an invitation to visit him at his town of Fungasar, on the north side of the island. It is situated on the next bay to that now called Massacre Bay, where De Langle was killed. The path across the island is a very difficult one to travel; it leads up through the valley, and across the dividing ridge, which is quite precipitous. The rain which had fallen made it very slippery, and the journey was fatiguing to those not accustomed to this kind of walking.

I was much struck here with the manliness and intelligence of the natives, and with their frank open expression of countenance. The colour of their complexion is rather darker than that of the natives of Tahiti. The outlines of face and figure are very like those we had left, their hair and eyes black, and their teeth good and white. Some of them had frizzled hair, but it was generally straight.

Just before arriving at the village, we were met by Toa, and some of his relations and attendants, who welcomed us to his village, saluting me by rubbing his nose with my hand; this is the usual custom.

He ordered a pig, taro, bread-fruit, &c., &c., for our entertainment. These were cooked in the universal Polynesian mode, by being covered up in a hole with hot stones. We were soon told that the feast was ready, but having had some experience of their cooking, we desired it might remain in the oven a little while longer. Their usual custom is to take it out the moment that the taro is cooked, and from daily practice they are well acquainted with the time required to cook it. This is scarcely sufficient to give the pig time to be warmed through. Our request prevailed, and in the course of half an hour we were summoned to the council-house or fale-tele, where strangers are always entertained. We were shown our seats, on one side of a circle, while Toa, with his family and friends, occupied the other. The mats, except one, were not very clean. The pig, which must have weighed one hundred pounds, was brought in, and laid with the taro and bread-fruit on banana-leaves. A butcher's knife was all that we possessed to carve it with. The whole village, old and young, men, women, and children, who were waiting in anxious expectation for their share, now surrounded us, and made it uncomfortable to eat, with so many hungry expectants; I made haste, therefore, to divide it, and with it they soon dispersed. The taro was exceedingly well cooked, dry, and farinaceous. The bread-fruit they said was too young, and not being considered good by them, they objected to giving us any of it, but did not hesitate to eat it themselves. A pig is a great treat to them, for although they have plenty, they prefer selling to eating them.

All kinds of provisions in these islands are enhancing in value, and will continue to do so. It is remarkable how the prices fluctuate. On some days provisions of all kinds will be exceedingly cheap, and almost any article will be taken in exchange; and then again nothing can be found to please the natives, or induce them to trade, although the quantity for sale is equally as great. It was not a little amusing to see the natives sitting whole days to obtain the price of their fowl or pig, and persisting in their refusal of the offer made; and this was sometimes done by a large number at the same time, all remaining true to each other until their poe or food became exhausted, when they would take the earliest opportunity of disposing of their different parcels.

In the grove near the village, we saw several piles of stones. I was told they were the graves in which they formerly buried the dead, just below the surface. On the top were placed stones, forming a high pile. Now they bury their dead in graves about three feet deep, and enclose them with the Dracæna, which grows rapidly, and forms a pretty and neat trellis.

Toa became quite communicative, and as he showed me about his village, he told me, through the interpreter, that before the missionaries came, the chiefs all had their "aitu" or spirits, which they worshipped, and that they felt themselves obliged to do every thing they commanded. His aitu were fresh-water eels, which he constantly fed in the brook near the village. I visited it, and requested him to catch one, which he attempted to do; but after a long search, turning over large stones, and examining holes, he was unsuccessful. He said there were many in it formerly, and quite tame; but since he had embraced Christianity, they had all been caught and destroyed. On farther questioning him, he told me that he had himself eaten them; and that formerly if any one had touched, disturbed, or attempted to catch one, he should have killed him immediately. He said his eels were very good to eat, and was sorry he could not find any more; and laughed very heartily when I spoke to him about eating his aitu. I mention this circumstance to show the powerful effect the Christian religion has had upon the ancient customs of this people.

After much persuasion, they were induced to sing some of their old war-songs. Mr. Drayton wrote one down as a specimen of their music; the words were written by one of the interpreters.

To the above they sing a kind of second, with very correct harmony. They do not seem to have any particular air among them, and in singing the above, they did not sound the same notes every time. All their music sounds alike, and the above will give a good idea of it. A translation of the song was made by the same interpreter, and is as follows.

A chief of Samoa attacks an enemy on another island and conquers. After the victors have embarked safely for their island, they sing as follows:

"Keep her away, and mind the helm."

And when they get home, the people sing,—

"We are glad you have come to your island of plenty,
We have waited a long time for our chief and canoes."

Toa, after his unsuccessful search for his favourite eels, went into the brook for a bath, which he told me he very frequently did during the day; and it was delightful to see the pleasure he took in it. The natives, indeed, are almost constantly in the water, and, consequently, very cleanly in their persons. Finding that it occupied too much of their thoughts on the Sabbath, bathing on that day has been forbidden.

This village contained about forty houses, of a large and commodious size, and about two hundred inhabitants, a number of whom were absent on a visit to Upolu.

Towards evening, we took our leave of Toa, thanking him warmly for his kindness; we were escorted to the outside of the village by his friends and relations, whilst Toa himself accompanied us to Pago-pago.

The natives have no fixed time for meals, eating whenever they feel hungry. Their food consists of pork, fish, bread-fruit, cocoa-nuts, bananas, &c., but principally of taro. All of these are produced in abundance. Water is their common drink, and, notwithstanding cocoa-nuts are so abundant, the milk is seldom used: the trouble of procuring them is too much for them. They use ava made from the Piper mythisticum, and it is the only intoxicating drink they have.* It is never used to excess, although old and young, male and female, are very fond of it. The taste, to one unaccustomed to it, is not pleasant, being somewhat similar to that of rhubarb and magnesia. Their mode of preparing it is the same as has already been described.

They sleep on the large coarse mats with which they always cover the floors of their houses. Over these they spread coloured tapas, some of which are also used for nets of protection against the numerous musquitoes. For a pillow they use a piece of bamboo supported on small legs. Their hair is frequently shorn close, and coral, lime, or ashes sprinkled over it to destroy the vermin, which are generated in great numbers in their tapas and mats.

According to old Toa, a native is in a comfortable condition when he has a good house; a well-made visiting canoe; a neat, handy, large and well-formed woman for a wife; a taro-patch with a good fence; cocoa-nut, and bread-fruit trees, with a reasonable number of pigs.

The women are now admitted to the same privileges as the men. The chiefs have still great power over the people, although the influence of the missionaries has tended greatly to diminish it. Most of the people look back to the days when polygamy existed with regret,

* The ava does not, according to the whites, intoxicate in the same manner as ardent spirits, but produces a temporary paralysis, tremors, and a confused feeling about the head, indistinctness and distortion of vision, somewhat resembling the effect of opium.

and cannot understand why they are restricted to one wife. They say, "Why should God be so unreasonable as to require them to give up all their wives but one for his convenience?" They pay just attention to their religious duties; morning and evening prayers are always said, as is grace before their meals, and with a devotion rarely to be seen among civilized men.

Their amusements seem to be few; their books are constantly before them, and a great portion of their time is employed over them. Old gray-headed men may be seen poring over the alphabet, and taught by some of the youngest of the family. The employment of the men is to cultivate and weed the taro, and to take care of the fences; they also make sennit for their houses, and canoes for fishing. The women are engaged in making mats, and the boys and girls play, and wait upon their seniors.

Next to study, fishing is their great employment. This is performed by driving the fish towards the nets in shoal water, where they are easily caught. The cast-net is also used.

The only amusement we saw, is a game called lafo-tupe, which is played with cocoa-nut shells, and resembles shuffle-board.

Mr. Murray is an amiable as well as a truly pious man, and the natives have imitated the example set by him. He studiously avoids any intercourse with them in the way of trade or barter, except so much as is necessary for the provision of his own family, and devotes his whole time to preaching and teaching the gospel. He is one of the missionaries engaged in translating the Bible, many parts of which are now completed, and extensively used by the natives, many of whom read and write well.

Their observance of the Sabbath is very strict; and it is impossible to get a native to do any thing whatsoever on that day, but perform his religious duties. They attend church regularly. In Mr. Murray's congregation there are about thirty communicants, and nearly one thousand attendants on public worship. They come from many of the surrounding villages. Mr. Murray has been here about three years, and the native preachers nine or ten; he is well acquainted with the difficulties of his station, but seemed to feel assured that his exertions were about being crowned with success. He represented to me that the natives were very tractable, and desired exceedingly to be taught; that they had much application, seemed to comprehend many things, and were certainly not surpassed in intelligence by any of the natives of Polynesia.

Polygamy, which formerly was practised to a great extent, still exists among those who have not been converted.

Circumcision is practised among them.

They carry their children in the same singular manner on the hip, as was shown in wood-cut of the low archipelago. They are early betrothed, without regard to age, the girl being saa, or tabooed, until of marriageable age. During the intervening time, all kinds of native property are accumulated, such as mats, &c., for the bridal day. Two days previous to it, the inhabitants of the district are gathered together for feasting and dancing. On the third day, the bride is produced before the assembled multitude, and the ceremony attendant on marriage that was customary among the Jews performed. After the marriage had been consummated, the dowry was exhibited, and each article being held up it was proclaimed by whom it was presented; the multitude, having consumed all the eatables, and exhausted their strength in rioting and debauchery, dispersed.

Infanticide has never been practised on this island.

I have seldom seen a more devout or attentive collection of people than I observed at times in the church meeting, which was held in the council-house at Pago-pago; the new church was undergoing alterations; for on its being completed, it was found it would not accommodate the congregation, when they determined to enlarge it.

Upon the conclusion of a long service, they were observed to divide themselves into three parties; one remaining in the church, and the other two repairing to different buildings. The object of this was, that they might listen to instructions from their native teachers explanatory of the sermon, and also receive exhortations to put away all that is unbecoming to the Christian character. The afternoon is employed in further explanations and examinations by the missionaries. The native missionaries have also meetings on Fridays.

Their mode of singing hymns is peculiar, the whole mass joining in some parts, with all the lungs they could muster. This exercise appeared to afford them great delight. The congregation were mostly dressed in tapas, or clothed in one sort of garment or other; but the person who attracted our attention most, was the consort of Pomale. From being the wife of the most influential personage, she had received more presents from us than any other; and she endeavoured, on this occasion, to display on her person the greater part, if not all, that she had thus acquired. These consisted of a red calico gown, four or five petticoats of different colours, woollen socks, green slippers, cap and bonnet, a large plaid blanket shawl, and a pair of polar gloves, the whole surmounted by a flaming red silk umbrella—and this with the thermometer at 87°! It was difficult to keep our eyes off her during the service, and before the end of it, all her finery became awry. The other natives also seemed to have the desire of exhibiting their acquisitions, though these consisted frequently of no more than a vest, or a pair of pantaloons, without shirt, or occasionally of a long-skirted coat, without either of the former garments, so that a small roll of tapa was needed to cover their nether parts.

Some unauthorized attempts were made to induce the natives to break the missionary laws, by offers of great value in their eyes; they were told the missionaries would not see them. On understanding which, they pointed to the heavens, and replied, "There missionary see." This was conclusive, and a just and severe rebuke.

The Peacock and Flying-Fish again joined us on the 18th of October, in eight days from Papieti. Orders were at once given them to proceed to Upolu, to commence the survey of that island. (See Appendix VI.) They did not sail, however, until the 20th, having been detained by the winds. The harbour of Pago-pago, though easy of access, is extremely difficult to leave, in consequence of the southeast trade-winds blowing directly in, and rendering it necessary to make short tacks. Indeed, a vessel no sooner gets headway on one tack, than it is found necessary to tack again. The sea is often heavy at the mouth of the harbour, and the shore is lined with a narrow coral reef all around it. I was glad to see the Peacock safe outside, after beating about four hours.

During our stay on this island, the whole was examined, the harbour surveyed, and the principal heights determined. Tide-gauges were kept on the north and south sides, and the observations for magnetic dip, variation, and intensity, made. The temperature during our stay of fourteen days varied from 73° to 88°; the mean temperature was 80·50°.

The climate of Tutuila is mild and agreeable, particularly at Pago-pago, where the temperature is lower than it is elsewhere on the island, in consequence of its generally being overshadowed with clouds that hang on the high land. There is usually a fine breeze, which sets in about ten o'clock, and continues until sunset. The nights being calm, much dew falls in fine weather. We had little fair weather during our stay, and the prognostication of the natives proved too true, respecting the difficulty of seeing the sun and stars. The wind at times was very strong, almost a gale, accompanied by light rain and mist. I was informed that there is a good deal of rain during the year, but seldom such a continuance of it as we experienced. There does not appear to be any particular rainy season, but they are liable to these high winds during the winter months, or from October to March.* I obtained from the pilot a register of the weather from January, 1839, till October of the same year, which will show more clearly the state of the climate. This will be found in Appendix VII.

In our explorations, nearly all the villages of this island vere visited by some of the officers of the squadron, and from their report they much resemble each other. Those of Fagaitua and Leone, on the southern coast, are the largest, and are more of the Devil's towns than the others. One of their customs is truly savage. They seldom use pork as a food, consequently it is a great rarity with them; but at intervals of several months the villagers assemble at a feast, at which thirty or forty hogs are killed, when they gormandize on them for four or five days, or as long as the food lasts. The whole is eaten, entrails and all. Fish and taro are the principal food, and large numbers of the natives may be seen fishing off the coast in fine weather. The kind of fish usually caught are mullet.

There is a large kind of worm which they esteem a great delicacy, and which is eaten with much relish. It is impossible to see them sucking down the entrails of the biche-de-mar, holithuria, and echina, without disgust. They also eat many of the shell-fish that are found on the shore.

The temperature found on the top of Matafoa, at the altitude of two thousand three hundred and fifty-nine feet, was at 4 P. M., 69·4°, whilst that on board the ship was 79·5°.

We made an endeavour here to search the reefs at night for shells, with flambeaux or torchlight, after the manner of the Chain Islanders, by which means it is said that many species of shells are taken, which are never seen by daylight. We cannot vouch for this being the case, our experiment not having succeeded. The leaves of the cocoa-nut were either too green or too wet to burn. If success really attends this method, it is a singular trait in the economy of mollusca, which are generally supposed to be partial to daylight. It was my determination to make another trial, under more favourable circumstances; but from our constant occupation and fatigue of the crew in the daytime, we were unable to renew the experiment.

A few days before leaving Pago-pago, Mr. Murray brought to my notice the account of a murder that was supposed to have been committed on a foreigner at the west end of the island, for the sake of the little property he had about him. The report, however, appeared to me to be too vague to authorize any delay for the purpose of making an

* During eleven days of our stay, the quantity of rain that fell was $4\frac{6}{10}$ inches.

examination into it; and finding the man was reported to be a runaway convict, I had no right to interfere in the affair, and therefore, took no steps to inquire into it.

On the 7th of November, 1837, this harbour exhibited one of those remarkable phenomena of the oscillation of the tidal wave. The observations made on it are extracted from the letter of a missionary resident at Pago-pago, to the Rev. Mr. Mills, of Upolu, who obligingly gave me permission to copy them. They will be found in Appendix VIII.

The weather during the preceding evening was boisterous, with frequent squalls from the east, which continued till 7 A. M., from which time the day was cloudy, with frequent light showers. After 5 P. M., it continued to rain until ten o'clock at night. On the 8th, the tide continued to ebb and flow in an irregular manner. The day was fine and very warm. This phenomenon does not appear to have been observed at any other place in the Samoan Group, but was experienced, as will be noticed hereafter, at the Group Hawaii.

The peculiar formation of the harbour of Pago-pago, would make it more likely to be observed there than elsewhere. The ordinary rise of the tide is no more than four and a half feet, and neither before, during the continuance, nor after this phenomenon, were any shocks of earthquakes observed in any part of the group where missionaries are settled.

NAVIGATOR CLUBS, ETC.

CHAPTER IV.

CONTENTS.

DEPARTURE OF THE VINCENNES FROM TUTUILA—HER NARROW ESCAPE FROM WRECK—APPEARANCE OF UPOLU—MESSAGE FROM CAPTAIN HUDSON—CASE OF TUVAI—COUNCIL OF THE CHIEFS—ARGUMENTS IN BEHALF OF TUVAI—CAPTAIN HUDSON'S REPLY—DECISION OF THE CASE OF TUVAI—HIS BEHAVIOUR—COMPLAINTS OF THE NATIVES AGAINST WHALERS—PEA'S VISIT TO TUVAI—OUTRAGES OF OPOTUNO—VISIT TO MR. WILLIAMS—INTERVIEW WITH MALIETOA—HIS DAUGHTER—REQUEST FOR A COUNCIL OF CHIEFS—THE VINCENNES ORDERED INTO PORT—SURVEYING PARTIES—LAKE OF LAUTO—ITS LEGEND—SUPERSTITION REGARDING IT—MOUNT TOFUA—SUBTERRANEAN STREAMS—GEOLOGICAL STRUCTURE OF UPOLU—STREAMS OF LAVA—CAVERNS—SWALLOWS AT SANGA—WATERFALLS—GIMBLET RELIGION—ATTEMPT TO CAPTURE OPOTUNO—FONO, OR COUNCIL, OF HIGH CHIEFS—ITS OBJECT—MR. J. WILLIAMS RECOGNISED AS CONSUL—REGULATIONS DISCUSSED AND ADOPTED—DEMAND FOR OPOTUNO—FONO ADJOURNED—ITS SECOND MEETING—SPEECH OF MALIETOA'S ORATOR—REWARD OFFERED FOR THE APPREHENSION OF OPOTUNO—TERRITORIAL DIVISIONS OF UPOLU—TYRANNY OF TAMAFAGO—WAR OF AANA—DESOLATION OF THAT DISTRICT—ISLAND OF MANONO—ISLAND OF APOLIMA—PROCEEDINGS OF THE PORPOISE AT SAVAII—DR. PICKERING AND LIEUTENANT MAURY LAND THERE—SURVEY OF THE COAST OF SAVAII—NATIVE MISSIONARY AT SALACHUA—FELIALUPO—ASAU—BAY OF MATAATUA—PECULIARITIES OF ITS INHABITANTS—SAPAPALE—DR. PICKERING'S JOURNEY IN THE INTERIOR OF SAVAII—CURIOSITY OF THE NATIVES—FISHING ON THE CORAL REEF—DESCRIPTION OF SAVAII—ITS PEAK—ITS INTERIOR—ITS WANT OF STREAMS—ITS CORAL REEF—ITS SOIL—THE PORPOISE PROCEEDS TO TUTUILA, AND THENCE TO UPOLU—REUNION OF THE SQUADRON—REVIEW OF THE MARINES—EARTHQUAKES.

EMMA MALIETOA

Engraved by F. Halpin from a sketch by A. T. Agate

CHAPTER IV.

UPOLU—MANONO—SAVAII.

1839.

The surveys of the island of Tutuila having been completed by tne 23d November, we made preparations for our departure, and on the 25th we weighed anchor. In leaving the harbour we had a narrow escape from wreck; the almost constant southeast wind, which is fair to a vessel entering the bay, and makes it easy of access, is ahead on going out, which renders egress difficult; it therefore becomes necessary to make frequent tacks, and a vessel must be well manœuvred to escape accident, for to miss stays would be almost certain to bring about shipwreck. When we beat out, the wind was light, and it failed altogether just as we reached the most dangerous part of the channel; we were in consequence brought within an oar's length of the reef, on which a heavy surf was breaking. The moment was a trying one, and the event doubtful; all were at their stations, and not a word was spoken. Of my own feelings on the occasion I have no very precise recollection; merely remembering that I felt as if I breathed more freely after the crisis had passed and we were in safety.

The afternoon was fine, and we sailed along the southern shore of the island, admiring its diversified surface, its luxuriant groves, and the smiling villages that crown its bays. Where the valleys come out from between the ridges to the shore, there is usually a level plain extending inwards for a couple of miles; these plains are occupied for the most part by groves of cocoa-nut and bread-fruit, beneath whose shade lie the dwellings of the natives. Many of the inhabitants were abroad in their canoes, employed in fishing; some of them scarcely seemed to notice the ship, passing them rapidly with all sail set, while others appeared to regard her with intense curiosity. In the evening we had much lightning, but no thunder.

The distance between Tutuila and Upolu, of thirty-six miles, was soon passed, and in the morning we were delighted with the view of the latter island as we ran down its coast to the westward. It appears much richer and more fruitful than the other islands of this group, and may be described as of moderate height, rising gradually in a succession of ridges from a low shore; here and there, broad and fertile valleys are seen, with numerous streams falling from the mountains in cascades. The eastern portion of the island is much more rugged than the western; the main ridge runs east and west, and ridges or spurs run back to it from the northern coast in a southeast direction. Between these lateral ridges are broad and fertile valleys, decreasing in width as they recede from the coast. The shore is lined with a coral reef, which is now and then interrupted by channels, and forms snug and convenient harbours.

At noon we descried the Peacock lying in the harbour of Apia, and shortly afterwards I received a message from Captain Hudson, saying that my presence was required on shore. In the hope that it was not a business of such a nature as to cause detention, I left the Vincennes in the offing, while I went ashore in my boat. On reaching the land, I found the chiefs engaged in the trial of a native called Tuvai, who had killed an American named Edward Cavenaugh, a native of New Bedford.

It appeared that on Captain Hudson's arrival the murderer was pointed out to him in the village, upon which he very properly determined to have the offender punished, and gave orders to have him arrested. He was in consequence seized in a house near the water, and carried on board the Peacock. Being taken by surprise, he offered no resistance to his capture. Captain Hudson then requested a conference with the neighbouring chiefs, who in consequence had assembled on the 27th.

The fono, as such assemblies are called, was held in the council-house, or fale-tele, where the chiefs were collected. The Rev. Mr. Mills acted as interpreter on the occasion. Captain Hudson, through him, stated that the object of his having requested them to assemble was to bring the accused to a trial before them, in order that if his guilt were established, he might be brought to condign punishment: he then pointed out to them the guilt and consequences of the crime of murder, and declared the course he had considered it his duty to adopt. The chiefs listened attentively to this address, and in reply, through the principal one, admitted that the man taken was in reality the guilty person, a fact known to every person upon the island. Captain Hudson then stated to them that it was absolutely necessary that Tuvai should be promptly punished, in order that others might be deterred from the commission of the same crime. He suggested, however, that in spite of the universal belief in Tuvai's having committed the crime, it was proper that he should undergo a trial, or at least an examination, in order that he might have the privilege of being heard in his own defence.

This suggestion being approved, Tuvai was brought on shore under a military guard, and placed in the centre of the building. He was an ill-looking fellow, of about twenty-eight years of age, and manifested no fear, but looked about him with the greatest composure.

The trial was simple enough: he was first asked by the chiefs whether he was guilty of the crime, to which he answered that he was; being next asked why he had committed it, he replied that he had done it in order to possess himself of the man's property, (clothes and a knife.)

The chiefs, among whom was Pea, of Apia, to whom the criminal was distantly related, made every effort in their power to save his life; stating that he was in darkness, and therefore unconscious of the guilt of the action, when he committed the murder; that as they had but just emerged from heathenism they ought not to be subjected for past actions, to laws they knew not; that these laws were made for people who occupied a more elevated station; that Tuvai was a poor man of no account, and was not a person of sufficient importance to be noticed by a great people like us; that faa Samoa (the Samoan fashion) did not allow men to be put to death in cold blood, but that after so long a time had elapsed, as in the instance before them, it admitted of a ransom.

Pea went on to say, that many bad acts had been committed upon natives by white men, with impunity, and asked whether the Christian religion sanctioned the taking of human life. He then appealed to our generosity to pardon the present crime, and assured us that no such offences should be committed in future.

Pea had one of those countenances which exhibits all that is passing in the mind. It was amusing to see him at one time exhibiting a picture of whimsical distress at the idea of being compelled to put his kinsman to death, and immediately afterwards laughing at something ludicrous which had occurred to him.

Pea was seconded in his endeavours by Vavasa, of Manono, one of the finest-looking of the chiefs, whose attitudes and movements were full of grace, and his manner exceedingly haughty and bold.

In reply to their arguments, Captain Hudson stated, that however freely other sins might be forgiven, in consideration of their late benighted state, even the darkness of Paganism could not extenuate the crime of murder. He told them that the Scriptures said, "Whoso sheddeth man's blood, by man shall his blood be shed;" that nothing but the life of the offender could satisfy the demands of justice, and that they must execute the criminal themselves.

This announcement caused much excitement; the chiefs again asserted that they knew no such laws; that by the customs of Samoa, the anger of the friends and relations of a person who had been killed was to be appeased by a present from the criminal or his relations, and by a form of submission, which consisted in knocking their heads three times on the ground. To this it was replied, that the guilt of the prisoner had been proved and admitted—he must die.

The chiefs, after much reluctance, consented, but expressed great repugnance to an immediate execution. They urged in the most strenuous manner, that the criminal should be carried on board ship, and executed there, or that he should be taken to some uninhabited island and left. These alternatives were refused by Captain Hudson, and the chiefs seemed in great distress.

At this point of the discussion, the Vincennes was announced as being in sight, and the proceedings were suspended. An officer was immediately despatched, who, as has been already mentioned, boarded that vessel off the harbour.

When I landed, I found the assembly anxiously awaiting the result of my arrival. Captain Hudson and myself had a private interview, in which he detailed all the facts, and stated that it had been his intention to compel the chiefs to make all the preparations for the execution, but before it was carried into effect to come forward and reprieve the criminal, at the same time requesting Mr. Mills to make an appropriate speech, stating the reasons for the pardon.

After a full discussion of the whole subject, we came to the conclusion, that it would be best to transport the criminal to some other island; for it appeared probable that this would have a better effect than even his execution, as it would be longer remembered, while to cause him to be put to death might naturally excite a desire of revenge.

This decision was at once communicated to the chiefs, with a statement, that in conformity with the laws of Tahiti in such cases, Tuvai should be transported to a desert island, where he would never again have an opportunity of killing a white man. The chiefs, although evidently relieved from the most intense part of their distress, were still much affected by this decision.

The prisoner was then ordered to be taken on board the Peacock whither he was followed by a crowd of natives, with many tears and lamentations, among whom his wife was the most affected. Among others, Pea, the chief of Apia, to whom, as has been stated, the prisoner was related, was very much distressed and excited. Unable to vent his rage and trouble in any other manner, he spent it upon the crowd around him, striking in all directions with a huge stem of a cocoa-nut leaf, by which he soon dispersed them. I felt a curiosity to see what effect the sentence would have upon the prisoner. Death he would have suffered without uttering a murmur; but when he heard he was to be taken from his native land, his firmness was overcome, and he was observed to shed tears. He made no resistance to his being removed on board ship, but after he got there he said he would rather be put to death and buried in his own native island, than banished to a desert one.

After this difficult business was arranged, they brought their own grievances before me, and particularly their complaints against the American whalers. They said that some of them had evaded their port-charges, and refused to pay for the provisions with which they had been furnished. To this I replied that I was ready to indemnify them for their losses, and should ask no other proof of them than their own statement. They appeared struck with the unexpected liberality of this offer; but, after consultation, as if to manifest a corresponding feeling, declined to accept it. I then informed them that their port-charges for the squadron should be paid, which gave much satisfaction, particularly to old Pea, who would derive the principal benefit from them. The fono then broke up in great good humour.

Pea and some of the other chiefs were very anxious to hear from me what sort of an island Tuvai was to be put upon. They asked many questions in relation to it, and always among the first, whether there would be any cocoa-nut trees, Nature's first and best gift to them, upon it. Wishing to make the intended punishment as terrible as possible to them, I always replied that there would be none whatever.

After Tuvai was again on board ship, old Pea paid him a visit, in the course of which the former melted into tears, howled bitterly, and begged that he might be taken on shore to be put to death, in order that his body might be buried in his native soil. It appeared from information that we received, that this was a part of a concerted plan to obtain a farther commutation of his sentence, and that this affecting interview was got up in order to excite our sympathies. Finding it did not produce the desired effect, old Pea went about the ship with a doleful visage, exclaiming, "Eoloisa-ia-tu Tuvai"—have compassion on Tuvai.

I was in hopes to find the surveys of Upolu nearly, if not quite finished; but the Flying-Fish, which was to have aided in performing them, had not yet been seen or heard from. This was no small disappointment, as it might compel me to bring the Vincennes into the harbour, and thus incur a serious delay.

Before I had decided upon this step, I learned that a chief of the name of Opotuno, whose capture had been considered so important by our government that a ship of war had been despatched for the express purpose, had again become troublesome, and was threatening vengeance upon all the whites who might fall in his power. I therefore determined to make an attempt to obtain possession of his person by stratagem. Lest, however, such an attempt should create disturbance in the island, or be productive of injury to the white residents, I determined, before putting my purpose into effect, to have an interview with the Rev. Mr. Williams, the principal missionary in these islands, both to consult as to the best mode of accomplishing this object, and to learn what effect it would be likely to have on the operations of the missionaries.* I accordingly set out for his residence at Fasetootai, about twenty miles to the westward of Apia, in the hope of seeing him. Mr. Cunningham, H. B. M. Vice-consul, was kind enough to accompany me.

We left the Peacock at sunset, and reached Mr. Williams's snug cottage about midnight. Nothing could be kinder than the welcome he gave us; and the pleasure he expressed at our visit soon made us feel at home. He gave us supper, and provided us with comfortable beds. Shortly after our arrival, another party was welcomed, consisting of three ladies and a gentleman of the mission, who were in like manner provided for, without apparent inconvenience.

Mr. Williams seemed to me exactly what a missionary ought to be, pious, cheerful, and meek, although resolute. His whole thoughts seemed to be directed to the welfare of those whom he had undertaken to enlighten. His views were pointed not only to the diffusion of the gospel, but also to the extension of the useful arts, and whatever could tend to elevate the condition and eradicate the vices of the natives.

After a long consultation, Mr. Williams came to the conclusion that there was no reason for fearing that the arrest of Opotuno would be the cause of any injury to the whites or missionaries. He said that Opotuno was a blood-thirsty fellow, and that it would be doing the islands a great service should he be removed; that there was not a

* Mr. Williams is the author of the well-known Polynesian Missionary Researches, and it will be our melancholy office hereafter, to speak of his falling a martyr in his efforts to propagate the gospel.

shadow of doubt that he had murdered twelve whites, of whom several were Americans; that he was a determined enemy to the whites, and in the habit of saying that he would omit no opportunity of killing all who might come within his power. Mr. Williams, however, doubted the success of any attempt to take Opotuno, unless it was made under disguise; for upon the approach of all men-of-war, and during their stay, he lived in the mountains of Savaii, where it was impossible to find him.

The situation of Mr. Williams's cottage is pretty; it stands within a few rods of the beach, and is surrounded by a nicely-dressed lawn, on which are several fine trees; the background is filled up with cocoa-nut, bread-fruit, and a variety of other trees. Near by is the tiny ship-yard of his son, Mr. John Williams, who was taken by his father to England, and there taught all the mechanical trades. He has returned thence within a few months, with his wife, and by the aid of a few natives has already built himself a vessel of about twenty-five tons burden, which he proposes to employ in trading among these islands.

The next day we returned to Apia. On our way we stopped at Sagana for the purpose of visiting Malietoa, the principal chief of the Malo or conquering party.

Sagana is a neat settlement, and is regularly laid out; it is situated on a small peninsula, across whose isthmus a stone wall has been erected, for the purpose of protecting the plantations within it from the swine. The village contains about six hundred inhabitants, and there is a school composed of about fifty scholars kept by Mr. Wilson, one of the missionary teachers, son of the missionary at Matavai Bay.

No preparation had been made to receive us, for I came unannounced; nor, indeed, had it been my intention to stop, but hearing that this was the residence of Malietoa, and that he was at home, we paid him a visit. He was well advanced in age, and it was generally remarked that he bore a striking resemblance to General Jackson. The resemblance is not confined to that of person only; for Malietoa possesses also not a little of the same energy of character.

I have rarely seen a place where more attention is paid to cleanliness than at Sagana. A similar regard to neatness prevails in the walks around the village, and in the cultivation of the taro, melons, and bananas, which is carried on in the immediate vicinity. The paths leading to these cultivated grounds pass through fine shady groves. The preservation of the broad walks and paths appears to be rather an amusement than a labour to the villagers.

Here Malietoa was seen in his domestic circle, with his wives and children around him. I found him in a small house, enjoying the afternoon breeze, with his daughter playing about him. She was about fifteen years of age, and decidedly the prettiest girl we had seen in this group; her name was Emma, and she was as intelligent as she was pretty.

The chief, whose hair was white with age, made us warmly welcome, and wished to go over to his fale-tele to receive us as became chiefs, but this I would not permit. His wives busied themselves in getting things in order, very much after the fashion of other parts of the world, when a stranger arrives unexpectedly. In a few minutes the fine mats were laid, the stools, calabashes, and straw put away. A clean shirt was slipped over the old man's head while my attention was called off to another object.

Malietoa's house was not larger than the others in the village, and exhibited no other difference from them than in containing a dais or platform, occupying about a third of it, and raised about a foot higher than the rest of the floor.

When the domestic arrangements were completed, large bunches of bananas and fresh cocoa-nuts were brought in and presented to us. Mr. Wilson was an excellent interpreter, and by his aid I had a long and agreeable talk with the old chief, who, when his wars were touched upon, appeared full of fire and animation.

I intimated my desire to have a conference with the ruling chiefs, for the purpose of transacting business, whereupon he readily assented to call a fono, and appointed the 4th of November as the earliest day on which he could possibly get the chiefs, a part of whom must come from Savaii, together. This day he named himself, after having made a reckoning of the six intervening days upon his fingers; I observed, however, that he found it necessary to repeat the count several times. Having transacted this business with him, and regaled ourselves on his hospitable fare, we took our leave.

On reaching the Peacock, I found that none of her surveying parties had returned, and the Flying-Fish was still missing; I thus became satisfied that I should be detained here for several days. I therefore sent orders for the Vincennes to make for the harbour, where she anchored in the afternoon, near the Peacock.

The next day, parties were despatched in various directions, so as to bring all parts of the island under examination at the same time.

One of these excursions was made across the island. On arriving at the highest point of the ridge, between Siuma and Siusinga, which has an elevation of two thousand and fifty feet, and just before the descent began, a clearing was found, in which were two mounds of earth, each about fifteen feet high, and one hundred and twenty feet in circumference; several stone walls were also seen. In respcct to these, there is a tradition that they were built by the warriors of Vavao, who invaded Upolu, and after their predatory warfare along the coast was over, occupied this commanding position for the purpose of cutting off the communication between the opposite sides of the island. The trees growing on these mounds are nearly two feet in diameter, and the missionaries have inferred from their inquiries that the invasion referred to occurred seventy or eighty years ago.

Messrs. Dana and Couthouy visited a lake called Lauto, which lies to the westward of this pass, and in the centre of an extinct crater. The edge of the crater was found to be two thousand five hundred and seventy feet above the sea, and the descent thence to the water of the lake is one hundred and twenty feet. These gentlemen succeeded in obtaining a line of soundings across the lake, by cutting down trees, and forming a raft of them. They found the depth in the middle nine and a half fathoms, decreasing thence gradually in all directions to the shore. The form of the lake is nearly circular, and it has a subterranean outlet. The hill in which this crater is situated is conical, and there is a low knoll at some distance to the south of it, which is the only other elevation in the neighbourhood, above the general height of the ridge.

The border of the crater is clothed with the usual forest foliage of these islands, which, however, exhibits here more than usual beauty, being decorated with the finely-worked fronds of the arborescent ferns, in widely-spread stars, and the graceful plumes of a large mountain palm.

The poets of the island have appreciated the beauty of the place, and allude to the perpetual verdure which adorns the banks of the lake, in the following line:

"Lauuto'o e le toi a e lau mea."
"Lauto, untouched by withered leaf."

There is a legend connected with this lake, that has more of poetic eauty and feeling than one would have supposed to exist among so rude a people. It is as follows.

Many generations since, during a war between Upolu and Savaii, a number of war-canoes from the latter island crossed over to attack Ulatamoa (or, as it is now called, Ulumoenga), the principal town in the district of Aana. At the time of their approach, two brothers, To'o and Ata, chanced to be paddling their canoes in the channel between the reef and the shore, and before they could reach the land were attacked by a party of Savaiians. After a valiant defence, Ata was overpowered and slain, while To'o narrowly escaped the same fate.

Overwhelmed with sorrow at the loss of a brother whom he tenderly loved, To'o retired to a neighbouring mountain, and burying himself in the darkest recesses of its forests, made them resound with his bitter lamentations. At length in his wanderings he came to the summit, where, stooping down, he scooped out with his hands a vast hollow, and, leaning over its brink, suffered his tears to fall in until it was filled. The lake thus formed has ever since borne the appellation of Lauu-to'o.

The regard of To'o for his brother's memory was further evinced by his adoption of Ata's name, conjoined to his own as his family title, and the appellation of Toomata, a contraction of To'o-ma-ata, is retained by his descendants, who are still chiefs of note in Upolu, and from whom the tradition was derived.

The lake of Lauto is regarded with superstitious dread by the natives, who believe it to be the abode of the spirits, who, in former times, were regarded with great veneration, and worshipped. These were supposed to inhabit the waters of the lake, in the shape of eels, as thick as a cocoa-nut tree, and two fathoms long. The attempt of our gentlemen to explore it was looked upon as such a profanation that their native guides left them, and regarded them as persons doomed to accident if not to destruction. The eels were represented as so savage and fierce that they would bite a person's leg off. No eels, however, nor any other fish, were seen in the lake.

In the neighbourhood of the crater no rock was observed in place, nor any light scoria. Only a few fragments of stone were scattered about.

The cone of the crater of Lauto, is flatter than the others of the same character that were visited, and particularly than that of Mount Tofua. This is the westernmost of them all, and lies behind Fasetootai. It rises so boldly, that it is seen distinctly from the sea. This, with all the other craters, are situated upon the central ridge, and the most conspicuous of those which remain, are Siusinga, which lies behind Sagana and Faliata. There is also one upon Mount Malata, in the rear of Fangaloa, and another on the southern side of the island, near Salomana.

The part of the ridge on which Tofua is situated, is much lower than the cone itself, and has gradually declined from its eastern end. The ascent from Fasetootai has, for the first three or four miles, an almost imperceptible rise; after this, the slope increases rapidly until it becomes quite abrupt. Even in the steepest parts, however, the rock was rarely visible, but is covered with a deep and fertile soil, arising from its decomposition, and mixed with vegetable mould. The whole declivity, from the very summit of the peak to the shore of the sea, is, like other parts of the island, clothed with a dense forest, which shuts out all view of the ocean, except from the top of the mountain. The ridge whence the cone rises was found to be one thousand one hundred and fifty feet above the level of the sea, and the angle of ascent thence upwards, was from 40° to 50°. The top of the peak, which forms the edge of the crater, is not more than fifteen feet wide in any place, and sometimes not more than half as much. It is of uniform height, and has a circular form; the cavity within it was estimated as having a circumference of about two and a quarter miles, and occupies the whole summit. The depth of the crater was found to be three hundred and sixty feet, and the whole of its interior is filled with lofty trees. The slope of the inner declivity was 60°. At the foot of this is an uneven plane, covered with earth and loose cinders or scoria. Upon this grow forest trees, many of which were more than one hundred feet in height, beneath which is a dense growth of shrubbery.

It was remarked, that although it had rained constantly for several days before this crater was visited, no water was found within it. This is accounted for by the fact that many of the brooks and streams on the island are subterranean throughout their whole course, while others are partly so. The former gush up near the sea-shore in large springs or fountains, forming natural, or feeding artificial pools, in which the natives bathe. According to our observations, such pools are so numerous on the western shore of the island, as to occur on the average, at intervals of a mile.

The rocks of this island are of the volcanic character that might be expected from the existence of so many craters. They are principally composed of a variety of basaltic lava, in which are found augite, felspar, albite, and chrysolite. Extensive currents of lava are seen, and are particularly abundant on the southern side of the island.

It appears as if these had flowed down towards the sea-shore in various directions, and that after their outer surface had cooled, the portion that remained liquid within had run out, and left a sort of tunnel. Such tunnels are numerous, and form the subterranean courses of the streams. Some of these natural tunnels are remarkable: among them, one visited near Sanga will serve as the type of all. It was found to be a cavern nine hundred and fifty-eight feet in length, extending in a southeast direction, and to have an average width of about fifteen feet; its roof was about eight feet in height. At the termination of this cave, there was a pool of water five feet in depth.

the temperature of which was observed to be 72°, while that of the air within the cavern was 76·6°, and that of the open air was 77·4°. These streams of lava are much more frequently seen on the southern than on the northern side of the island; they are also larger on the former of these sides, where they were in some places four miles in width.

The path from Vivimanga to Sanga is called the Stone Road, formed in part of blocks of lava. There are also in this neighbourhood many caves in the streams of lava. The cave at Sanga is the largest of these, and to describe it will convey an idea of those of smaller dimensions.

The cave at Sanga was dedicated to the god "Moso," who was supposed to reside in it. The entrance was found to be closed by an artificial wall built across it, about three feet thick; it is four feet high, and six feet wide. The sides, roof, and floor of this cavern are comparatively smooth, and are covered with stalagmites of a light yellowish colour, which are not calcareous.

It had been reported that this cave was frequented by a peculiar kind of swallow, which never ventured into the light of day. Mr. Peale, who was one of the party that visited it, found swallows in abundance, which made a bat-like noise, or rather one having a resemblance to the rattling of pebbles. So far, however, from their being a peculiar species, as they had been represented, they were the common species of the islands, and instead of shunning the light of day, they were continually passing in and out of the cavern, which was merely a place selected by them for breeding. On the ledges of the sides and roof their nests were found, composed of pieces of moss glued together. The eggs were white, and of a large size in proportion to the bird, and no more than one was found in each nest.

In traversing the island of Upolu, many deep gorges were seen, in which there were waterfalls. One of these cascades was measured, and found to be seven hundred and fifty feet in height, so that the whole of the water was dissipated in spray before it reached the bottom. These glens are wild in the extreme, and beautiful, from the great variety and peculiar character of the foliage with which they are clothed.

The south side of Upolu, like that of Tahiti, is much more luxuriant than the northern, which is owing to a like cause, namely, that, it receives more moisture from the prevailing winds.

The wild orange grows every where in great abundance, and in some places the road was literally strewed with the fruit, which here equals the cultivated variety in size.

In the different jaunts across the island, many of the "Devil's" or unconverted towns were visited, where our parties were always treated with great hospitality. At the town of Siusinga the chief who entertained our party was a priest of the *Gimblet* religion. This new faith has made some progress among these islands, and has the following singular origin:

A native of Savaii, by name Seeovedi, was taken from that island by a whale-ship, and did not return for several years. During his absence he visited several ports, where it would seem he obtained some notions of the forms and ceremonies of the Roman Catholic Church. Possessed of considerable natural shrewdness, he founded on this knowledge a plan to save himself from labour for the future, by collecting followers at whose expense he might be maintained. During his absence, and while on board the whale-ship, he had received, as is usual in such cases, instead of his native name, that of Joe Gimblet; and this cognomen is now firmly attached to the sect of which he was the founder.

Having formed the plan of founding a sect, he did not scruple as to the means of carrying it into effect; for he boldly claimed a heavenly mission, professing to hold converse with God, and asserting that he possessed the power of working miracles, raising the dead, &c. He soon gained many proselytes, and had attained great consideration and authority, when unfortunately for him he was called upon to exert his pretended power of raising the dead, by restoring to life the favourite son of a powerful chief called Lelomiava, who had been murdered.

Joe did not hesitate to undertake the accomplishment of this miracle. He in the first place directed a house to be built for the reception of the body, and when it was finished he required that it should be supplied with the best provisions. In conformity with this requisition, the choicest articles of food that could be obtained were regularly handed to Joe for the use of the defunct, upon whom he alone waited, while every other person except the chief and himself was excluded from the building.

The food thus regularly supplied as regularly disappeared, and Joe assured the chief that his son had eaten it, and under this bountiful allowance would soon recover his strength, and walk forth. In this way time wore on, until the patience of the old chief began to show symptoms of being exhausted. This somewhat alarmed Joe, but as he was a fellow of infinite resources, he contrived to evade inquiry and procrastinate, hoping, no doubt, that some lucky incident might turn up, by which he should be enabled to extricate himself from the dilemma. Unfortunately for him, however, after another month of anxious suspense, the old man's pigs and taro fell short, notwithstanding the chief's dependants had for a long time been restricted from using them. All of them were in fact much reduced by their compulsory fast, with the exception of Joe, whose rotundity of form seemed to indicate that he at least ran no risk of starvation. Whether it were owing to the suspicions which his jolly appearance excited, or that he began to entertain doubts of Joe's supernatural powers, is not known; but one day old Lelomiava determined to satisfy himself of the progress making in the restoration of his son. With this design he entered the house, and was shocked with the sight of his son's body in a state of loathsome putridity. He immediately summoned Joe, and informed him that it was time that the promised miracle should be accomplished, adding, that it must be done by the morrow's dawn. Joe immediately redoubled his exertions, and prayed hastily to all the saints of his calendar. He, however, knew full well what would be his fate if he remained to encounter on the morrow the anger of the savage chief. He therefore effected his escape during the night, and made his way to his native island. There he remained for some time incog., but now ventures to appear openly, practising his impositions boldly, and is the worst antagonist the missionaries have to deal with.

This story was related by the old chief himself, who, instead of finding his son restored to life, was compelled to bury his body, which he did, with the exception of the head. This he put in a box, and suspended beneath the peak of the roof of his house, where it remains, a witness of his credulity and of the gross imposition that was practised upon him.

While the party remained at Siusinga, a sick native was brought from the coast to a neighbouring house, and their host, the Gimblet priest, was called upon to pray for him. This afforded them an opportunity that might not otherwise have occurred, of learning some facts in relation to the ceremonies of this sect.

On this occasion, the priest approached the house where the sick man lay, and when upon the stone platform in front of it, he drew forth a book from the folds of tapa in which it had been carefully enveloped. He then called upon Jehovah, returning thanks for the many blessings which had been conferred on his people, and asked for a continuance of the same, invoking the name of Jesus. He ended by inquiring the Divine pleasure concerning the sick man, and begging mercy for him.

The nature of the book could not be distinctly seen, as it was again carefully enclosed in the tapa as soon as the ceremony was over; but so far as it was visible, it bore an unquestionable resemblance to a blank note-book!

The proselytes of this sect, in case of sickness, confess their sins to one another, and have a number of fast-days, which are rigidly kept. Their Sabbath occurs only once a month, and is celebrated by the firing of guns and the puerile mummery in which their worship consists.

In pursuance of the resolution I had adopted, Captain Hudson set out on the 30th of October, with the boats of his ship, for the purpose of attempting the capture of Opotuno. This noted chief of the neighbouring island of Savaii, had, as has been stated, committed several murders and other outrages. Among other acts, he had taken possession of two boats, sent on shore by the whale-ship William Penn, Captain Swain, of Nantucket, killing the chief mate, and the two boat-steerers. The third officer of the vessel was also wounded, and left for dead upon the beach; he was, however, picked up by some females, who removed him to a hut, where, through their kind attentions, he recovered. He did not, however, rejoin his ship, but remained for some time on the island.

The most surprising part of the history of this transaction is, that Captain Toby, of the ship Swift, of New Bedford, afterwards purchased these boats from Opotuno, although he knew that chief had obtained them by murdering this captain's own countrymen.

Captain Hudson fell in with the Flying-Fish, on his way to Savaii, and took her with him, to aid in carrying on the stratagem by which the watchfulness and suspicions of the wary chief were to be lulled to rest.

On their arrival off the part of the island where Opotuno usually resides, they made for the shore under pretence of surveying, and reached the village of Setipetea, which adjoins that where he dwells. We afterwards learned that no sooner had the boats got within the reef, than he prepared for his flight to the mountains. The news of the capture of Tuvai, and the reappearance of boats from a vessel (the Peacock) which had passed about ten days before, served to put him on the alert. He had, however, become so daring that he did not at once fly, but awaited more decided indications of hostility; and when Captain Hudson, accompanied by only two men, passed through his village, having left his boats only a mile distant, he entertained the intention of shooting him. He had actually cocked his gun for this purpose, when one of his followers advised him not to fire, as he would bring great trouble on the island if he shot a chief. When the boats' crews afterwards entered Opotuno's village, the inhabitants showed much alarm, but the chief was missing. It was therefore considered advisable to make no hostile demonstrations; as no good purpose could have been effected by following him to the mountains, where it would have been impossible to apprehend him.

The boats therefore returned, and although without succeeding in the main object of the expedition, something was gained in reviving his apprehensions of being captured. His village was not destroyed, because to do so would have been no injury to him, but only distressing to its poor inhabitants. He would have laughed at the idea of his being punished by the burning of their habitations, as it is said he did so when an attempt was made, during a previous cruise of the Vincennes, by her commander, who visited his village, and burned two or three of his houses.

The impunity he has hitherto enjoyed has served to render him audacious, and it is not long since he put to death an American seaman, who had been left sick in his charge.

Opotuno is detested by his brother chiefs, not only for his aggressions upon foreigners, but on his countrymen also. Only a short time before our arrival, he seduced and carried off the wife of Vavasa. This act was considered so outrageous, and was so deeply resented, that we were informed a war was only prevented by the near relationship of these two chiefs. The Samoans regard with horror the idea of those connected by ties of consanguinity, fighting against each other.

Opotuno is not only related to Vavasa, but is the adopted son of old Pea of Manono, a connexion which was not without its effect in averting hostilities.

On the 4th of November, a fono was held, according to the appointment made with Malietoa, in the fale-tele of Apia. All the officers who could be spared from the ships were ordered to attend. Old Pea, the chief of Apia, seemed to be the master of ceremonies on the occasion. Clean mats were spread for the chiefs, and chairs and benches borrowed from the missionaries' houses were placed for us opposite to them. All the highest chiefs of the "Malo" party were present, except Pea of Manono, and two minor chiefs of Savaii. Malietoa presided. His whole demeanour was dignified, composed, and thoughtful. His personal appearance has already been spoken of, and the form of his head, his white hair, and dignified bearing, again reminded us of General Jackson. He is slender and tall, although somewhat bent by age. It was to be regretted that his dress was ill chosen, and rather detracted from the respect he would have inspired had he appeared in his native garb; he wore pantaloons, a round jacket, and a pink and white striped cotton shirt.

Tooa, the nephew of Malietoa, who acted as spokesman, and whose countenance betokened the interest he felt in the business, attracted attention in the second degree. Then came Mole, the son of Malietoa, Maletau, their general, the most renowned leader in the war of Aana, and Tai-ma-le-lagi, Malietoa's brother. There were also present a number of chiefs of less distinction, among whom was old Pea of Apia; although he was compelled to take his place, yet he did not fail to be conspicuous, not merely by his personal appearance, but by his officiousness.

The proceedings were conducted with great ceremony, but there was a marked difference between this fono, and the solemnity of our Indian councils. The Samoan assembly appeared more quiescent, the proceedings exhibited more refinement, and the customs partook of an Asiatic character.

In all such meetings, a rigid order of precedence, that seems well understood by every one, is established; all conversation is carried on in a whisper; no one is seen standing in the presence of a superior, and sitting with outstretched legs is considered indecorous. Articles were never passed over a person, and no native ever ventured to come in contact with a chief.

The background on the side of the natives was filled up with inhabitants from different parts of the island.

On the opposite side of the building, the officers of the squadron and the missionaries formed a numerous group. Among the latter was our friend Mr. Williams and his son, whom I had appointed to act as consul until the pleasure of the government of the United States was known, and whom it was intended to present in this capacity to the meeting, in order that he might be recognised formally by the chiefs, Messrs. Heath, Mills, and Wilson were also present; and Mr. Heath, who was believed to be best acquainted with the Samoan language, was kind enough to officiate as our interpreter.

The object I had in view, in requesting the fono to be called, was to procure the formal enactment of laws and regulations which might secure to our whale-ships a certainty of protection and security, and at the same time to prevent impositions being practised by them upon the native government, of which, as has been stated, complaint had been made. To the breach of these laws, it was intended that the penalty of a fine should be attached, in order to secure obedience to them.

The meeting being organized, I in the first place presented Mr. John Williams, as the consul of the United States, whom the chiefs recognised as such with great willingness and satisfaction.

We then entered upon the discussion of the proposed regulations. which were adopted in a form which promises to be mutually beneficial, being highly advantageous to them, and at the same time insuring a certainty of security to American vessels that may visit the islands they could not before enjoy.*

One of the articles referred to the redress of injuries committed by the natives, and provided for the punishment of those who had been guilty of crime, by giving them up.

Wishing to rid these islands of a pest both to natives and foreigners, I now, as authorized by the spirit of this article, made a demand for the murderer, Opotuno, and stated that a compliance with this would settle all disputes between us. This demand produced a great sensation among the chiefs, and much excitement prevailed in the meeting. Malietoa, in reply to it, expressed himself strongly in detestation of the character of Opotuno, and stated that his capture by us would give him satisfaction, but argued that the regulations now enacted could not apply to his past misdeeds, and that he would only come within its operation should he be again guilty of like crimes. He in short pleaded that the law could have no *ex post facto* bearing.

He next argued, that the inevitable consequence of any attempt on their part to seize Opotuno, would be to involve the whole group in a civil war, for he was not only a powerful chief himself, but connected with others still more so; and that a civil war was that which he most desired to avoid. He however went on to say, that so far as he was concerned, no opposition would be made to any steps on our part to secure one whom they knew to be guilty of great outrages; but he could not in any way assist.

In conclusion, he stated that the islands had, until within the few years that had elapsed since he obtained the command, been the seat of continual wars; that they were now aware of the advantages of peace, and had a just sense of the benefits they in consequence enjoyed; and declared that he should do all in his power to preserve the blessings of peace, and maintain the unwonted state of prosperity. For these blessings he ascribed high acknowledgments to the missionaries, saying that he hoped the Samoan people would in due time profit by the lessons taught them, and adopt all the improvements of the Papalangis.

Few persons have ever inspired me with more respect than this old chief, and his sentiments were delivered by Tooa in an impressive manner.

It was not my object to drive them to extremities, or to press for an

* A copy of these regulations will be found in Appendix IX.

instant decision. I also wished to give them time to reflect upon and canvass the regulations just adopted, and perceived that they began to be fatigued with the length of the conference. I therefore proposed that before they gave me a final answer in relation to Opotuno, they should take time for consideration and reflection, for which purpose I suggested that the meeting should be adjourned until the next day, which was accordingly done.

On the 5th November we again met, when the arguments urged the day before were a second time brought forward, and the necessity of their taking measures that should effectually prevent outrages upon the persons and depredations on the property of white men, strongly set before them. They met these arguments with complaints against the white men who had come to the islands or been left upon them, saying that many of them were bad fellows, and had caused much trouble. I at once told them that if they would bring these turbulent persons to me, I would take them away from the islands, and that the laws they had now assented to, were such as would secure their punishment for any future offences.

In this state of the proceedings we were favoured with a set speech from the official orator of Malietoa, an old blind chief, who stood up, supporting himself by leaning with both hands upon a long stick. In this attitude he poured forth such a torrent of words as few of us had ever before heard; and if eloquence be composed of elocution and a ready flow of language, he was fully entitled to the praise of possessing it.

As we learned from the translation of this speech, its object was to urge the necessity of going to war, in order to secure the murderer, Opotuno, for the purpose of delivering him up. This, however, was intended only for effect; for these, as we well knew, were not the real sentiments entertained by Malietoa.

This speech was made up of short and distinct sentences, was spoken in a loud voice, and contained many repetitions.

However contrary this speech may have been to the cool determination of Malietoa, it seemed to meet the popular feeling; and there is no saying what might have been the consequence, had not the missionaries contrived to check the outburst. It was now proposed that the fono should receive and publish a document, offering a large reward for the seizure and delivery of Opotuno, dead or alive. This proposition was a new source of excitement, and old Malietoa exclaimed with emphasis, 'Give me the paper!—I will put it upon my house; where all the world shall see it."

A copy was then nailed on the pillars of the council-house, which Pea was made responsible for, and others were prepared and distributed to the several chiefs.

The meeting was then dissolved, and every one present evinced the greatest satisfaction that the whole of the business before it had been concluded in so satisfactory a manner.

The island of Upolu is divided into three districts, viz., Atua, Tua-Masanga, and Aana. Each of these was formerly governed by a separate and independent chief, styled Tui. Atua occupies the eastern end of the island, which extends as far as the town of Lauli; Tua-Masanga is the middle division, and includes the towns of Siuma and Safata, on the southern shore; Aana lies west of this, and comprises the remainder of the island. The first of these districts is of the greatest extent, the second is at present the most powerful, and the third is the most fertile. The union of these districts under one general government, in which the island of Savaii is also included, is a late event. Previous to 1830, this island had suffered from the usurpation of a chief of Manono, called Tamafago, who was a great tyrant, but who had contrived to cause his person to be considered as sacred, and to impress on his countrymen the idea that it would be sacrilege to disobey, hurt, or even to touch him. After the conquest of a rival district in Savaii, he assumed the style of king of that island, "O le Tupu o Savaii," a title which Malietoa now enjoys, but without deriving from it any power.

Tamafago not only ruled at Savaii with royal and divine attributes, but obtained a complete ascendency over Upolu, where he compelled all to give up their property to him, and to yield the women of all classes to his desires.

Finally, his tyranny and excesses exceeded the bounds of patience, and the people of Aana rose against him, conquered, and put him to death. From this arose the war of Aana, which will be again spoken of; for the chiefs of the other islands considered themselves bound to avenge the death of Tamafago. The people of the other districts of Upolu were not united in the support of their neighbours of Aana, who had made themselves almost universally odious by their haughty bearing. The war was a bloody one, and resulted, after a continuance of two or three years, in the entire defeat of the people of Aana, by those of Manono, who expelled them from their district, and forbade their return to it on pain of death.

This fertile region remained entirely unoccupied until the arrival of the missionaries; but when the Christian influences of their preaching began to be felt, the decree that condemned Aana to solitude was annulled, and the few of its former inhabitants who had escaped slaughter, were permitted to return to their ancient homes.

The island of Manono, whose inhabitants exerted such an influence in the closing scenes in the war of Aana, is situated within the sea-reef of Upolu. It contains eleven hundred inhabitants, and is the residence of the chief Pea, who must be distinguished from the inferior personage of the same name who resides at Apia. This island is covered with forests throughout its whole extent; its circumference is about four miles; and it is the station of one of the English missionaries.

In spite of its small extent and scanty population, Manono is identified with the political history of all the other islands of the group; for, during the reigns of the two Tamafagos, it held supremacy over them. The reason of its acquiring and exercising this political supremacy, is principally to be ascribed to the possession by its inhabitants of the small island of Apolima, which they used as their "olo" or citadel. To this retreat, inaccessible except at a single point, the inhabitants of Manono were in the habit of retiring when pressed by too powerful an enemy, and when his rage had spent itself, they thence returned to their home with undiminished numbers.

This natural fortress lies between Manono and Savaii, and soundings extend to it both from the shores of Upolu and Savaii. The coral reef attached to it is but small.

Drawn by A.T. Agate — Engraved by R. Holmes

APOLIMA.

Apolima, on the most cursory examination, is evidently the crater of an extinct volcano. Perpendicular cliffs rise from the sea around its whole circuit, except at a single point on its northern side. Here the lip of the crater is broken down, and admits the water of the sea into a small bay, which affords a safe harbour for boats. The entrance to this is so narrow as to admit no more than one boat at a time, and is dangerous whenever there is any surf. It may, therefore, be easily defended. There is only one other point on the island where it is possible to effect a landing, namely, at a small height to the westward of the bay, and here it can only be done when the water is perfectly smooth. But an enemy landing here would have made no progress, for before the interior can be reached from this point, the steep and precipitous rocks remain to be climbed.

The highest point of Apolima is on its south side, where it is four hundred and seventy-two feet above the sea. The perpendicular cliffs which face the sea are of course bare of vegetation; but with this exception the whole surface is covered with cocoa, bread-fruit, and other trees, or with plantations of taro, yams, &c.

In the centre of the island is a village of about twenty houses, and the permanent population consists of no more than about seventy-five persons. The people are evidently jealous of the maiden reputation of their natural fortress, and showed much concern when we visited it, which the women even manifested by shedding tears.

It can be readily understood from this description of Apolima, that whatever party held it would be able to maintain possession of it against great odds, and thence to take advantage of any weakness or want of watchfulness on the part of their enemies.

While we were engaged at Tutuila and Upolu, the survey of the island of Savaii was performed by Lieutenant-Commandant Ringgold, in the Porpoise. It has already been mentioned that this vessel had been detached for that purpose, and that Dr. Pickering, from the Vincennes, had gone in her. The brig first touched at Sapapale, the residence of the Rev. Mr. Hardie, who gave them a cordial welcome, although much surprised at so unusual an arrival.

Many of the natives collected to view the white men, of whom so many had never been seen together on the island. In their remarks, they, among other things, praised our people for their beauty.

Dr. Pickering and Lieutenant Maury were landed here, to remain upon the island while the brig was employed in surveying it; the former to examine its productions, the latter to observe the tides. Mr. Hardie kindly afforded them accommodations in a new house he had just been erecting.

Lieutenant-Commandant Ringgold, after landing Dr. Pickering and Lieutenant Maury at Sapapale, proceeded around the island for the purpose of surveying it. He began with the examination of the large bay of Paluale, near the eastern point of the island. Here there is a missionary station, under the superintendence of Mr. M'Donald, who had resided there for about six months, with his wife and children. The natives are peaceable, but are described as inquisitive and rude. The village is prettily situated, and is approached through a boat-passage in the reef.

The south side of the island was found rocky and iron-bound, with a heavy surf breaking on it. Towards the western end of the island, the rocks around the points were worn into cavities, and the sea rolling into them produced innumerable spouts of water.

When the brig was abreast of the deep inlet of Salealua, a native missionary came off in a whale-boat. He tendered every possible civility, and was very desirous that a trade might be opened with the village of the same name. This is situated at the head of the bay, upon a sandy beach, and has around it and upon the sea-shore a considerable extent of level plain, filled with groves of cocoa-nut and bread-fruit. Upon examination, no shelter was found for vessels in this bay, and the urgency of the duty required that the brig should pass on without farther intercourse with the shore.

Near the northwestern point of Savaii is the large and beautiful village of Felialupo, with a snug little cove for boats. This place is under the charge of a Tonga missionary. The natives were friendly, and disposed to exchange their poultry and fruit, for tools, cloth, &c.

The next inlet on the north side, was that of Asau. This was supposed to be the only place where there was any probability of finding a harbour. But the hope of such discovery was frustrated, for there is only a small and shallow entrance through the reef, and within the reef the shore forms an extensive flat.

Many canoes from this village visited the brig, and before they had taken leave a theft was discovered. A commotion immediately took place, and the native missionary at once gave himself up as a hostage, until the article (a whetstone) should be brought back. A canoe was then despatched to the chief, and in the course of an hour he came on board, bringing the article. Both the chief and the native missionary expressed great mortification that such an occurrence had taken place. Several small presents were made to them, and they returned to the shore highly pleased.

Proceeding on the survey, the brig arrived off the north point of the island, and reached the bay of Mataatua, which was examined, and found to afford a good anchorage. The brig was anchored here, and the harbour surveyed. This is the only harbour in the island where a vessel can anchor with safety, and here supplies of hogs, poultry, and vegetables, may be had in abundance; wood and water are also easily obtained, the latter from copious springs near the beach.

A great difference in form, physiognomy, and manners, from those of the adjacent villages, was observed here, as well as a change in the character of many articles of manufacture. The war-clubs and spears were of uncommon form, and neatly made.

This bay is surrounded by a white coral beach. The natives appeared harmless, but manifested great curiosity. The women are more gracefully formed than at the other islands.

The native missionaries appeared to exercise much influence over them, having put a stop to many of their former evil practices.

On the 24th, the brig again arrived off Sapapale, after an absence of nine days. Here they were joined by Dr. Pickering and Lieutenant Maury, and found the old chief Malietoa and his son Mole, who were extremely courteous. On the former being presented with some articles, he remarked, that "our property was very good, but our good-will better."

Dr. Pickering engaged natives to accompany him into the interior, and to visit the Mu or burnt district. Preparations for the journey were made in advance, and among other things, it was stipulated that there should be only two meals a day,—one early in the morning, and another in the evening. The first day, however, was to form an exception.

Mr. Hardie accompanied the party for a few miles, and they soon after their departure met a native who was styled "the Lord of the Forest." The party were desirous that this man should accompany them, for his appearance promised more than that of the others, and it seemed it was necessary to obtain his permission before they could enter the forest. In times of scarcity, his domains become of great value, in consequence of the quantity of wild yams they yield. This person agreed to accompany them, and they proceeded along a good path through cultivated grounds of taro, dracæna, &c. Mr. Hardie, before leaving the party, endeavoured to make the natives understand the nature of Dr. Pickering's errand; the latter was unable to make himself understood by them. They had not proceeded far before they came to an uninhabited house, where the natives stopped for the purpose of preparing dinner, the cooking of which occupied three hours! The day was in consequence well advanced before they again started, and at about 4 P. M. they reached an open shed, about two miles from the last stopping-place, where the natives concluded to halt for the night. The occupants, who consisted of two elderly women and a young man, were dispossessed, and the shed was enclosed by hanging up leaves of the Heliconia, which resemble those of the banana. They then prepared some excellent cocoa-nut pudding, and heated some cocoa-nut milk in the shells. This beverage is usually taken by them every morning and evening; the natives all saying grace before their meal, and prayers before they went to rest. It was late the next morning before Dr. Pickering could get the party in motion, and pursuing their route, they soon overtook the Lord of the Forest, who had preceded them, and was employed in cutting a path through the woods, although that already made might have been easily passed through. No inducement could make them change their purpose, and they continued to work at their turnpike, lopping off large branches, beating down ferns, &c. After some time, they reached a rising ground, which they found to be on one side of a crater, about a thousand feet above the sea, and seven miles inland.

Dr. Pickering now concluded that it was a hopeless task to attempt to penetrate into the interior with such guides, and determined to return, which he accordingly did. He found the rest of his party a mile in advance of their previous encampment, where they had built for themselves a fine house, and each man had collected two large baskets of yams for provisions. This was their stopping-place for the night, and among other cares for the Doctor's comfort, they constructed for him a native pillow, formed of a piece of bamboo, with legs lashed to it about three inches high.

The natives were in high spirits during the evening, talking and laughing immoderately. They succeeded in getting off by nine o'clock the next day, and reached the coast about noon.

During the stay of Dr. Pickering and Lieutenant Maury on this island, they were objects of great curiosity; and whenever they walked out they were followed, not only by boys, but grown men, who did not, however, offer to molest them in any way. When they passed through the villages, all the inhabitants, not excepting the scholars from the schools, came out to look at them. The latter, however, did not abandon their books, but retained them in their hands; for all, whether young, middle-aged, or old, are anxious to learn, and their perseverance, as in other parts of the group, is astonishing.

Dr. Pickering here witnessed the taking of fish in a different mode from that practised on the other islands. Application was made to the chief, and through his influence a meeting of the head men of the town was called, and a fishing expedition agreed upon. The net, if it could be so called, was prepared, and in the course of two days every thing was ready. The net was a kind of cheval-de-frise, made of the leaves of the cocoa-nut tree, split and wound round a line, and was little less than half a mile in length. It was more formidable in appearance than in reality. This net was taken out at high water to the coral reef, in three pieces, then fastened together, and thus made to enclose a large extent of water. This space was gradually contracted by doubling up the net, which answered the same purpose as the drawing of a seine. The fish did not attempt to pass it, and were thus driven towards a certain point, where a sort of sack of matting had been placed for them to enter. As the fish were gradually enclosed by the mat, and the tide fell, the scene became an animated one. Men, women, and boys, to the number of two or three hundred, were eagerly engaged in picking up or catching the stragglers as they were seen leaping up; the whole area seemed alive with fish, jumping in every direction, some over the heads of the natives, and thus escaping, while others leaped into hand-nets. About a canoe-load was caught, comprising thirty different kinds of fish, some of which were six or eight pounds in weight, but the majority were smaller. The haul was considered an unsuccessful one, which was attributed to some misunderstanding and mismanagement among the natives, by which a large stone fell on the net, and allowed many of the fish to escape.

Savaii is the most western island of the Samoan Group, and is also the largest, being forty miles in length and twenty in breadth. It is not, however, as populous, or as important, as several of the others. It differs from any of the others in its appearance, for its shore is low, and the ascent thence to the centre is gradual, except where the cones of a few extinct craters are seen. In the middle of the island a peak rises, which is almost continually enveloped in clouds, and is the highest land in the group. On account of these clouds, angles could not be taken for determining its height accurately, but it certainly exceeds four thousand feet.

The interior of the island is rarely entered, even by natives, and has never been penetrated by strangers. The only settlements are upon the shore, along which the natives always journey, and there are no paths across it.

Another marked difference between Savaii and the other large islands, is the want of any permanent streams,—a circumstance which may be explained, notwithstanding the frequency of rain, by the porous nature of the rock (vesicular lava) of which it is chiefly composed. Water, however, gushes out near the shore in copious springs, and when heavy and continual rains have occurred, streams are formed in the ravines, but these soon disappear after the rains have ceased.

The coral reef attached to this island is interrupted to the south and west, where the surf beats full upon the rocky shore. There are, in consequence, but few places where boats can land, and only one harbour for ships, that of Mataatua; even this is unsafe from November to February, when the northwesterly gales prevail.

The soil is fertile, and was composed in every part of the island that was visited, of decomposed volcanic rock and vegetable mould.

The Porpoise, having taken Dr. Pickering and Lieutenant Maury again on board, set sail for Tutuila, for the purpose of joining the Vincennes, and beat to windward along the south side of Upolu. During this passage many of the crew became sick, which rendered it necessary to stop for a few days at Pago-pago, in order to recruit them. Here they all speedily recovered, except one man, named David Blodget, who died. The disorder was attributed to the dampness of the vessel.

The delay in the arrival of the Porpoise at Apia caused me to send the Flying-Fish to Tutuila, whence they both returned to Apia.

Previous to sailing, at the pressing instance of the chiefs, I ordered the marines and small-arm men of the squadron, in all about one hundred and fifty, to be sent on shore, with their music, for exercise. They had been well drilled to act on shore should occasion require, and were provided for the occasion with blank cartridges. The natives from far and near were collected to witness the review, and few scenes that occurred during the voyage were as amusing as this. The old and young were equally delighted, and it was ludicrous to see them endeavouring to imitate the soldiers, in their marches and countermarches. They were not satisfied unless the drummers were constantly beating, and were particularly delighted with the bass-drum. The firing occasioned some alarm at first, but when they saw it did no harm, they became reconciled to it, although even to the last they would scamper off to a distance at each discharge.

The review left an impression on their minds of the superiority of our arms that will not soon be forgotten.

The men were embarked at sunset, and had many jokes to relate of the conduct of the natives, and particularly old Pea, who on this, as on other occasions, acted as master of the ceremonies.

During our stay in this group, we experienced two slight shocks of earthquakes; their occurrence here is not unusual, but there is no account of any damage having been done. Their motion is generally tremulous and horizontal; one, however, has been experienced of a wavy description. They are said by the foreigners often to produce the sensation of sea-sickness.

On the 10th of November, the whole squadron was assembled in the harbour of Apia, after having been actively engaged since the 8th of October in examining the different islands, and making surveys of their coasts and harbours, &c. This work was all expeditiously and well done, with the exception of the south side of the island of Upolu, which was imperfect in some respects; it was consequently re-surveyed in the following year, and the charts finished. Besides the surveys, full series of experiments were made in magnetism, and extensive collections obtained in natural history, botany, &c., the islands being traversed by parties in several directions for this purpose. For the results in these departments, the reader is referred to the Reports of the Naturalists; and to the Hydrographic Atlas, for the charts.

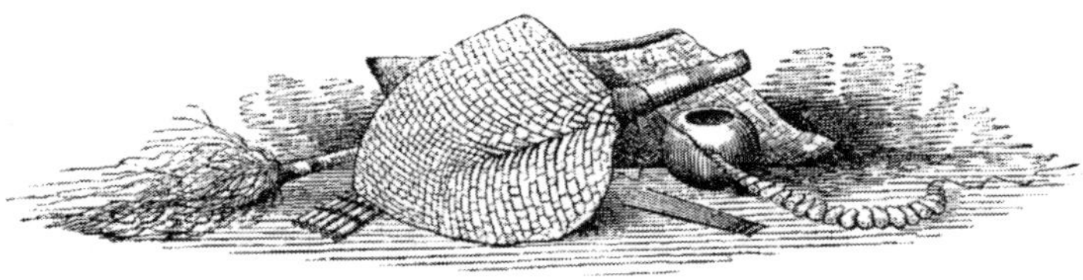

FAN, BASKET, NECKLACE, COMB, ETC.

CHAPTER V.

CONTENTS.

GEOGRAPHICAL POSITION OF THE SAMOAN GROUP—ITS HARBOURS—ITS TIDES AND CURRENTS—ITS CLIMATE—SIZE OF THE ISLANDS—SOIL AND PRODUCTIONS OF THE GROUP—ITS CULTIVATION—ITS QUADRUPEDS—ITS BIRDS—ITS REPTILES—ITS FISH—ITS PRODUCTS AVAILABLE FOR COMMERCE—LANGUAGE OF ITS NATIVES—THEIR DISEASES—THEIR GENERAL APPEARANCE—THEIR NATIONAL CHARACTER—LABOURS OF THE MISSIONARIES—NATIVE MISSIONARIES—SELECTION OF THEM FOR THE NEW HEBRIDES—POPULATION OF THE ISLANDS—RELIGION OF THE HEATHEN—THEIR IDEA OF THE CREATION—OF A FUTURE STATE—THEIR OMENS—THEIR SUPERSTITION—THEIR DANCES—THEIR MUSICAL INSTRUMENTS—THEIR AMUSEMENTS—PAPALANGI SHIP—THEIR BIRTHS—THEIR MARRIAGES AND COURTSHIP—THEIR ADOPTION OF CHILDREN—THEIR BURIALS—THEIR MOURNING—THEIR MANNERS AND APPEARANCE—THEIR DRESS—IMPROVEMENTS IN THE ANCIENT DRESS—TATTOOING—THEIR MANUFACTURE OF TAPA AND MATS—SAMOAN CANOES—BOAT-SONG—HOUSES OF THE NATIVES—THEIR LIGHTS—THEIR FOOD—THEIR HABITS—THEIR MALANGAS—THEIR PUNISHMENTS FOR CRIMES—THEIR WARS—THEIR OLOS—THEIR PEACE-MAKING—CLASSES OF SAMOAN SOCIETY—ALLOTMENT OF LANDS—MODE OF GOVERNMENT—DESCENT OF CHIEFTAINSHIP—CEREMONIES AT THE FONOS.

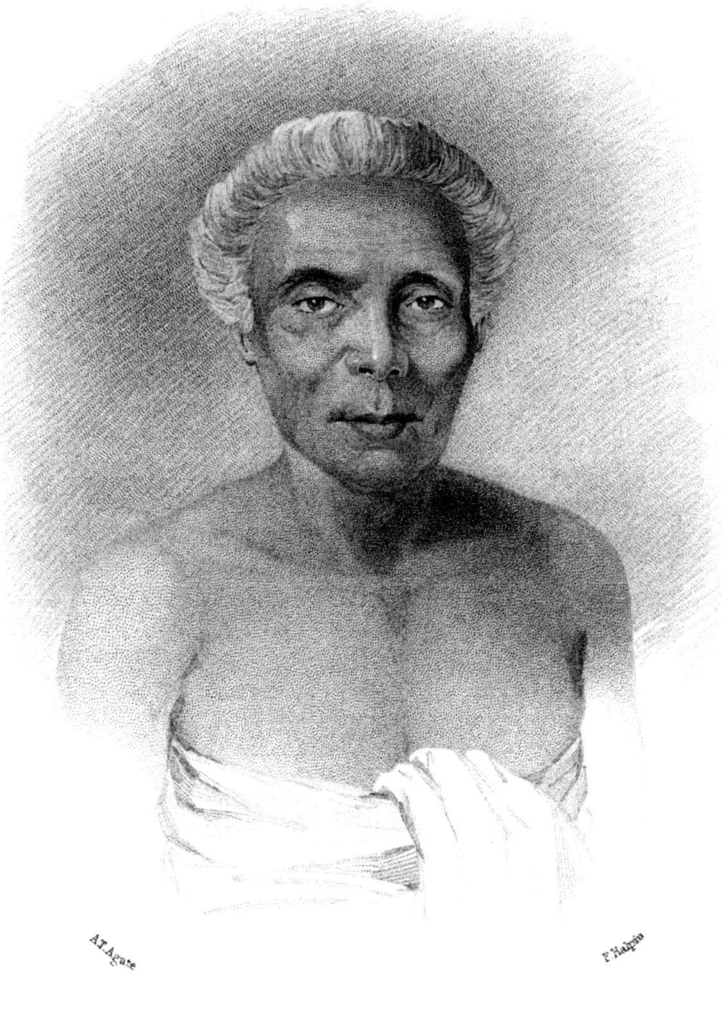

MALIETOA

CHAPTER V.

SAMOAN GROUP.

1839.

During the time that the squadron remained in the Samoan Group, all the islands of which it is made up were visited; not only were the examinations, spoken of in the two preceding chapters, made, but their shores were minutely surveyed by boats; the meteorological instruments were duly registered; astronomic and magnetic observations made, and a full record of the tides kept. We have thus obtained a large amount of information, which will be more easily intelligible in a condensed form, together with a great number of facts in relation to the aboriginal population, which may be made more interesting when applied to give a general view of the habits, character, and state of civilization among the natives, than if dispersed in isolated remarks in the accounts of the separate tours in which it was obtained.

The group lies between the latitudes of 13° 30′ and 14° 30′ S., and the longitudes of 168° and 173° W. The islands, as has been seen, agree in the general character of being of volcanic structure, and having coral reefs; differing, however, in the modifications of these formations, which have been from time to time described. The harbours are usually situated within the reefs, but Tutuila is an exception to this rule, by the possession of the deep land-locked basin of Pago-pago. This is, of all the ports, the one best adapted for the refitting of vessels; but Apia, in Upolu, in the latitude of 13° 48′ 56·6″ S., and longitude 171° 41′ 09″ W., is not so difficult of egress, and in consequence of its proximity to the fertile district of Aana, the most convenient for vessels seeking only a temporary anchorage and refreshment.

The approach to Pago-pago, and the other harbours of the Samoan

Isles, is not difficult; and as the soundings extend in some places for a distance beyond the reefs, vessels may drop an anchor in case of necessity.

The flood tide among these islands sets to the westward; beyond its influence, on the southern side of the islands, a current generally prevails to the eastward, while it runs westward on their northern side. Vessels, therefore, when beating to windward, would find it to their advantage to keep on the southern side of the group, where there is not only a favourable current, but where the winds would be found more regular, and calms less frequent.

Tidal observations were made contemporaneously at Tutuila, Upolu, and Savaii; these show a regular difference of one hour in the tidal wave between Tutuila and Upolu: the tide at Savaii appears from the record to have been more irregular than at the other islands, which may in part be attributed to the extent of the reef, and the difficulties that were encountered from the want of aid in making the observations.

The climate of these islands may be termed variable, and there is much bad weather, particularly during the winter months, when long and heavy rains, attended at times with high winds and northerly gales, are frequent. Destructive hurricanes also occur, and of these one is still recollected which blew down the bread-fruit trees, and destroyed many of the houses.

The air is more moist than that of the Society Islands, and the vegetation in consequence more luxuriant. Thunder and lightning are often experienced, but during the summer months light winds and calms are the prevailing characters of the climate.

Some of our gentlemen made the remark, that, to judge from the time at which the bread-fruit was gathered, there must be a great difference between the seasons of this island and Tahiti; for when we arrived at Tutuila, that product was ripe and in abundance, although when we left Tahiti, only a few days before, it was unripe and not to be had. The same remark was made in relation to the vi-apple (Spondias dulcis.) But, by comparing the voyages of Cook and Wallis, it would appear that the time of the year at which the bread-fruit is in season at Tahiti is not constant, for both these navigators found it in perfection, although they visited that island in different months. If there be a difference between the time of the ripening of the bread-fruit in the Society Islands and this group, the greater moisture and higher mean temperature of the Samoan climate will account for it.

The temperature of the air at Apia varied from 77·7° to 80·2°; that of the water from 81·25° to 83·75°.

The mean height of the barometer in the group was 30·128 in.

The islands of the Samoan Group contain two thousand six hundreu and fifty square miles, which are divided as follows, viz.:

Savaii	700
Upolu	560
Tutuila	240
Manono	9
Apolima	7
Manua	100
Oloosinga	24
Ofoo	10

The soil of all the islands is rich, and arises chiefly from the decomposition of volcanic rocks. At Tutuila, it was remarked that the vegetation was luxuriant, and the trees of large growth. At Upolu the forests seemed more sombre than those of Brazil, although the same kind of growth appeared to prevail.

The trees do not branch out until near the top, which renders it difficult to obtain botanical specimens. The trunks are covered, and even the summits of the trees sometimes overgrown, with the leaves of the scandent Flagellarias and Freycinetias, a climbing Piper, and other vines, as Hoyas, Convolvulus, &c. The lower part of the trunks are enveloped with ferns, of which there are many varieties, and with some species of Pothos, which give the whole ground a matted or woven appearance.

The woods in the interior of the islands are very thick, and often composed of large and fine trees; among them are, tree-ferns, a species of banyan, pandanus, and several species of palms. Among other plants a species of Cerbera was observed, with beautiful clusters of large and odorous white flowers, which yielded a quantity of white viscous sap, that our botanist, Mr. Rich, thought might be manufactured into caoutchouc. On the whole, the species of trees are much more numerous than at Tahiti, and the vegetation in consequence richer and more varied. The woods, however, are not enlivened by showy flowers, and the few of these that are seen are of a white or grayish hue, which is to be ascribed to their being but little exposed to the rays of the sun, in consequence of the umbrageous foliage. Many of the flowers seen on the ground were unknown to our botanist, as were several fruits.

Among the trees which have been named, that which struck us as most remarkable was the species of banyan (Ficus indica), called in these islands Ohwa. Some of these were seen, whose pendant branches had taken root in the ground to the number of thousands, forming stems from an inch to two feet in diameter, uniting in the main trunk more than eighty feet above the ground, and supporting a vast system of horizontal branches, spreading like an umbrella over the tops of the other trees. For the sketch of one of these I am indebted to Mr. Peale.

OHWA TREE.

The bread-fruit is the most abundant of all the trees, and grows here to a large size; the vi-apple, the cocoa-nut, and the wild orange are also found in great numbers; and at Tutuila a large lime-tree was seen in full bearing, which was said to have been planted before the arrival of the missionaries.

Among the most singular of the vegetable productions is the stinging tree, of which the natives are much afraid; for if its leaves be touched an eruption is produced, particularly if the skin be wet. Its leaf is cordate, but quite smooth.

The arborescent ferns are not as numerous as at Tahiti, but grow to a larger size. The palms give a character of luxuriance to the country, from the variety of their foliage. Rattans ninety feet in length were seen running over the trees.

Bamboos and the wild sugar-cane were very common; the latter is used in thatching houses: the wild ginger also abounds.

Of the wild nutmeg (Myristica,) two species were seen, which are small trees, and likely to be passed without notice, were it not for the peculiar manner in which branches grow out of the trunk, which is in whirls, at regular intervals, like the white pine (Pinus strobus) of our Northern States.

It was remarked that the character of the vegetation approached more nearly to that of the East Indies than of the Society Islands, and the leafless acacias were the type of those we afterwards saw in New Holland; but there are some plants which appear peculiar to these islands.

Many of the trees we have named, as well as other plants, are objects of cultivation; but the ground cleared for this purpose does not extend far from the coasts, near which all the villages are situated.

To clear the land, the bark is burnt off the trees, after which they are permitted to stand until they become dry, when they are cut down and used as fuel.

The cultivated plants and trees are, bread-fruit (of which they have twenty varieties), cocoa-nut, ti (Dracæna), bananas, taro, paper-mulberry, tacca, from which arrow-root is made, and of which they have several sorts; sugar-cane, which is not made into sugar, but used only for thatching; coffee, ava (Piper mythisticum), sweet-potato, pine-apple (Anana), brought by the missionaries from the Society Islands, yams, the papaya, and tobacco in small quantities. The agave has not been introduced; but in a few years lemons and sweet oranges will be produced in great quantities from trees which have recently been planted.

To the cultivation of the tacca they pay little attention, yet the quality of the fecula (arrow-root) made from it is said to be superior.

The missionaries are endeavouring to teach the natives the best mode of cultivating the sugar-cane and manufacturing it, and it is said that a few persons have adopted the new methods. At present they find a substitute for sugar in the root of the ti plant, which is baked in ovens, and yields a large quantity of saccharine juice resembling molasses.

Great attention is paid to the cultivation of the yam. They are planted in October, and are ripe in February and March. The vines run up the trees, and when they die, the root is known to be ripe. To plant them, they are cut, like the potato, into pieces containing *eyes*, which are laid in heaps and covered up until the sprout appears. The pieces are then set out at distances of about three feet from each other.

Hearing that there were some extensive savannas in Upolu, overgrown with the wild sugar-cane, I directed Assistant-Surgeon Whittle and Mr. Couthouy, to proceed to the east end of the island, where they were said to grow. They, however, saw nothing of the kind except a few small patches of that plant.

There are no traces among these islands of any native quadruped, nor any other of the mammalia, except a species of bat (Pteropus ruficollis), which is very destructive to the bread-fruit. Swine have now become abundant, and the missionaries have introduced cattle, which are rapidly increasing, and will in a few years be in sufficient numbers for the supply of vessels. Horses have also been brought to the islands.*

The first large quadruped ever seen by these islanders was a mule. With it they were much astonished, and it was considered so great a curiosity that it was carried around the island of Upolu for the purpose of gratifying the natives with a sight of it. They gave it a name, signifying—the hog that travels over the ground.

Poultry of all descriptions is plentiful, and pigeons abound, which, however, are considered sacred, and not used as an article of food. Of the latter bird (Columba oceanica), between sixty and seventy specimens of different varieties were obtained, but it is remarkable that of all these, none were the same as those found in the Society Islands. To the Zoological Report I would refer for further information on this subject. There are but few birds of game, and none of the hawk genus. A philomel was pointed out by the missionaries as the principal singing bird, and the woods of Tutuila were filled with warblers. The note of the philomel, although much praised, did not appear agreeable to me.

The pigeon is commonly kept as a plaything, and particularly by the chiefs; for this purpose they are fastened to a stick by a thread about twelve feet in length. They are taught to fly from and return to the stick, and when well tutored to this feat, the possessor of the bird exhibits it with much pride and satisfaction. One of our officers unfortunately on one occasion shot a pigeon, which caused great commotion, for the bird was a king-pigeon, and to kill it was thought as great a crime as taking the life of a man. The people were not to be pacified until the interpreter told them that the officer belonged to "man-of-war," which intelligence, together with a small present, satisfied them, and the matter was settled.

To justify their regard for them, we were told that when the inhabitants of Aana were driven away, about eight years since, by the people of Manono, the pigeons abandoned the district, but that upon their return to their homes, the pigeons again made their appearance in their former abodes.

Snakes were found in Upolu, and sea-snakes are reported to have been seen off the islands.

Fish are taken in the neighbouring waters in great abundance and variety. Besides other modes of taking them, they are caught on the reefs by women, who place baskets near the holes where they are accustomed to take shelter. They are also speared by torchlight, and

* On Upolu there are now twenty head of cattle, and seven horses.

taken in deep water by the hook. Among the sea-fish, mullets are very numerous, and are frequently seen leaping from the water in immense shoals.

One of the modes in which fish are caught by the Samoans, was witnessed at Samatau. About a dozen canoes formed themselves into a ring around what appeared to be a dark circular spot in the water, about six feet in diameter, and which was moving along with a slow and unequal motion. This was a shoal of the small fish called lou, which is about two inches in length. The shoal being thus surrounded, the circle of canoes was gradually lessened, until the fish, finding themselves enclosed on all sides, ceased to move forward. At this moment, the head fisher, who was seen standing up in the canoe with a net in his hand, threw it dexterously over the shoal, upon which all the other men dove at once from the boats, and remained for several seconds under the water, where they secured the sides of the net. On reappearing, all regained their canoes except four, who remained to take charge of the net, which with its prize they conveyed to the chief.

These islands furnish abundant supplies for the refreshment of vessels, but as yet there are few articles which can be rendered available in foreign commerce. Tortoise-shell, of which a little has at times been procured at Savaii, cocoa-nut oil, and arrow-root, are nearly all that can be procured in quantities beyond the immediate wants of the visiters. Caoutchouc, gum Arabic, castor beans, orris-root, ginger, and coffee, might however be easily added to the list of exports. In return for what they can furnish, the natives now look to objects of real utility; beads, jews-harps, &c., once so much in request, are now scarcely prized; and cotton cloth, writing-paper, and hardware, particularly needles and other small articles of utility, are the kinds of manufactured goods which are most sought after.

The Samoan language is soft and smooth, and is the only one of the Polynesian dialects in which the sound of s is found. The letters that the missionaries have found necessary to adopt in order to write it, are only fourteen in number, viz.: A E F G I L M N O P S T U V. In attempting to sound the words of other languages, they use L instead of R, S for H, and P instead of B. The G has a nasal sound, as in *ong*.

It has nearly the same construction as the Tahitian, nevertheless the Samoan is far from being understood by the natives of the Society Islands. The Samoans say that they never can acquire it—"their jaws are too stiff." The missionaries also have great difficulty in speaking it, and are liable to make many mistakes which appear absurd to the natives.

We have seen that it possesses the sibilant sound of s, and every one of the words terminates with a vowel.

A separate dialect is appropriate to the chiefs, all of whose actions, the parts of their bodies, &c., have different names from those of the common people. The Philological Report is referred to for further information upon this subject.

Many of the Samoans reach the age of seventy or eighty years. There is, however, a great mortality among the young children, which is probably owing to their exposure to the weather. Those who survive, grow up robust and healthy.

Among the diseases which afflict the adults, one of the most usual is a spinal affection, which results in caries and produces humpback. This is no doubt owing to the peculiar manner in which the children are carried. Catarrhs and bronchial disorders, occasioned by the exposed life of the natives, are prevalent, and a white resident died of phthisis during our stay. The dysentery, as an epidemic, is unknown, but sporadic cases of it occur, occasioned by imprudence in diet.

There is an eruptive complaint, called ilamea, which covers many of the children under the age of ten years with sores, and which seems more particularly to attack the face and head. The mode in which it is treated is singular: the child is rubbed with the husks of the cocoa-nut, until all the scabs are removed; a soft preparation of the bread-fruit is then applied, after which they are washed. This operation is undergone every time they bathe, which is daily. When the bread-fruit is not in season, a decoction of the husk of the cocoa-nut is used in its place.

The elephantiasis prevails to a great extent among men who are past the middle age; and some of the cases are truly frightful. There are also many instances in which women are affected by it. It does not appear to cause the least degree of pain. Among the reasons that have been assigned for the frequency of this disease are, the habit of eating their food without salt, and the use of cocoa-nut water; to which may be added exposure at night, and want of sufficient exercise. The latter cause, whether it be capable of producing this disease or not, unquestionably exists; for they are in the habit of sitting for hours with their legs bent under them, which must cause a stagnation of healthy circulation. Laziness, however, cannot be ascribed to them as a part of their national character, for they are disposed to exertion, and willing to be employed. When, therefore, they have received sufficient instruction, and civilization has taught them new wants, they will probably become an industrious and thriving people.

Ophthalmia, which is supposed to arise from the reflection of the sun from the sandy beaches near which all their villages are built, is so prevalent, that, to speak within bounds, not less than a fifth part of the population is affected with it.* In most cases it was observed to begin on the inner corner of the eye, whence it extends gradually over the pupil, until the sight is completely lost. As the disease advances, the thickness of the film increases, and when it has covered the eye, that organ becomes enlarged and appears to project. From appearances it would not be difficult to remove the film, and thus cure the disorder; but the natives have not made any attempt of the kind. Several cases of total blindness arising from this disorder were seen.

The venereal disease does not exist at Tutuila, and is hardly known in the other islands. This serves to prove how great a superiority this island possesses over Tahiti in the chastity of its females, who in general observe their marriage vow with strict fidelity.

Fevers are rare, and those of a remittent and intermittent type are unknown; in fact, the geological formation of these islands is by no means favourable to the generation of the miasmata that cause them.

No means of medical assistance are attached to the English mission, and the missionaries, therefore, can do but little in alleviating the maladies of the natives. Even their slight knowledge of remedies affords some alleviation, and their practice is far preferable to that of the natives, who always abandon to their fate those who are very ill.

Among the few curative means that the natives do employ is a sort of shampooing. This is performed by rubbing the body and limbs with the hands, at first gently, and gradually more and more roughly. These manipulations are applied as a restorative after fatigue, and to alleviate pain. For the former purpose they are effectual, and often abate, if they do not remove, the latter.

Among all the Polynesian islanders, the men of Samoa rank, in point of personal appearance, second only to the Tongese; and many specimens of manly beauty are to be seen among them. As much cannot be said of the women, who are rather ill-formed and stout. When very young, however, some of them are pretty, and their colour is light, being little darker than that of a brunette or South American Spaniard. The girls are lively, have a good expression of countenance, and, what is rare in Polynesia, have some degree of bashfulness.

The average height of the men is five feet ten inches, and some of

* It is so common at Savaii, that at least one case of blindness, in one or both eyes, is to be seen in every family.

the chiefs, whose limbs are well rounded, would be called fine-looking men in any part of the world. Their features are not in general prominent, but are well marked and distinct, and are all referable to a common type. The nose is short and wide at the base; the mouth large and well filled with white and strong teeth, with full and well-turned lips; the eyes black, and often large and bright; the forehead narrow and high; and the cheek-bones prominent. It was observed that some of them had the eye turned up at the outer corner like the Chinese. Of beard, they have but little, but their hair is strong, straight, and very black; instances, however, were observed, where it had been turned to a carroty red, by washing it with lime-water for the purpose of destroying the vermin (Pediculus humanus).

Little difference was perceived in the shape of the heads of the two sexes, for observing which there is a good opportunity among those who have embraced Christianity, who shave off their hair. The general form of the skull is broad and short, and is highest near the crown.

When the islands were first visited, the natives were represented as ferocious and treacherous. This arose in a great degree from the bloody conflict they had with the boats of La Perouse's squadron; and the opinion was kept up by the just resentment they in some cases manifested for wrongs committed on them by lawless visiters. The instance of Opotuno, however, shows that this idea of their character is not entirely without foundation. Viewed in a more favourable light, they are, as we found them, kind, good-humoured, intelligent, fond of amusements, desirous of pleasing, and very hospitable. Both sexes show great kindness and love for their children, and age is so much respected that only old men are admitted to council. As a shade on this picture, they are indolent, covetous, fickle, deceitful, and little reliance can be placed upon them. To illustrate these features of their character: the first question asked when a chief receives a visiter is, "What present will you take?" for they consider it incumbent upon them to bestow some token of regard, and a neglect to offer it would be indecorous. This custom was always complied with, when any of our officers visited them, and although it was evident they did not wish to part with any thing valuable, their choicest possessions were exhibited as if for the choice of the stranger. On the refusal of their offered presents, great joy was always to be observed in their countenance and manner, showing that they rejoiced in an escape from loss, while they had at the same time performed the prescribed rites of hospitality. This risk being over, they were too happy to supply us with cocoa-nuts and fruits. In spite, however, of the apparent liberality with which these were furnished, they do it in expectation of a full return. In pursuance of this hospitality, it is the custom when a stranger passes through a village without showing an intention to stop, to follow him and offer food.

The Samoans are usually very inquisitive, and it was amusing to excite their curiosity. Among other things mentioned for the purpose was, that white men often wore false teeth and wigs. The latter practice in particular seemed strange to them, and they called it "thatching the head." A terrestrial globe was also shown to some of them, whereon the position of their islands and their small relative importance was pointed out. This excited great surprise, for until within a few years they had no idea that there was any country except their own.

If the chiefs are liberal in their tenders of presents to their visiters, they on the other hand do not hesitate to ask for whatever they see. They may, in fact, be styled sturdy beggars. One of the most persevering in his mendicancy, was no less a person than Vavasa, the proud and overbearing chief of Manono. They usually began with begging from the humblest individual, and ended with the highest; and when they had obtained all they could, would go over the side of the ship ridiculing our folly for giving so much.

Old Pea, by way of excusing himself when charged with being a great beggar, said he did not keep any thing he got for himself; that it was the Samoan fashion always to ask for every thing he saw. It mattered not if his request was refused, he was as content as if he obtained what he desired, but he said he should have blamed himself if he had not asked.

The beneficial effects of the labours of the missionaries are more evident among the Samoans than at Tahiti. The spread of the gospel has not been opposed by evil habits of the same inveterate character, and the natives of this group have been more easily reclaimed from their vices than those of the Society Islands. The greatest obstacle to the success of the missionaries has arisen from the presence of a few abandoned white men, who attach themselves to the heathen chiefs. Their opposition, although injurious to the missionary cause, yields little benefit to themselves, for of every thing they acquire, the chief under whose protection they are, takes half; and although no opposition is ever made to their departure from the islands, they are not permitted to take any thing with them. The vices of these men excite the disgust of the more well-disposed of the natives, who often express their astonishment at their ignorance of sacred subjects, and ask if it be possible that such men can have been brought up in a civilized community.

The first attempt to introduce Christianity is related to have occurred in the following manner. Some years before the arrival of the missionaries, a vessel was wrecked upon the island of Upolu, and her cargo seized upon by the natives, many of whom, even to the present day, regret that they did not then understand what riches were thus placed at their disposal. Their mode of treating the prize was farcical in the extreme: pipes were made out of candlesticks, clothing was thrown away as valueless, and many injured themselves with the fire-arms. The crew were well treated, and fed for a long time, although the natives were greatly astonished at the quantities of pigs required for their support, and entertained fears lest they should breed a famine in the land. The captain advised his crew to turn missionaries, and set them the example himself. He met with much success, and succeeded in building several churches, until, upon the arrival of the English missionaries, he was compelled to relinquish his assumed occupation. It is not probable that even the captain was deeply versed in religious knowledge, and very certain that the crew could not have been; but their success appears to have arisen from the great veneration with which white men were at first regarded by the Samoans. They looked upon them as a sort of spirit, whom it was impossible to hurt or to kill; and the ships first seen off the coast were considered as heavenly messengers, prognosticating some dreadful calamity. The bad conduct of their nautical visiters has destroyed this reverence, and foreigners generally no longer meet the kind welcome they formerly received; this observation does not apply to the missionaries, who receive all the honour that is due to their good intention, of which the natives are fully aware.

The Wesleyan missionaries, and those of the British Board, reached these islands about the same time, or the former were perhaps the first to arrive. The influence of the Wesleyan tenets, and the number of their followers, increased rapidly under the superintendence of the Rev. Mr. Turner. Difficulties, however, arose between the two parties of missionaries, which were finally adjusted between the two boards in London, and the Wesleyans abandoned this field for that of the Feejee Group. This arrangement was amicably made, and I heard of only one individual on either side, who showed an uncharitable spirit towards his fellow-labourers of the other party. In spite of the removal of the Wesleyans, there is still a large number of the natives who adhere to the tenets and forms taught them by Mr. Turner, and still retain a strong attachment to him.

The missionaries were from the very first taken under the protection of the most powerful chiefs, and have never received either insult or injury from any of the natives. They have established schools in many of the villages, but have found a difficulty in obtaining native teachers.

A printing-press has also been established at Upolu, and rapid progress is making in the translation of the Scriptures, of which some portions are already published. Many publications have issued from this press: among them I regretted to observe a small tract containing a violent attack upon the Roman Catholics. The sight of this surprised me, as it contradicted the opinion I had formed, from my intercourse with the missionaries, of their liberality and freedom from intolerance. The sole object of the tract was to prepossess the minds of the natives against the missionaries of the Papal Church, in case they should visit these islands. This struck me as being at variance with the first principles of our religion; and I could not refrain from expressing an opinion that the tract was calculated to do much harm.

The labours of the English missionaries have been much aided by native assistants, who have been both industrious and successful; and among them, those of Raratonga have the merit of having led the way. They have acted under the direction of Mr. Williams, and he was loud in praise of their exertions. I witnessed a most interesting meeting of these native missionaries, for the purpose of selecting nine from their number to accompany Mr. Williams to the New Hebrides, which has perhaps left a more deep impression on my mind from the melancholy result of that attempted mission.

Great anxiety was exhibited by the candidates; and I have never seen a more proper state of feeling, or listened to more correct sentiments than were expressed on this occasion. All appeared devoted to their calling, and some of them were quite eloquent. After the choice was announced, those upon whom it had fallen manifested a cheerful but not unbecoming triumph, while the rejected candidates were evidently grieved and disappointed. The former were now invested with new apparel, which, although no more than a striped cotton shirt,* gave them an air of consequence among their brethren, which was amusing to us who could draw comparisons between this simple garment and prouder kinds of canonicals.

Each of the resident missionaries now delivered a long harangue,

* This garment is the only remuneration that they receive during each year from the missionary funds, and with it they feel themselves well requited.

I have to acknowledge the obligation under which I feel myself to the missionaries, both individually and collectively, for their kindness and attention. They did all in their power to further the objects of the Expedition, and to them the squadron is mainly indebted for a great part of the facilities we enjoyed of becoming acquainted with the manners, habits, and customs of the Samoans.

which was replied to by one of the selected. The subjects of these discourses were, on the one hand, advice in reference to the duties about to be entered upon, and on the other a recognition of the weight of the responsibility incurred by the successful candidates.

Most, if not all, of those selected for the new mission were fine-looking men, and they were chosen out of many applicants, for their steady habits and strict moral conduct. The term of their engagement on the new duty was three years, after which they were to return to their wives and children, who were not to accompany them.

The extent and influence of the labours of the missionaries may be best understood by a comparison between the whole population of the islands, with the numbers of those who have embraced Christianity, and attend the schools.

The entire population of the group is estimated at 56,600, of whom 14,850 have embraced Christianity, and 12,300 attend the schools. These numbers are thus distributed:

ISLANDS.	POPULATION.	PROFESSORS OF CHRISTIANITY.	PUPILS.
Eastern Group . .	2,000	150	150
Tutuila	8,000	2,200	1,900
Upolu	25,000	8,000	6,200
Savaii	20,000	4,000	3,700
Manono	1,100	400	230
Apolima	500	100	120
Total,	56,600	14,850	12,300

The whole number of foreign missionaries is eleven, of whom one resides in Tutuila, six in Upolu, three in Savaii, and one in Manono.

The number of native teachers is one hundred and thirty-eight, of whom five are in the Eastern Group, thirty-one in Tutuila, fifty in Upolu, thirty-six in Savaii, twelve in Manono, and four in Apolima.

Besides those counted as having actually embraced Christianity, it is said that two-thirds of the whole population belong to the Christian party.

Of those who attend the schools, about ten thousand read, and this newly introduced habit has of course made a very great change in the habits of a majority of the people, but the number of heathen still left is sufficient to furnish an idea of their original manners and customs, which will in a few years be either entirely lost, or so modified by the spread of the gospel as to change their character entirely. The rapidity with which this change is going on, rendered it desirable to obtain as much information as possible in relation to the pristine manners of this people.

As respects their ancient religion, we have obtained the following particulars of the heathens. They acknowledge one great god, whom they call Tangaloa-lagi, but pay less worship to him than to their war-gods, Tamafaiga, Sinleo, and Onafanua. The first entices them to war, the second leads them to it, and the third is a female goddess, who encourages them to fight.

Mafuie is their god of earthquakes, who was deemed to possess great power, but has, according to the Samoans, lost much of it. The way in which they say this occurred is as follows. One Talago, who possessed a charm capable of causing the earth to divide, coming to a well-known spot, cried, "Rock, divide! I am Talago; come to work!" The earth separating at his command, he went down to cultivate his taro-patch. His son, whose name was Tiitii, became acquainted with the charm, and watching his father, saw him descend, and the earth close after him. At the same spot, Tiitii said, "Rock, divide! I am Talago; come to work!" The rock did not open, but on repeating the words, and stamping his foot violently, the earth separated, and he descended. Being a young man, he made a great noise and bustle, notwithstanding the advice of his father to be quiet, lest Mafuie would hear him. The son then asked, "Who is Mafuie, that I should be afraid of him?" Observing smoke at a distance, he inquired the cause of it. Talago said, "It is Mafuie heating his oven." Tiitii determined to go and see, notwithstanding all the persuasions of his father, and met Mafuie, who inquired who he was. "Are you a planter of taro, a builder, or a twister of ropes?" "I am a twister of ropes," said Tiitii; "give me your arm, and I shall show you." So taking the arm of Mafuie, he twisted it off in a moment. Such a practical illustration of his powers soon made Mafuie cry out, "Na fia ola, na fia ola!"—I desire to live, I desire to live! Tiitii then took pity upon him, and let him go. The natives, on feeling an earthquake, exclaim, "Thanks that Mafuie has but one arm! if he had two, he would shake the earth to pieces."

The god Salefu supports the earth. They have likewise Mesua, Faana, Tinitini, Lamamau, who are gods of lightning, rain, whirlwinds, &c. These gods are said to reside on an island to the westward, from which quarter their bad weather usually comes.

They had, likewise, many inferior gods, who watched over particular districts. These various gods owned certain animals, reptiles, fish, and birds. In some few districts inanimate objects were worshipped, thus: a branch of bamboo, with a bunch of cocoa-nut fibres tied on the top, was worshipped in Manono. They also had carved blocks of wood and stone erected in memory of dead chiefs, which they worshipped.

The account they give of the creation of their island is as follows:

Tangaloa, their great god, who lives in the sky, sent down the bird Tuli (a kind of snipe), his daughter, to look what was below. She reported to her father that she saw nothing but sea. Tangaloa then rolled a stone from heaven, which became the island of Savaii, and another which produced Upolu, and the same for the others.

This did not suit Tuli, who returned to ask for inhabitants. He gave her orders to plant the wild vines (fuefue), which after growing were ordered by him to be pulled up and thrown into heaps, from which worms were produced. Then it was desirable that they should become human. Spirits were accordingly sent to them by Tuli, and the worms became man and woman.

Their notions of a future existence are quite vague. They believe, however, in a happy future state, where every thing good is provided. Some say that it is on their own island, others on distant islands, and for the chiefs at the residence of the gods on Pulotu, an island to the westward. They also believe that the spirit goes there immediately after death; that in these places it never rains; that they eat and drink there without labour, and are waited upon by the most beautiful women, who are always young, or as a chief expressed it to one of our officers, "whose breasts never hang down."

The spirits, according to their belief, often come down to wander about at night around their former dwellings; some spirits are believed to die, while others are immortal; some dwell in subterranean abodes, and are eaten by the gods. Some persons believe that after death they become "aitus," or inferior gods.

They believed in many omens, which were carefully watched. If the black stork, called matuu, flew before them on a war expedition, in the direction they were going, they deemed it betokened success; but if in any other direction, it was an ill omen. If a dim moon, or very bright starlight, or comet, were observed, it always indicated the death of a chief; and a rainbow was a sign of war.

The squeaking of rats was an unfortunate omen. Sneezing was also considered unlucky; if any one of a party sneezed on a journey, their further progress was postponed.

I was told that the Samoans have a great dread of being abroad in the dark, and that when obliged to pass about their villages by night, they use flambeaux made of the dried stalks of the cocoanut-leaf to light them on their way. This fear is partly owing to superstition, which makes them fearful of encountering some spirit or aitu, with which their imaginations people the groves, springs, rocks, trees, &c. They are in the habit of occasionally making a feast for the king's aitu, when a number of pigs are prepared, and a quantity of taro, fruit, &c., is gathered. The portion for the aitu is placed near his supposed dwelling-place, and the dependants and others enjoy themselves on the remainder.

They were formerly in the habit of presenting their first fruits to the aitus and chiefs. This custom still continues among the heathen, but the Christian party present theirs to the missionaries. The ceremony usually takes place in January or February. In drinking ava, the first cup was always presented to the gods.

There is an account of a large lizard which dwells on the south side of the island, and is worshipped as an aitu. The description given of it makes it two fathoms long and as large round as a cocoa-nut tree, with huge scales, and a mouth filled with sharp teeth. It is said to dwell in a stream near Safata, into which the natives frequently throw meat. Some of them declare that they have seen him, and that he has dwelt there upwards of fifty years.

It is not remarkable, however, that they should have this tradition; and this circumstance affords an additional proof that they have had frequent intercourse with the Tonga, or Friendly Islands, where a similar tradition is spoken of in Mariner's Tonga Islands.

Among their other superstitions is that of a malignant spirit that resides in the vicinity of Apolima, in the shape of an enormous eel, of from six to ten fathoms long, and large in proportion, which attacks canoes and drags them down.

A story is told that is said to have happened only a few years ago. While two natives of Manono were swimming across the channel in the reef, they were drowned in the sight of many others; immediately a large canoe was manned, and went in quest of them; the crew of this canoe encountered the monster, and wounded it. The canoe was upset, and although a few saved themselves by swimming to the shore, the greater part of them were destroyed. When asked if it was not a shark, (of which they have two kinds, the tanifa, or great white shark, and the masi, or small blue one,) they replied, it was a monstrous *pasi*, which is the name applied to the murœna or conger eel.

Their dances and other amusements are in a great degree abolished, but they are still practised in the heathen villages, and even the

Christian women may still be induced to exhibit the former, which they call *siva*. The mode of performing it differs from that of the Tahitians, but is like it lascivious, and neither of them would be called dances in our sense of the term. The dance is usually performed by young girls, who stand up before the audience, throwing their arms, legs, feet, and hands, in numerous strange attitudes, which are any thing but graceful. The others who are present sing amusing words, in two or three parts, while a third or fourth part is kept up in a coarse grunt or guttural sound, in the bass clef. The words are comprised in short sentences, each of which finishes suddenly with a staccato note, and a violent gesture. The music of one of the dances at Tutuila was as follows:

The females, unlike those of Tahiti, have not many musical voices among them, but, in common with other uncivilized races, have a perfect knowledge of time.

The men, on the contrary, produce round rich sounds, rather below tenor, but as wild as nature would have them to be.

The dance of the girls at Upolu consisted entirely of motions of the body, and was so indelicate as to produce disgust. The chaunt which accompanied it was sung with a high voice, and three or four women were employed in beating time on the mats with short sticks, in which most of the spectators joined with their hands. In all cases they kept time with the greatest accuracy.

The Samoan drum is made of a part of a tree, hollowed out; they have also an instrument, formed of a loose slat fitted into a board, on which they beat time with two sticks. Their flute, if it may be so called, is made of bamboo, as are also their pipes, which resemble those of Pan.

The dances of the men are by no means indecorous. Those who perform them vary in number from two to a dozen, and are divided into two parties. These parties alternately advance and retreat, which gives an appearance of animation. Clapping their hands, swinging them to and fro, or clasping them over their heads, they follow each other in a circle, leaping up and down, and turning suddenly around, keeping time to the music. The dances continue a considerable time, and end with a sudden clap of the hands and a simultaneous shout.

Drawn by A.T. Agate. A. Jones Sc.

SAMOAN DANCE.

The music to which they danced in Upolu, was as follows:

The song is usually extemporaneous, relating to some recent occurrence. The following is a translation of one of them, obtained by Mr. Couthouy through one of the interpreters.

The Papalangi has come to Samoa,
The Papalangi has come to Vaiusu,
Let us all go down to the spring.
The Papalangi is fond of the Siva.
Where is the pig? Where is the fattened fowl?
The Papalangi cannot join in the Siva.
Kindle up a bright blaze! Where are the virgins?
I am going to get some cocoa-nuts.
Look at this Samoan, how finely he dances!

These dances are usually performed in the fale-tele, where strangers are entertained. The inhabitants and their guests occupy different ends of the building, and alternately keep up the dancing and singing. Through the latter all the news is made known, occurrences related, and inquiries made and answered.

Many of the nights are spent in this way; and much of the daytime in eating, bathing, and sleeping.

Besides these dances, there are various games. One of these, called "lupe," is played by two persons, who sit opposite each other. One of them presents his closed fist to his opponent, and then rapidly holds up one, two, three, or all the fingers and the thumb, striking the back of his hand on the mat at the same time. If his opponent fails of instantly holding up a like number of fingers, he loses a point, and ten points finish the game.

"Lafo litupa" is also played by two persons, who place about fifty beans of the Mimosa scandium before them; then taking up four at a time, they throw them up in the air, and catch them on the back of the hand; the player who catches a hundred soonest is the winner.

Tuae-fua: this is played by five or six persons. It resembles the sport of the Chinese jugglers with iron balls. The first player sometimes takes as many as eight oranges, throwing them successively into the air, and endeavours to keep the whole in motion at once. They are very dexterous at this: if they miss three times, the game is lost.

Tui-muri affords the natives much amusement. Any number of persons may play at it. They seat themselves in a circle, and divide into two parties. An orange is suspended from above, about two feet from the ground, and each person is supplied with a small sharp-pointed stick. The orange is swung round, and as it passes, each one endeavours to pierce it, some with great eagerness, others quite calmly, and others again with a wary coolness, all of which affords much amusement to the bystanders. The party wins who first succeeds in fairly hitting the orange fifty times.

It is played for mats, trinkets, &c., but more generally for a baked pig, which is eaten when the play is over.

Litia: this is a general sport, sometimes whole villages playing against each other; it is in fact an exercise in spear-throwing. Two parties furnish themselves with light sticks of the Hibiscus tiliaceus, about eight or ten feet long and as thick as a finger. The bark is stripped off, which makes them very light. The two parties arrange themselves in a line, and strive to throw these as far as possible; the party which succeeds in throwing fifty the farthest wins the game. The usual distance to which they are thrown is about forty yards, and one would conceive it almost impossible for them to be thrown so far. A grand feast usually terminates the sport, which the losing party pays for.

"Lafe," is a game confined to the chiefs, who play it for pastime. Four persons sit at the corners of a mat, ten or twelve feet long, in whose centre is placed another of ten inches square; the persons at opposite corners are partners; each party is provided with five circular pieces of cocoa-nut shells, from two inches in diameter to half a cocoa-nut. The first player lays his smallest piece on the little mat, and his opponent tries to knock it off, and leave his own in its place. Each in his turn endeavours to knock his opponent's pieces off. The party which first succeeds in knocking its opponent's pieces off one hundred times, wins the game. The pieces of cocoa-nut are finely polished and carved with a variety of devices.

Sham club-fights and wrestling-matches are common, and frequently end in broken heads, limbs, and teeth knocked out.

PAPALANGI SHIP.
SAMOA.

An odd amusement of the natives was seen in the forest, in one of the clearings near one of the heathen villages, and at a short distance from Apia, (the vignette gives a good idea of it.) A fine large tree had been lopt of its branches (except at the very top), for a mast; around this a framework of timber, after the model of a vessel, was constructed; all the timbers were carefully fastened together with sennit, and with the requisite curvature; from the bow a large and long piece of timber projected, and at the stern a rudder was contrived, with its tiller; but instead of its ordinary movements as with us, it was intended to act vertically, in the way to which they are accustomed in managing or steering their large canoes with an oar; vines and creepers were used for the rigging; ballast had likewise been placed in the hold.

This afforded them great amusement, and showed an ingenuity in the construction of this Papalangi ship, as they called it, which had cost them much time and labour.

There is no ceremony at births, or indeed any inconvenience. The mother generally proceeds immediately to the spring, bathes and washes her infant, and at the same time her usual occupations are resumed. The naming of the child frequently takes place some time before its birth, for sex makes no difference in the names, which are given indiscriminately to males and females.

The mothers often suckle their children until they are six years old; and I was told of an instance where a woman gave nourishment to three children of different ages at once, the eldest removing the youngest sometimes by force from the mother's breast.

It is their practice to wash the children frequently in the fresh-water streams.

When a native wishes to get a wife, the consent of the chief is first obtained. Then he takes a basket of bread-fruit, and offers it to the girl of his choice. His suit is considered as accepted if she partakes of it. He must then pay her parents a certain price for her, which varies with the station and ability of the parties. A chief's daughter is valued high, viz., at half a dozen hatchets and as many fathoms of cloth.

Another mode of courtship is to go to the house of the object of attachment or desire, and be entertained. If the family show a friendly feeling towards the young man and eat with him, his addresses are favourably received. The formal offer is made by a large present to the family of the female, which being accepted, the match is made, and if refused, the courtship is at an end. The parents expect their children to abide by their decision. The "Malo" party have been in the habit of taking wives from their conquered enemies when they thought proper. At a marriage ceremony a great feast is made, particularly if it be a chief's.

A man is at liberty to repudiate his wife and marry again on certain conditions, but the woman cannot leave her husband without his consent.

Adultery was formerly punished with death, and is very seldom committed. Among single women, intercourse with a Samoan before marriage, is a reproach, but not with transient foreigners.

It is a common practice for parents to make a present of their children to chiefs or others, who adopt the child as their own, and treat it ever after as such. After it is grown up, one-half of its earnings goes to its adopted parent. This custom gives the chiefs many adopted children of both sexes, who continue to live with them, and are in all respects treated as their own; and spreads their connexions far and wide.

In their burials at Upolu, they have but little ceremony. The body is enveloped in many folds of tapa, and deposited, as has already been described at Tutuila, with the ti planted around. No utensils, arms, &c., are deposited with the bodies; for, according to their belief, they have these things provided for them in their Elysium. A feast is made for the attendants, consisting of pigs, taro, bread-fruit, &c.; presents are made by all the relatives to the family of the deceased, and if the family can afford it, a small canoe is procured for a coffin. After the body has lain in the grave some time, they take up the skull and place it in a box in their houses. The reason assigned for this is to prevent their enemies from possessing themselves of it, for it was a custom in their wars to violate the sanctity of the grave. We heard that a few of the bodies of chiefs had been preserved by oil and heat; and the missionaries informed me that they had seen the bodies of those who died thirty or forty years before, preserved in this manner.

Their mode of showing their grief is to burn themselves to blisters, (forming indelible marks,) with little rolls of twisted tapa, which, on being lighted, soon produced a coal. They also scratch their bodies. The females are said (in token of affliction for deceased friends) to have pricked holes in the corpse, and sucked out the fluids. All these practices may be now said to be passing away, and are almost obliterated.

There is already a very great difference, not only in dress but in appearance, between those who have adopted Christianity, and those who adhere to heathenism. The latter have a wild look, to which their long hair, tied in a bunch behind, adds not a little; and when going to war they let it hang down in wild confusion, which increases their savage appearance.

DEVIL MAN.

On the other hand, the Christians crop their hair short,—a fashion which was introduced by the missionaries.

The hair of the children is cropped close, except a lock on each side of the head. The manners of the people in the Christian and heathen villages are as different as their appearance. In the latter no schools are seen, nor any of the incipient marks of civilization. Their reception of strangers in the Christian villages is always kind and hospitable, although, as has been stated, a return is looked for. Among the heathen, the manner of reception cannot be counted upon with certainty, for they at one time welcome their visiters with cordiality, and at another are rude, insolent, and anxious to obtain all the strangers possess. When in good humour, they entertain their guests with the lascivious dances we have described, performed by native girls. Their whole manner and conduct are so different from those of villages within a short distance of them, that the effect produced on the latter by the instruction of the missionaries, appears almost miraculous.

In the heathen villages the dress of the Samoans is to be seen in its primitive simplicity. It is no more than the titi, which is a short apron and girdle of the leaves of the ti (Dracæna), tied around the loins and falling down to the thighs. The women besmear themselves with cocoa-nut oil mixed with turmeric, which gives them a shining yellow tint, that is considered as a beauty; on each breast is a spot of reddish brown, of a singular shape, and of various sizes, from that of a dollar to that of a dessert-plate. They do not show the least sign of feminine bashfulness, while those of the Christian villages cover their bosoms, and exhibit as much modesty as those of any country.

SAMOAN GIRL.

During the last ten years the dress of the natives has undergone much change; the titi has been increased in length, and extends all round the body; it has a neat and pretty effect when first put on, but requires renewing often, as the leaves wilt in a few days; this garment is well adapted to the climate, being cool, and the necessity of frequent change insures cleanliness.

The Wesleyan missionaries from the Friendly Islands have introduced the siapo, of Tonga, which has now come into common use. It is soft, pliable, and not glazed, and is principally used as a wrapper, after the manner of the pareu of the Tahiti Islanders. A piece of cotton cloth is usually worn by the chiefs as a siapo.

The maro is worn when engaged in active exercise, or in war, as being less cumbrous. The women often wear a beautifully white shaggy mat (ie sina), hanging from the neck to the feet. It is woven by hand from the fine threads of the hibiscus; they also sometimes wear wrappers of the siapo form, and the tiputa, a kind of poncho, of the same material, after the old fashion of the Tahitians, which is more becoming than the loose gown introduced into that island by the missionaries.

There is another kind of mat, of very fine texture, worn on great occasions, and used in their dances as a kind of cloak. It is ornamented with a border of red feathers. This is the most valuable property they possess, for they cost much pains to the manufacturers, and are often a year or eighteen months in their hands.

In the way of ornaments they use but few. The men usually wear a shell (the ovula) suspended around the neck by a string.

Their hair formerly claimed much of their attention, as it does still that of the heathen, who, as has been seen, wear it long and have it nicely combed and twisted up in a knot on the top of the head. The females frequently used to wear a wreath of flowers, which gave them a picturesque and pleasant appearance; but the use of flowers as ornaments has been interdicted by the missionary teachers.

Tattooing, if not in reality, at least in appearance, may be said to form a part of dress. It is performed by persons who make it a regular business. The age at which it takes place is from fourteen to eighteen, and is usually considered the initiation to manhood. The usual colouring matter is obtained from the kernel of the candle-nut. Tattooing is here called ta-tatau, and is tastefully drawn. The natives are very fond of it. It is expensive to the family, for the operator always receives a high price for his labour, consisting of the finest mats, siapo, and other property, as agreed upon before the operation is begun. The instrument used is made of bone, sharp like the teeth of a comb, and requires but a slight blow to enter the skin. The part tattooed on the males is from the loins to the thighs, but the women have only a few lines on their hands and bodies.

The articles of which their dress is composed are manufactured by the females, who are exceedingly industrious. The common cloth or tapa is made of the inner bark of the paper-mulberry, which is cultivated for the purpose in nurseries. It is cut when the stem is about one and a half inches in diameter; the inner bark is separated and washed in water, which deprives it of some of its gum; it is then beaten until the adhesion of the fibres forms many of the strips into a single mass. The mallet used for this purpose is about two inches square, and about fourteen inches long, with a handle at one end; two of its faces are grooved and the other two smooth; the bark is laid on a board, and struck with the mallet in a direction at right angles with its fibres; the grooved sides are used to spread out the fibres, and the smooth ones to knit them together. The grooves also give a thready appearance to the surface.

This method differs from that practised at Tahiti, where the bark is beaten with a smaller mallet, upon a spring-board; and the tapa made here is of inferior quality. The tapa is often printed with colours in patterns. This is performed in a mode similar to that practised in Europe before the introduction of copper rollers. Instead of engraved blocks, they form tablets, about as thick as binder's boards, of pieces of large cocoa-nut leaves, by sewing them together. One side of the tablet is kept smooth and even, and upon this cocoa-nut fibres are sewed so as to form the required pattern, which is of course raised upon the surface of the tablet. These tablets are wet with a piece of cloth well soaked in the dye, after which the tapa, which for this purpose is well bleached and beautifully white, is laid upon them and pressed into close contact. The dye is made from herbs and roots, and is of various colours.

The women also manufacture the mats. Some of these have been mentioned in describing the dress of the natives: the finest kinds are made of the inner bark of the paper-mulberry; those of coarser texture of the leaves of the pandanus, which are nicely scraped and bleached. The mats are all made by hand, and by interlacing the fibres; one of the finest description will require the industrious labour of a year.

Among the mats are some of as fine a texture and as soft as if made of cotton. These are rarely or never manufactured at present, and are solely possessed by the chiefs, in whose family they are handed down from father to son, as heir-looms. They are considered as their choicest treasures, and are so much coveted, that wars have been made to obtain possession of them. For the latter reason they are called Fala-taua.

There are several distinct trades among the men besides that of tattooing; among the most esteemed is that of canoe-building, in which there is no little skill displayed.

The usual fishing-canoe is made of a single tree, with a small outrigger to balance it. They have no large double canoes, such as are seen in Tonga and Feejee.

The largest canoes are from thirty to sixty feet long, and capable of carrying from ten to twelve persons. They are formed of several pieces of plank, fastened together with sennit. These pieces are of no regular size or shape. On the edge of each plank is a ledge or projection, which serves to attach the sennit, and to connect and bind it closely to the adjoining one. It is surprising to see the labour bestowed on uniting so many small pieces, where large and good planks might be obtained. Before the pieces are joined, the gum from the bark of the bread-fruit tree is used to cement them close and prevent leakage. These canoes retain their form much more truly than one would have supposed, and I saw few whose original model had been impaired by service. On the outside, the pieces are so closely fitted as frequently to require close examination before the seams can be detected. This perfection of workmanship is astonishing to those who see the tools with which it is executed. They are now made of no more than a piece of iron tied to a stick, and used as an adze. This, with a gimlet, is all they have, and before they obtained these iron tools, they used adzes made of hard stone or fish-bones.

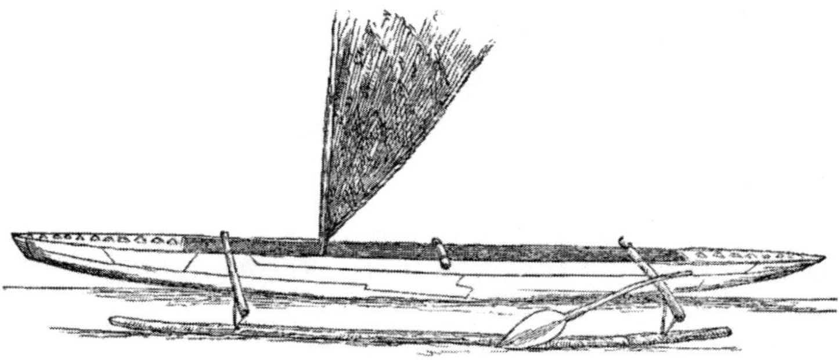

SAMOAN CANOE.

These canoes are built with a deck forward and aft. They are long and narrow, and their shape is elegant. They are paddled by natives, who sit two abreast, and are guided by a steersman. The seat of honour is on the forward deck, in the centre of which is a row of pegs, to which the large white ovula shell is attached by way of ornament. The natives find no difficulty in occupying this place, as they manage to sit in almost any position with ease to themselves; but a stranger who attempts it, and is for any time confined to one of these places of honour, will repent of the distinction he enjoys before many minutes are over. One of our gentlemen was treated with this distinction, and will long recollect the words of the song they sing.

"Lelei tusilava le tau mua,
Leango tusilava le tau muri."

"Good above all is the part before,
Bad above all is the part behind."

The uneasiness, from his account, does not only proceed from the small place left to sit upon, but also from the constant apprehension of being precipitated into the sea. This faa Samoa, or Samoan fashion, is any thing but agreeable.

Having both a prow and stern, these canoes cannot be manœuvred without tacking; consequently the out-rigger, that constitutes their safety, is, in using their sail, alternately to leeward and windward, and does not, when to leeward, add much to the stability of the canoe. They carry less sail than the canoes of the other natives of Polynesia, and to guard against the danger of upsetting, the natives rig a sprit or boom (suati), projecting from the opposite side to that on which the out-rigger is fitted. This boom is secured with guys to the top of the mast. When the wind blows fresh, some of the men go out upon it, and thus balance or counteract the force of the wind. Those on the other side of the canoe are kept ready to go out on the out-rigger when that becomes necessary. The sail is made of a mat, of a triangular shape, with its apex below: some of these are ten feet high.

None of the canoes we saw at the Samoan Group are calculated for long voyages. Those used in their intercourse with the Tonga Islands, are the large double Feejee canoe, of which I shall speak when I treat of those islanders.

In their trips from town to town, they are generally on parties of pleasure, termed *malanga*, and are frequently to be met with singing their boat-songs.

These songs have but little variety, are destitute of melody, and have small pretensions to harmony. They consist, for the most part, of two short strains, repeated alternately, the first by a single individual, and the second by several. Their voices are loud, and have generally a tenor character; the strains are mostly in the minor scale, and sung in the key of two or three flats.

The following boat-songs will give an idea of them:

Another:

"Cook* tells you pull away,
I will do so, and so must you."

The work in which the Samoans show their greatest ingenuity, is in the construction of their native houses, and particularly of their faleteles or council-houses, some of which are of large dimensions. They are built of the wood of the bread-fruit tree, and there are two modes in use, their own, and that borrowed from the Friendly Islands. The true Samoan house is slightly oval; those of the Friendly Islands are oblong. They may be said to consist of three parts, the centre and two ends; the former is erected first. For this purpose the three centre posts, which are twenty-five or thirty feet high, are usually first raised; on these rests the ridge-pole. A staging or scaffolding is now erected, nearly in the form of the roof, which serves for ladders and to support the roof temporarily. The roof is commenced at the

* All the natives have some knowledge of Captain Cook, derived from their communication with the Friendly Islands.

ridge-pole, and is worked downwards. The cross-beams are lashed in at different heights, connecting the centre portions of the roof together, and are fastened to the upright centre posts. The rafters are made of short pieces, placed at equal distances apart, and form the curve that is required to construct the roof. Between the largest rafters are smaller ones, about one foot apart. Across the rafters are placed and fastened many small rods, about an inch in diameter. The whole is neatly thatched with the sugar-cane or pandanus-leaves, and the rafters are terminated by a wall-piece, made of short pieces of wood, fastened together and to the rafters, so as to form the ellipse required for the roof. The end portions, of similar small pieces, are made to correspond to the required curvature of the roof and the ellipse of the wall-plate. Posts are now placed in the ground, about three feet apart, to receive the wall-piece, which is fastened to their tops. There is no fastening used but sennit, made of cocoa-nut fibres. The rafters are generally made of the hibiscus, which is light and strong. The eaves extend about a foot beyond the posts. The smaller houses generally have permanent sides; the larger ones are open all around, but mats are hung up as curtains by the occupants, and any part may be used as a door.

SAMOAN FALE-TELE.

After the whole is finished, the interior has the appearance of an extensive framework, from the number of cross-beams, which are used as depositories for their property, tapas, mats, &c.; and in some cases the favourite canoe of the chiefs is placed on them. After a full inspection of one of these fabrics, one cannot but view these natives not only as industrious, but as possessing great skill and ingenuity. The thatching lasts four or five years. There is no floor to the house, but the ground is covered with stones about the size of a small egg. There is usually a paved platform on the outside, about three feet wide. In some cases this is raised a foot, and serves to keep the house dry, as the stones allow a free passage to water. On the pavement are laid coarse mats, and the finer ones are spread above, covering about half the area.

These fine mats are rolled up until required. Many baskets hang here and there, with some cocoa-nut shells to contain water, and the ava-bowl. Mats are suspended about as screens. At night, each sleeper is usually supplied with a musquito-curtain, called tai-namu, which, forming a kind of tent, by being passed over a ridge-pole or rope, and falling on the ground, answers all the purposes required.*

On one, and sometimes on both sides of the centre-post of the houses, is a small circular hearth, enclosed by stones of larger size; this is the place for burning the dried leaves of the cocoa-nut, which serve them for light at night. Although these do not give out much smoke, yet as they burn for a long time, the house gradually becomes filled with soot, for there is no outlet above for its escape.†

As they always use the flambeau to light them on their return from their feasts, it produces a singular and pretty effect to see an assembly breaking up, and the different parties winding through the groves with torches, throwing the whole into bold relief. A rude lamp is also used, made of a cocoa-nut shell, with a little oil in it, and a piece of vine-stalk for a wick, and likewise the nut of the Aleurites triloba, or candle-nut, several of which are strung on a thin stick.

Many white-washed houses are now to be seen, for the natives have been taught the use of lime by the missionaries, and are beginning to use it in their dwellings. All the missionaries' houses have plastered walls, and board floors, and are very comfortable. There is a great quantity of fine timber on these islands, for building purposes. The timber of the bread-fruit tree and hibiscus, are alone made use of by the natives. The missionaries have their planks or boards sawed by hand, and generally by foreign carpenters.

The food of the Samoans is prepared in the way practised at Tahiti, and generally consists of bread-fruit, bananas, taro, sweet-potatoes, and yams. Fish is supplied in quantities from the reef, and they also eat the large chestnut, vi-apple, and arrow-root, the fecula of which they begin to manufacture in some quantities. Although it would scarcely be supposed necessary, where every thing is so bountifully supplied by nature, yet they make provision for times of scarcity and for their voyages, of the bread-fruit, made when green into a kind of

* Musquitoes are exceedingly annoying to strangers, but I did not remark that the natives were troubled with them. Their bodies being well oiled is a great preservation against the bites of these insects.

† The prevalence of sore eyes is said to be owing to the smoke of the lamps.

paste, and rolled in banana-leaves. This undergoes a partial fermentation, and is called mahi. It is not unlike half-baked dough, and has a sour unwholesome taste. They eat birds, &c., but a large wood-maggot which is found on the trees, is looked upon as the most delicious food they have.

They have much variety in their cooking, and some of their dishes are exceedingly rich and agreeable to the taste. They practise several modes of cooking the taro-tops; one, by tying them up with cocoa-nut pulp and baking them, in which state they resemble spinach cooked with cream, but are sweeter. Another dish is called faiai, made of the scraped and strained cocoa-nut pulp boiled down to the consistency of custard. It is eaten both hot and cold.

The habits of the Samoans are regular. They rise with the sun, and immediately take a meal. They then bathe and oil themselves, and go to their occupations for the day. These consist in part of the cultivation of taro and yams; building houses and canoes. Many fish; others catch birds, for which purpose they use nets affixed to long poles. They generally find enough to employ the mornings, in getting their daily supply. After this is done, they lounge about, or play at their various games, eat about one o'clock, and again at night, retiring to rest about nine o'clock. The men do all the hard work, even to cookery.

The women are held in much consideration among this people, are treated with great attention, and not suffered to do any thing but what rightfully belongs to them. They take care of the house, and of their children, prepare the food for cooking, do all the in-door work, and manufacture the mats and tapa.

They are cleanly in their habits, and bathe daily; after which they anoint themselves with oil and turmeric. This custom, I have no doubt tends to preserve the health by preventing the excessive perspiration which the heat of the climate naturally brings on. It is, however, at times offensive, for the oil is apt to become rancid.

The Samoans are of a social disposition, more so, indeed, than the other natives of the Polynesian islands, and they are fond of travelling. The reasons they have for taking these journeys are various: thus, when there is a scarcity of food in one part, or a failure of the crops, they are in the habit of making a "faatamilo," or circuit, around a portion of these islands, so that by the time they return, (which is at the expiration of three months,) their own taro has grown and the bread-fruit season come around. They are now in their turn prepared to afford the same hospitality and accommodation to others. The old people are usually left at the village to take care of it, whilst the younger portions are gone on one of these malangas, or journeys. During these expeditions, a sort of trade is frequently carried on. The different portions of the inhabitants are each celebrated for a particular staple. Some excel in making mats; others in building canoes; the districts in which the seaports are, obtain a variety of articles from ships, which are subsequently distributed over the whole group.

It may readily be supposed that there are many circumstances which make this mode of communication inconvenient, particularly when the travelling party is a large one, in which case it absolutely breeds a famine in its progress.

I have before stated that every village has its "fale-tele," which is the property of the chief. In this their "fonos" or councils are held, and it is also the place where strangers are received. The mode of receiving visiters is attended with much ceremony. A party enters the village without inquiring where or how they are to be entertained, and take up their quarters in the "fale-tele." In a short time the chief and principal personages collect and visit the strangers, telling them in a set speech the pleasure they enjoy at their arrival, and their delight to entertain them. This is mostly said in what they term "tala-gota," the speech of the lips, and much complimentary language ensues. The Samoan language abounds in phrases adapted to this use, and worthy of a refined people.

After this interchange of compliments, the young women assemble to treat the strangers to "ava." This is prepared after the usual mode, by chewing the Piper mythisticum. During this time the young men are employed collecting and cooking food. This is all done with great despatch. The pigs are killed; the taro collected; the oven heated; and baskets made to hold the viands. In the feast they are well assured of sharing, and therefore have a strong stimulus to exertion.

The strangers, on receiving the food, always return part of it to the entertainers. Thus all the village is occupied with the entertainment, and a scene of frolicking ensues until the strangers see fit to take their departure.

Among the heathen, dancing during the evening always follows this feast; but the Christian villages have abolished all dancing.

These visits are not always paid or received in a spirit of hospitality. The chief of a powerful district takes this mode to exact tribute from his less powerful neighbours, and they are on such occasions extremely overbearing and insolent to their entertainers.

For crimes, they have many forms of punishment, among which are: expulsion from the village in which the offender resides; exposure of the naked body to the sun; flogging; cutting off the ears and nose; confiscation of property; and the compulsory eating of noxious herbs.

When a murder has been committed, the friends of the person slain unite to avenge his death; and the punishment does not fall upon the guilty party alone, but on his friends and relatives, who with their property are made the subjects of retaliation. If any delay in seeking redress in this manner occurs, it is received as an intimation that the injured party, whether the family, the friends, the village, or whole district to which the murdered person belonged, are willing to accept an equivalent for the wrong they have sustained. The friends of the murderer then collect what they hope may be sufficient to avert retribution, and a negotiation is entered into to fix the amount of compensation. When this is agreed upon, it is offered to the nearest relative of the deceased, and the parties who present it perform at the same time an act of submission, by prostrating themselves before him. This closes the affair.

For some crimes nothing but the death of the offender could atone. Among these was adultery; and when the wives of chiefs eloped with men of another district, it generally produced a war. This was one of the causes of the wars waged by Malietoa.

There existed, however, means by which the code was rendered less bloody, in places of refuge for offenders, such as the tombs of chiefs, which were held sacred and inviolate.

Wars were frequent among the Samoans before the introduction of the gospel, and scarcely a month passed without quarrels being avenged, and with blows. The last and perhaps the most bloody war that has ever occurred on these islands, was about the time of the first visit of Mr. Williams, the missionary, in 1830, when the inhabitants of one of the finest districts, that of Aana, in the western part of Upolu, were almost exterminated. This war continued for eight months, and only those were saved who escaped to the olos, or inaccessible places of refuge, or were protected by the "Malo," the ruling or conquering party.

When the missionaries arrived, in 1836, and for upwards of a year afterwards, Aana was without a single inhabitant; but through their influence upon the Malo party, it was agreed at a large "fono" to restore the exiles to their lands. Aana is again (in 1839) the finest part of the island, and will be in a few years quite a garden.

These wars, like those of all savage people, were attended with great cruelty, and neither old nor young of either sex were spared. It is related that after the last battle of Aana, a fire was kept burning for several days, into which hundreds of women and children were cast.

Their wars were seldom carried on in open fight, but stratagem was resorted to, and all enemies that could be attacked were killed, whether in their houses, or when accidentally met with at their work in the taro-patches.

Their arms consisted of clubs and spears, made of the iron-wood (Casuarina), bows and arrows, and of late years, the musket. The man who could ward off a blow and at the same time inflict a wound on his adversary, was considered the best warrior. Each village had its separate commander, and there was no general, their operations being from time to time decided in council. Their spears were pointed with the sting of the ray-fish, which, on breaking off in the body, caused certain death.

The olos, above mentioned, were usually on the top of some high rock, or almost inaccessible mountain, where a small force could protect itself from a larger one. One of these olos, or strongholds, of the people of Aana, during the late war, was on a high perpendicular ridge, which forms the western boundary of the bay of Faleletai, and it was the scene of many a bloody contest. The Manono people, coming by night, would land at the foot of the hill, and attempt its ascent, while those on the top would roll and hurl down stones, generally overcoming them with ease, and driving the invaders back with great slaughter. The latter, however, took a fearful and truly savage revenge for their various defeats. Laying in wait until the women came down to fish on the reefs, they set upon them, and massacred them all. The burning of houses, the destruction of the bread-fruit, cocoa-nut trees, taro-patches, and yam-grounds, &c., were the ordinary features of these conflicts.

Prisoners were sometimes spared, but they were usually held as subjects of retaliation, in case of any of the adverse party being killed.

Upon the occurrence of a cause of war, the parties sent to their respective friends in the different towns to solicit their aid. Such solicitations usually resulted in the whole district, and sometimes the whole of the island, being engaged in a civil war.

On going to war, they were accustomed to cast their hair loose, or to tie it up in various forms, and, to add to the fierceness of their appearance, they wore large bunches of false hair, which also in creased their apparent height.

In making peace, the conquered party was required to make submission, by bringing loads of stones, fire-wood, and green boughs, and to bow down very abjectly in the presence of the chief. They were also required to pay a large amount of tapa, mats, and other property.

The government of the Samoans is more refined in principle than could well be expected. The rule of hereditary chiefs is acknowledged, and the distinction of the several classes well defined. Great respect is paid to the chiefs, and particularly to the "Tupu," or highest class. To this belong Malietoa, Pea of Manono, &c. The second class consists of the near relatives of the first, and of others who have large possessions; the third, of the petty chiefs of villages; next come the tulafales, who are a well-defined class between the chiefs (alii) and common people. These tulafales are proprietors of the soil, and householders; they possess considerable influence, and act as advisers of the chiefs, and the executors of their orders. Like the chiefs, they derive their rank from descent. There is no distinct name for the common people as a class, but the chiefs in speaking of them always apply some opprobrious epithet. The son of a low-born woman by a chief ranks as a chief, although he has no authority, and the son of a noble woman by a man of mean birth, may be either a chief or a commoner.

The lands are allotted and distinguished by known boundaries. The natural heir of the former owner succeeds, and is the feudal chief or leader in war, but all his dependants are free to cultivate it. Lands may be sold, which is done at public meetings, and the bargain is made binding by sticking their staves into the ground, or digging a portion of it up.

The whole power lies in the high chiefs of the "Malo" or conquering party. They assemble in fono, and determine the general laws and rules of action. At the head of this is Malietoa, who is now considered the head chief of Atua, and is supposed will shortly acquire that of Tui of Aana. Each of these districts formerly had a separate chief, bearing the same title of Tui, but in their wars with Manono, nearly all the descendants of these princes were killed off. To obtain this title requires the consent of the chiefs of Manono, and part of Savaii, which belongs to the ruling party.

The fono may levy what contributions it pleases, particularly on those they have conquered. The present "Malo" or government is designated "Malo-to-toa"—the gentle government.

Although there is no supreme authority acknowledged in any one individual, yet there are instances of chiefs of districts assuming and maintaining it. The late Tamafago, of whom some account has already been given, was one of these. He assumed the attributes not only of a king, but of a god, and after conquering a rival district on Savaii, he took, as has been stated, the name, "O le Tupu o Savaii"—the King of Savaii. After he was killed, Malietoa succeeded to the same title; but it now confers no power, and is considered merely as complimentary.

Each district and town has its own government. An elderly chief generally presides, or is considered as the head of the village, town, or district. In these primary fonos or meetings, the affairs are generally discussed by the alii (chiefs) and tulafales (landholders), and what they determine on is usually followed. The great fono, or general assembly, is seldom called, except on matters affecting the whole of the island or district. The subject is calmly debated, and most thoroughly discussed; the final decision, however, is not by vote, but is adopted after consultation, and is governed by the opinions of the most influential chiefs. It thus appears that these assemblies have little influence upon the course the chiefs may have determined to pursue, and serve chiefly to insure the united action of the district in carrying the designs of the chiefs into effect. The tulu-fano or decree, promulgated by the council, is to be obeyed, and those who fail are punished by the Malo, being plundered by them of their lands, &c.

In the descent of the office of chief, the rule of primogeniture is not strictly followed, but the authority and title always remain in the same family.

It is the custom at the fonos to compliment the head chiefs, and invoke blessings on them in prayers, that their lives may be prolonged and prosperous. I was informed that these assemblies were conducted with much ceremony, but I was much disappointed in the one I witnessed. The forms of proceeding may, however, be different when strangers are not present. The fonos generally begin at an early hour in the morning, and last until late in the afternoon. One of the most pleasing of the ceremonies is that in which the chiefs are supplied with food during the time the meeting is in session. After the food is prepared and dished in fresh banana-leaves, the wives and daughters of the chiefs attire themselves in their best dresses. They then enter the fale-tele, and approach their fathers, husbands, and brothers, &c., before whom they stop, awaiting their instructions as to whom they shall hand the viands. When they have obeyed their directions, they retire. The whole duty is conducted with the utmost decorum, and while it is going on, no conversation is permitted except in a low voice. I learned from the missionaries who had attended some of their meetings, that the manner of speaking was good, and the self-possession of the orators remarkable. The speakers generally have persons near them who act as a sort of prompters, and remind them of the subjects it is desirable they should speak of. The whole proceedings are conducted with the utmost quiet, and no disturbance is allowed.

SAMOAN PET PIGEON.

CHAPTER VI.

CONTENTS.

DEPARTURE FROM THE SAMOAN GROUP—WALLIS ISLAND—DISPOSITION MADE OF TUVAI—CATHOLIC MISSIONARIES AT WALLIS ISLAND—HOORN ISLAND—MATTHEWS' ROCK—REMARKABLE THUNDER-STORM—BALL'S PYRAMID—PORT JACKSON ENTERED—ARRIVAL AT SYDNEY—VISIT TO THE GOVERNOR—FORT MACQUARIE—FATE OF MR. WILLIAMS—DESCRIPTION OF THE TOWN OF SYDNEY—ITS STREETS—ITS RESEMBLANCE TO AMERICAN TOWNS—RELIGIOUS SECTS—PREVALENCE OF INTOXICATION—GOVERNMENT-HOUSE—DRIVE TO SOUTH HEAD—PUBLIC GROUNDS—MR. CUNNINGHAM THE BOTANIST—HIS MELANCHOLY FATE—COUNTRY AROUND SYDNEY—THE NATURALISTS LEFT IN NEW SOUTH WALES—GENERAL DESCRIPTION OF THAT COLONY—ITS FORESTS—ITS GEOLOGICAL STRUCTURE—COUNTRY SOUTHWEST OF SYDNEY—LIVERPOOL MOUNTAINS—COUNTRY NORTH OF THE LIVERPOOL MOUNTAINS—ILLAWARRA—DROUGHTS AND FLOODS—REMARKS OF MAJOR MITCHELL—RIVERS OF NEW SOUTH WALES—ITS MINERAL PRODUCTS—ITS WATER—THE HEIGHT OF ITS MOUNTAINS—ITS CLIMATE—ITS TEMPERATURE—PREVAILING WINDS—ITS VEGETATION—MONOTONY OF ITS SCENERY—ITS HORTICULTURE—NATIVES OF AUSTRALIA—THEIR NUMBERS—THEIR PHYSICAL TRAITS—THEIR CHARACTER—THEIR WANDERING HABITS—THEIR TIMIDITY—THEIR CONFLICTS—THEIR CORROBORY DANCES—ACCOUNT OF ONE OF THEIR BATTLES—THEIR MODE OF SETTLING PRIVATE QUARRELS—THEIR WEAPONS—THEIR MODE OF CLIMBING—THEIR CANOES—THEIR SOCIAL SYSTEM—THEIR CUSTOM OF MAKING YOUNG MEN—THEIR MARRIAGES—BURIAL OF THEIR DEAD—THEIR RELIGIOUS OPINIONS.

Drawn by A. T. Agate.

Eng.d by Rawdon, Wright & Hatch.

SYDNEY.

CHAPTER VI.

NEW SOUTH WALES.

1839.

On the 10th of November we weighed anchor from Apia, and made all sail to the westward; and on the 11th had lost sight of Savaii. Officers were stationed for the three following nights to look out for the periodic showers of meteors, but the nights were cloudy, and none were seen.

On the 12th we made Uea or Wallis Island, and at 3 P. M. were off its southern end, which is situated in latitude 13° 24′ S., longitude 176° 09′ 22″ W. Instead of a single island as might be expected from the name, there are nine separate islands, varying in circuit from one to ten miles, and enclosed with one extensive reef. The land is, in general, high. We made a running survey of this group.

While off Wallis Island, we were boarded by a canoe, in which was a native who spoke a little English. I had thus the means of communicating with the shore, and resolved to take advantage of it by landing the prisoner Tuvai. I conceived that this would accomplish all the ends I had in view in removing him from his native island, particularly as the course of the wind is such, for the greater part of the year, as to prevent canoes proceeding from Wallis Island to the Samoan Group, and there is in consequence no communication between them. His fate would of course remain a mystery to his countrymen, and the impression I had hoped to produce on their minds would be effectually made. My original intention had been to land him at Hoorn Island, which is two days' sail further to the south; but a similar opportunity might not perhaps have presented itself there.

Having decided on this course, I committed him to the charge of the person who had boarded us, and gave particular directions that

O

he, with his rolls of tapa, should be immediately taken and presented to the chief. The customs of the islanders promised that this would insure him good treatment, by giving him at once a protector; or at least that he would be only robbed by a single person, and not exposed to the pillage of the whole population, who would in all probability have stripped him of his property the instant he landed, if not restrained by the authority of a chief.

Tuvai seemed delighted at being released from his confinement on shipboard, and took his leave by shaking hands with the sentry. Thus, while the culprit has not been exposed to any unnecessary severity of punishment, I feel satisfied that I fully accomplished my object of convincing his countrymen that they could not hope to commit murders upon their white visiters with impunity.

These islands appear to be well wooded, and we saw many large native houses upon them. As we drew near, we perceived upon a rocky flat a few natives waving a white flag. The native who came on board informed me that the inhabitants were numerous, and that among them there were ten white men.

It is said that the Catholic missionaries who were expelled from Tahiti were landed on this island, when, the moment they reached the shore, they were stripped of all they possessed. They, notwithstanding commenced their good work, and are reported to have performed it effectually.*

The entrance to the lagoon is on the south side of the group, and the pilot, if so he may be called, informed me that there was ample room for the ship to pass within the reef. Wood, water, and refreshments may be obtained here.

Towards evening we stood on our course with a strong breeze, regretting that time did not permit of landing and obtaining a more full account of this little-known land. But the season for operating in high southern latitudes was rapidly approaching, and I was aware that, to say nothing of the extent of sea that was to be traversed, I must spend a considerable time at Sydney in making the necessary preparations for a long and arduous cruise.

Hoorn Island was made the following day. It was discovered in 1616 by Schouten and Le Maire. Its highest point is two thousand five hundred feet above the sea; on its northern side many rocks are visible, and the whole surface appears bold and precipitous, affording, as far as we could perceive, little soil for cultivation. Cocoa-palms in

* While in the Feejee Group, I learned that a Catholic mission had already been established there; that it was prospering, and that it had already been the means of saving an English vessel from capture, by a timely notice to the crew.

considerable numbers, were, however, observed upon a low point projecting from its southern side.

This island is inhabited, and I have been informed that an unsuccessful attempt to establish a mission upon it was made by the Catholics in 1840.

Taking our departure from Hoorn Island, we made all sail to the southward, passing about sixty miles to the westward of the Feejee Group, which was to be afterwards a subject of close examination. On crossing the meridian of 180° we dropped the 14th of November, in order to make our time correspond to that of the Eastern Hemisphere, to which our operations were for some months to be confined.

On the 18th, we saw Matthews' Rock, whose height we ascertained to be one thousand one hundred and eighty-six feet. It is of a conical shape, about a mile in circumference, and principally composed of conglomerate. A dike of basalt was observed occupying about a third of the width of the island. In order to obtain specimens, a boat was despatched to endeavour to effect a landing: the undertaking proved difficult, but was accomplished by Dr. Fox and Midshipman Henry, who swam through the surf. They brought off some specimens of porphyritic rock, and a few small crystals of selenite. Patches were seen on the northern side of the island appearing as if covered with sulphur. As has been so often mentioned in speaking of other uninhabited islands, great numbers of birds were seen upon and around it. This island is in latitude 22° 27′ S., longitude 172° 10′ 33″ E.

For several days preceding the 18th, a current had been perceived setting southwest; it was tried here, and found to set in that direction at the rate of three fathoms per hour. The wind began here to haul to the northward and eastward.

We had the misfortune on this day to lose one of our Six's thermometers, after having made a cast of two hundred fathoms with it. The difference between the temperature at the surface and at that depth, was 14°, the former being 76°. The following day (19th), a cast of six hundred fathoms was made by the Peacock. The temperature below was 50°, while that at the surface was 73°.

On the 24th, we had a remarkably severe storm of thunder and lightning; the ship appeared filled with the electric fluids; the points of the conductors, the mastheads, and yardarms were illuminated with Corpo Santos; and several of the officers declared that they had felt electric shocks. The gale blew violently, beginning from the northwest, and then shifting to the southwest. During its continuance the thermometer fell seventeen degrees.

For the two following days we had head winds, and a heavy cross sea.

On the 26th November, we made Ball's Pyramid, which appears to be a barren rock rising abruptly from the sea.

On the 24th, 25th, 26th, and 27th, we experienced a current setting at the rate of twenty-four miles a day to the northeast. On the 28th it set east-northeast at the rate of twenty-five miles per day. From the latter date the current began to set strongly to the southwest, showing that we had entered the stream which sets in that direction along the coast of New South Wales.

At sunset on the 29th November, we made the light-house on the headland of Port Jackson. We had a fair wind for entering the harbour, and although the night was dark, and we had no pilot, yet as it was important to avoid any loss of time, I determined to run in. I adopted this resolution, because, although we were all unacquainted with the channel, I was assured that the charts in our possession might be depended upon, and I stood on under a press of sail, accompanied by the Peacock. At 8 P. M. we found ourselves at the entrance of the harbour. Here a light erected on a shoal called the Sow and Pigs, since the publication of the charts, caused a momentary hesitation, but it was not long before it was determined where it was placed, and with this new aid, I decided to run up and anchor off the Cove. In this I succeeded, and the Peacock, directed by signal, followed the Vincennes. At half-past 10 P. M. we quietly dropped anchor off the Cove, in the midst of the shipping, without any one having the least idea of our arrival.

When the good people of Sydney looked abroad in the morning, they were much astonished to see two men-of-war lying among their shipping, which had entered their harbour in spite of the difficulties of the channel, without being reported, and unknown to the pilots. Their streets were speedily alive with our officers and men, who were delighted at finding themselves once more in a civilized country, and one where their own language was spoken.

The Porpoise and Flying-Fish arrived the next day.

The morning of the 30th was beautiful, and the scene that broke upon us was totally unlike any we had hitherto witnessed during our voyage. In particular, the strong resemblance of all that we saw to our own homes, and the identity of language, gave us indescribable feelings of pleasure.

Our consul, J. W. Williams, Esq., came early on board to welcome us. He communicated the information that the Relief had arrived safely, and landed all our stores, which were ready for us and close at hand; after which, and about ten days before our arrival, she had sailed for the United States.

Our arrival was duly announced by an officer, and through him I was informed that the governor, Sir George Gipps, would be happy to receive me at eleven o'clock. In compliance with this intimation, I had the honour of waiting upon his Excellency at that hour, in company with Captain Hudson, and our consul. I made my apologies for having entered the harbour in so unceremonious a manner, and stated the reasons why I could not tender the customary salutes.

The reception I met with was truly kind: every assistance which lay in his power was cordially offered; and I was assured that I had only to make my wants known to have them supplied. The use of Fort Macquarie was immediately granted me for an observatory, a position which, being within hail of my ship, gave me great facilities for conducting my experiments, and at the same time superintending my other duties.

Fort Macquarie is situated on Bennilong's Point, which forms the eastern side of Sydney Cove; it covers about half an acre of ground, and is twenty feet above high-water mark; it has a few guns mounted, but they are in no condition for service.*

A few days before our arrival, it had been debated in council, whether more effectual means of fortification were not necessary for the harbour. The idea of this being wanted was ridiculed by the majority; but the entrance of our ships by night seems to have changed this opinion. Had war existed, we might, after firing the shipping, and reducing a great part of the town to ashes, have effected a retreat before daybreak, in perfect safety.†

I may in this place acknowledge the open-hearted welcome we met with from all the government officers, military and civil, as well as from the citizens. Our reception was gratifying in the extreme, and cannot be too highly appreciated. The Australian Club‡ was thrown open to us by its committee, and parties, balls, &c., were given in our honour; in short, all our leisure time was fully occupied in the receipt of these hospitable attentions.

The day after we anchored at Sydney, the brig Camden also arrived. By her we learned the melancholy intelligence of the death of the Rev. Mr. Williams, from whom we had parted so short a time before at the Samoan Group. He was then, as will be recollected,

* I understand that since our visit to Sydney, Fort Macquarie has been demolished.

† Since our visit, however, several new fortifications have been erected.

‡ At the Australian Club, I had the pleasure of seeing Count Strezleski, well known in the United States, who was travelling in New South Wales.

about setting forth to propagate the gospel among the savages of the New Hebrides, and was in full health and high spirits, in the ardent hope of success in his mission. My information in respect to this sad event, was derived from his associate, Mr. Cunningham. They had placed native missionaries at Rotuma and Totoona. Mr. Williams then landed at Tanna, which they found in a high state of cultivation, and where they were hospitably received by the natives. These were Papuans, and spoke a language much like that of the Hervey Islanders. At Tanna, Samoan missionaries were also left, and they thence proceeded to Erromango. Here they found a barren country and a different race of men, black, with woolly hair, who did not comprehend a word of any of the languages known to the missionaries.

The natives, although apparently suspicious, exhibited no symptoms of actual hostility. Mr. Williams, with Mr. Harris, Mr. Cunningham, and the master of the vessel, landed, and were strolling about, amusing themselves with picking up shells. While thus engaged, they had separated from each other, and Messrs. Harris and Williams were in advance of the others. On a sudden the war-shout was heard, and Mr. Harris was seen running, pursued by a crowd of natives. He was soon overtaken by them, and killed. Mr. Williams then turned and endeavoured to reach the boat, but he had delayed too long, and although he reached the water, he was followed into it and slain also.

Mr. Cunningham and the captain escaped, although with difficulty, and after some fruitless attempts to recover the body, left the island. Mr. Cunningham was of opinion that the attack had not been premeditated, but arose from a sudden desire to obtain possession of the clothes of the persons who were on shore; he was also satisfied that a single loaded musket in the hands of those left in the boat, would have been the means of saving these two valuable lives.

I had, in a conversation with Mr. Williams at Upolu, expressed my belief that the savage inhabitants of the New Hebrides would not be safely visited without the means of defence. He had in reply declared himself averse to the use of fire-arms or any other weapon in the propagation of the gospel; being of opinion that it would be more easily and effectually disseminated without them.

The missionary cause has sustained a great loss in Mr. Williams's death; for in him were united a true spirit of enterprise and fervent zeal, with great perseverance and a thorough knowledge of the native character. I still think with melancholy pleasure of the acquaintance I had the good fortune to form with him.

The town of Sydney may, for convenience of description, be considered as divided into two parts; the line that separates them coincides nearly with that of George Street, the Broadway of Sydney. The old town lies on the east side of this line, and occupies the eastern promontory of the Cove; it is the least reputable part, and is almost filled with grog-shops and brothels, except at its extreme eastern quarter, where there are a few genteel buildings, in agreeable situations. The streets to the south and west of George Street are well laid out, and are rapidly filling up with good houses.

George Street extends in a direction nearly north and south for two or three miles, as far as Brickfield Hill, which is also nearly covered with buildings. On George Street are many of the public buildings, among which are the barracks, the markets, the post-office, and the banks. The remaining buildings which front upon it are chiefly occupied as shops, in which almost every description of European manufactures may be procured, and some of them at moderate prices. Several streets run parallel, and others at right angles to George Street.

The houses of Sydney are for the most part well built and commodious. On the western side of the town are many handsome buildings and extensive public grounds; towards the eastern side is a large square, called Hyde Park, upon which are situated the offices of the colonial government, the church of St. James, and the Catholic cathedral.

Sydney contains about twenty-four thousand inhabitants, which is about one-fifth part of the whole population (120,000) of the colony; and about one-fourth of this number are convicts. In truth, the fact that it is a convict settlement may be at once inferred from the number of police-officers and soldiers that are every where seen, and is rendered certain by the appearance of the "chain-gangs." The latter reminded us, except in the colour of those who composed them, of the coffee-carrying slaves at Rio; but the want of the cheerful song, and the apparent merriment which the Brazilian slaves exhibit in the execution of their tasks, was apparent.

When viewed from the water, Sydney appears to great advantage. It lies on the south side of the harbour, and covers two narrow promontories, separated and bounded by coves. The ground rises gradually, and thus exhibits its buildings to great advantage, giving it the air of a large commercial city. It is chiefly built of a drab-coloured sandstone, resembling that employed in the new public buildings at Washington, but of a lighter hue. Red brick is also used in building, and the suburbs contain many neat cottages and country-seats. The sandstone is a beautiful material, but is not very durable. The view of the town is diversified with the peculiar foliage of Australian trees, among which the pines of Norfolk Island and Moreton Bay are most conspicuous. At the time of our arrival, the trees were infested with locusts (Cicada), which made a noise absolutely deafening. The sound this insect produces is the same as that made by the analogous species in the United States, but is continued here during the heat of the day, and ten times more deafening.

Handsome equipages abound; and the stage-coaches are numerous. These, with the costume and demeanour of the more respectable part of the population, struck us as being more like what is seen in our towns than in those of Europe. Every thing has a new look about it, and the people manifest more of the bustle and activity of our money-making and enterprising population than are to be seen in old countries. The acquisition of wealth seems to be the only object of all exertion here, and speculation was as rife as we had left it in the United States. Cutting down hills, filling up valleys, laying out and selling lots, were actively going on. There are in truth many particulars in which the people of Sydney resemble those of America. This is observable, among other things, in the influence of the public press. In Australia, however, it is more licentious than any except the lowest of our newspapers; taking unwarrantable liberties with private character, and is far from being remarkable for discrimination.

All the religious sects of the British Islands have their representatives here. Each has its ardent advocates, who appear to be in continual war with those of the others. The contest between them had risen to a great height at the time of our visit, which is probably to be ascribed to the agitation of a question in relation to the distribution of the school-fund.

In one particular, a most striking difference is to be observed between the scenes to be witnessed at Sydney, and in the cities of the United States. This consists in the open practice of the vice of drunkenness, which here stalks abroad at noonday. It is not rare at any time, but on holidays its prevalence surpasses any thing I have ever witnessed. Even persons of the fair sex (if they may be so called) were there to be seen staggering along the most public streets, brawling in the houses, or borne off in charge of the police. However highly coloured this picture may be thought, it is fully corroborated by the police reports of the Sydney papers on Monday mornings. The police-officers themselves are among the venders of the intoxicating liquid.

The facilities for indulgence in this vice are to be seen every where in the form of low taverns and grog-shops, which attract attention by their gaudy signs, adapted to the taste of the different orders of customers, as "the King's Arms," the "Punch-Bowl," the "Shamrock," the "Thistle," the "Ship," the "Jolly Sailors." Of these, two hundred and fifty are licensed by the government, or more than one to each hundred souls. Among them a small shop was pointed out, which from the extent of its custom, yielded the enormous amount of £200 for rent to its owner annually, a sum far beyond the apparent value of the whole property. The quantity of rum which is consumed in the colony may be estimated from the facts, that the revenue derived from its importation was in 1838 £189,450, and that the supply amounts nearly to eight gallons annually for every individual in the colony.

This state of things arose, of course, originally from the habits of the abandoned persons who formed the nucleus of the population. It might, therefore, have appeared to be the duty of the successive governors to restrain the vice, or even to render its commission impossible, by prohibiting importation. So far as penalty goes, this has been attempted, and a fine of five shillings is levied on all who are convicted of drunkenness before a magistrate; but, on the other hand, rum was actually at one time the only circulating medium, and in it the prices of land, labour, and food were estimated, and for it they were freely exchanged. Even for the charitable purpose of erecting a public hospital, Governor Macquarie granted to four individuals, who defrayed the whole expense, the monopoly of the right of purchasing all the spirits imported into the colony, and of landing them free of duty, for several years, with the additional consideration of a quantity of rum from the king's stores.*

The old Government-House, where I had the honour of seeing Sir George Gipps, is a low, cottage-shaped building, which has no pretensions to beauty, and appears to have been built at different times, having been enlarged as often as additional accommodation was needed. During the summer months the Governor resides at the Government-House at Paramatta.

A new palace or government-house is at present building, in the public grounds which lie to the eastward of the old one, from which a road extends through them towards the South Head of Port Jackson. This road is the usual promenade and drive of the citizens of Sydney. After leaving the government domain, it enters Wooloomoloo, a region covered with the country-seats and cottages of the higher classes, which although originally little more than a barren rock, has been

* It is related, that a highly respectable individual transmitted complaints against Governor Macquarie to the home government; and that, by way of answering these expostulations, the reply of the Governor was: "There are but two classes of persons in New South Wales, those who have been convicted, and those who ought to be."

brought into a high state of cultivation by its occupants. The drive in this direction may challenge comparison for beauty with any part of the world. It presents innumerable and picturesque views of the noble bay, and of the promontories that jut into it, occupied by mansions and ornamental grounds. On reaching the South Head, a view of great beauty is also seen. The point thus named, is a bold headland, about two hundred and fifty-four feet in height, on which stands the light-house, a fine tower, with a brilliant revolving light.

The public grounds are in part occupied by a Botanical Garden, which was laid out by Mr. Cunningham, the botanist of the colony, to whose memory a monument is about to be erected in the garden, which is itself a memorial of his fine taste, and his successful cultivation of the science he professed. Mr. Cunningham perished by a melancholy death, which is still spoken of with regret. He had, in his capacity of botanist, accompanied Major Mitchell, the Surveyor-General of the colony, on a tour of exploration in 1835. In the pursuit of his researches, he wandered from the party, and did not return. As soon as he was missed, the native guides were sent in search of him, but returned without having succeeded in finding his traces. Major Mitchell then instituted a fresh search, in which the tracks of Mr. Cunningham's horse were found, and followed for ninety miles. Within this space three places were seen where he had stopped and encamped. From the last of these, the tracks of the horse were again followed, until the carcass of the animal was found dead through fatigue and starvation, with the whip tied to the bridle, and all his accoutrements about him. Retracing their steps to his last encampment, they ascertained, on close examination, that he had there killed his dog for food, and his footsteps were seen as if making rapid strides for the bed of a river, which he had followed to a pool, into which he had plunged. Farther down the river. some shells were found near the remains of a fire, which had evidently been kindled by a white man. Here all further traces of him were lost, and the search abandoned in despair.

Some months afterwards, a second search was made by Lieutenant Vouch. In the course of this, some natives were taken near the Brogan river, in whose possession a part of Mr. Cunningham's clothing was found. They stated that a white man had come to them in a state of great exhaustion; that he was hungry, and they fed him, but that during the night they had become afraid, and killed him. The body was never found.

Lieutenant Vouch inferred that Mr. Cunningham had become deranged by the severity of his sufferings, and that this had caused him to wander about at night, which, with other suspicious movements, had alarmed the natives, who, under the influence of their terrors, had murdered him.

Thus ended the useful life of one who had raised himself to eminence by his own exertions, and had, by his virtues and scientific acquirements, gained the esteem of all the pure and good of the colony, by whom he will be long affectionately and honourably remembered.

These grounds have many pleasant shady walks, and afford an agreeable promenade for the inhabitants of Sydney; and one of them encircles the whole, with occasionally a rural seat and arbour.*

The aspect of the country around Sydney is sufficient to prove that New South Wales is very different, in its general features, from other parts of the globe. This is chiefly owing to two causes: the aridity of its climate, and the prevalence of sandstone rock. This rock may be readily examined at the Heads of Port Jackson, and on the shores of the many coves that surround this beautiful harbour. Its colour is pale yellow or drab, and it lies in beds nearly horizontal and of various thickness, whose upper surface, except where broken by ravines and water-courses, forms a table-land. The average elevation in the neighbourhood of Sydney is from three hundred and fifty to four hundred feet. At this level it extends in gentle undulations to a great distance inland.

This arid soil yields but a scanty growth of vegetable products, which, consisting of burnt pasture, and thinly-scattered trees and shrubbery, give to the whole region a look of desolation. The grass does not every where conceal the bare rock, and the thin soil supports only a few gum trees (Eucalypti), and bushes. Throughout the wide plain there is little to relieve the eye, except here and there a small cultivated spot.

As I did not consider it necessary that any of the naturalists should accompany the squadron on its southern cruise, they were left at Sydney, with orders to visit such parts of the country as might appear to offer the best opportunities for making collections in their respective departments.† This enabled me to obtain much information in rela-

* At the end of the walk around the government domain, the following inscription is calculated to excite a smile: "Be it recorded, that this road round the inside of the government domain, called Mrs. Macquarie's Road, so called by the Governor on account of her having originally planned it, three miles and three hundred and seventy-seven yards in length, was finally completed on the 13th day of June, 1816."

Governor Macquarie has literally put his mark on the town of Sydney, where hardly a single street, square, or public building can be passed, without seeing his name cut in stone.

† For orders, propositions of officers, and letters respecting their employment, see Appendix XI.

tion to the interior of this interesting country, its productions, and its original inhabitants. The narratives of several of these journeys will be given hereafter, but so much of what they learned as is general, together with such additional information as was gained from other sources, will form an appropriate introduction to the account of their tours.

The interior of the country, for a distance of sixty or eighty miles to the north and south of Sydney, presents the same characters which have just been described, except that deep gorges are from time to time met with, and that some parts of it are of a more undulating character.

On proceeding inwards from the coast, the country at a distance seems to be traversed by ridges, but on approaching their apparent position, they melt away into rounded elevations, of very gradual inclination. Still farther to the westward, the undulating region is bounded by inaccessible declivities and lofty mural precipices. These are the edges of the Blue Mountains, which are seen from Sydney, skirting the horizon like low hills, which have so little appearance of elevation that it at first seems to be difficult to conjecture how they came to be called mountains, when seen only from the coast. This ridge runs north and south, and rises at some points to the height of three thousand five hundred feet.

It is not many years since this ridge was considered as inaccessible, and the deep gorges which intersect its sandstone rocks as impassable. Its peaks rise in many places abruptly, and present such difficulties, as to have deterred travellers from attempting to scale their summits, or from seeking a passage through the ravines, which in the season of rains are swept by impetuous torrents.

The same description will apply to the mountains which bound the Illawarra district to the west, where sandstone also occurs, broken into precipitous heights, and deep gorges. At the Kangaroo Pass, the Illawarra Mountain is nearly two thousand feet high; its rapid acclivity is covered with a dense vegetation, until within three hundred feet of the summit; whence upwards a perpendicular face of rock is exposed. The path through this pass winds among the narrow breaks of the rock, and is toilsome to both beast and rider.

In one of the gorges which open upon this pass is a beautiful waterfall. The deep narrow glen opens abruptly upon the passenger, and exhibits its bare rocks, and the tiny stream is seen leaping from one projection of the rocky shelves to another, which break its headlong course, until, lost in spray, it reaches the bottom, where its waters collect, at the depth of two hundred and fifty feet below its upper edge, in a limpid pool. This gorge opens to the westward, and looks out upon a mountain range.

Seven miles further, a descent by a similar path leads into the Kangaroo Valley. This valley is nearly twenty miles in length, and has an average breadth of about three miles; it is surrounded on all sides by vertical precipices, from one thousand to one thousand eight hundred feet in height.

In consequence of the aridity which has been mentioned as a character of the soil about Sydney, and which is also a prevailing character throughout the rest of the country, there are many continuous miles of waste lands, which by the inhabitants are called "forests." These are very different from what we understand by the term, and consist of gum trees (Eucalypti), so widely scattered that a carriage may be driven rapidly through them without meeting any obstruction, while the foliage of these trees is so thin and apparently so dried up as scarcely to cast a shade. Thus miles may be traversed in these forests without impediment. A few marshy spots are occasionally seen, covered with thickets of brush; and in other places there are tracts so dry that even the gum tree will not grow upon them, and which receive the direct and scorching rays of the sun.

The exceptions to this general character are found in the occasional rising of basalt in conical peaks. The productions of the soil where this rock prevails, are in striking contrast to those of the arid lands of the sandstone formations; and the geological character of a basaltic ridge can be detected at a distance by the luxuriant vegetation with which it is clothed. These ridges become more and more frequent as the distance from the coast increases, and are occasionally interspersed with granite.

The latter rock is first seen in the Clwyd Valley, near Mount Victoria, and about eighty miles from Sydney. This valley lies in the western mountain range, which separates the waters that flow towards the east and west. The land falls gradually to the westward, until, in the Darling Valley, at a distance of four hundred miles, it is only about four hundred feet above the sea.

For some distance beyond Mount Victoria, granite characterizes some extensive ridges, and basaltic mountains are occasionally combined with those of granite.

Beyond Bathurst, about one hundred and twenty miles to the west of Sydney, a compact limestone, in which there are many caverns, occurs between ridges of granite and basalt; but, according to Major Mitchell, the sandstone reappears on proceeding further west, towards the Darling Valley, and is accompanied by the same sterility as upon the coast.

Before reaching the western barrens there are many fine and fertile valleys, among which, besides Bathurst, is that of Wellington, distant about two hundred miles from Sydney. Both of these are already settled.

As to the more remote parts of the interior of New Holland, no positive knowledge has yet been obtained. The prevailing opinion appears to be, that an extensive desert extends throughout it; and this opinion is supported by citing the dry and scorching character of the winds which blow from the west. The greatest distance to the westward which has been explored, is only four hundred and fifty miles, which is not a fourth part of the distance to the western coast. It will thus be seen that a vast field of discovery is still open, which will no doubt be ere long explored, under the auspices of the British government.

To the southwest of Sydney the same compact limestone seen at Bathurst makes its appearance at Argyle, also about one hundred and twenty miles distant from the former place. This stone yields lime of good quality, and is also a valuable material for building.

According to the best accounts, the range of granite appears to begin in Van Diemen's Land, and after being interrupted by Bass's Straits, runs through New South Wales in a broad belt. Near Bass's Straits it rises into a lofty group of mountains, called the Australian Alps, the only snowy ridge known in Australia, and continuing thence northward, it forms the dividing range of the waters.

The basaltic ridges of this southern region are said occasionally to reach a height of four thousand feet, and a limestone similar to that of Argyle and Bathurst, which contains many fossils, extends to the "Limestone Plains," where it is succeeded by the usual sandstone. How far this limestone extends to the southward has not been ascertained. The finest districts in this southern section are those of Port Philip, Argyle, Bass, and Bong-Bong.

To the northward, beyond the Hunter river, the country is intersected by basaltic ridges, which increase in number until they merge in the Liverpool Mountains, of which many of them are spurs. Between these ridges, and to the north of the Goulburn branch, sandstone again prevails, and forms a great extent of barren country; but the smaller valleys being generally bordered by ridges of basalt, are for the most part fertile.

The Liverpool range of mountains, although it has been traced for many miles in an east and west direction, is said by Major Mitchell to be a prolongation of the range which runs parallel to the coast. According to him, at the distance of one hundred miles inland, the range trends to the northward, and thence pursues a course to the northeast.

To the northward of the Liverpool range, plains of considerable extent spread over the country, and form the district of New England, which affords fine pasturage. These plains lie at an altitude of between two and three thousand feet, and from that circumstance enjoy a much cooler climate than Sydney, although five degrees nearer the equator.

The most remarkable part of New South Wales is the district of Illawarra, situated on the coast, about sixty miles to the south of Port Jackson. This is a narrow strip, that seems to be formed by the retreat of the sandstone cliffs from the sea, to a distance which varies from one to ten miles. The cliffs or mountains vary in height from one thousand to two thousand feet. This region is extremely fruitful; its forests are rich with a great variety of foliage, and of creeping plants which twine around the trees. The great size and number of the trees served to remind the gentlemen who visited it, of the vegetation of the tropical islands, luxuriant with tree-ferns, bananas, banyans, &c. This luxuriance is in part owing to a rich and light soil, composed of decomposed basalt and argillaceous sandstone, mixed with vegetable mould, but more to the peculiarity of its climate. The high cliffs which bound it to the west, keep off the scorching winds which reach other parts of the coast from that quarter, and the moisture of the sea-breeze intercepted by them, is condensed, falling in gentle showers. For this reason, it is not subject to the long and frequent droughts that occur in other parts of New South Wales.

These droughts are sometimes of such long continuance, that we at one time read of the whole country having been burnt up for want of rain, a famine threatened, and the sheep and cattle perishing in immense numbers.

These have been succeeded by long-continued rains, which have raised the rivers thirty or forty feet, flooded the whole country, deluged the towns and villages, and completely destroyed the crops. Such floods carry with them houses, barns, stacks of grain, &c., drown the cattle, and even the inhabitants are in some cases saved only by being taken from the tops of their houses in boats.

The year of our visit, 1839, added another instance to the list of disasters of the latter kind; and the published accounts state that twenty thousand sheep were lost in the valley of the Hawkesbury by the floods. Such evils indeed appear to be of frequent occurrence and the settler in New South Wales has to contend with the elements in an unusual degree.

Such disasters are equally injurious to the husbandman and the wool-grower; for the same cause that destroys the crops, also carries off the stock, so that it is only the large capitalist who can successfully struggle against or overcome such adverse circumstances. It is some recompense for this state of things, that one or two favourable years will completely repay all former losses; and it is due to the perseverance and industry of the inhabitants of New South Wales to say, that they have already, in spite of the difficulties they have had to encounter, made it one of the most flourishing colonies on the globe.

What these difficulties are, may be better understood by quoting some remarks of Major Mitchell, the Surveyor-General, who has had greater opportunities than any other person of examining the country, every accessible portion of which he has visited.

"Sandstone prevails so much more than trap, limestone, or granite, as to cover six-sevenths of the whole surface comprised within the boundaries of nineteen counties, from Yass Plains in the south, to the Liverpool range in the north. Wherever this happens to be the surface, little besides barren sand is found in the place of soil Deciduous vegetation scarcely exists there; no turf is found, for the trees and shrubs being very inflammable, conflagrations take place so frequently and extensively in the woods during summer, as to leave very little vegetable matter to turn to earth.

"In the regions of sandstone, the territory is in short good for nothing, and is, besides, generally inaccessible; thus presenting a formidable obstruction to any communication between spots of a better description."

The information obtained from other sources does not, however, sustain so very unfavourable a picture; it may, indeed, be true, when applied to the labours of husbandry alone, but there is reason to believe, on the other hand, that the excellence of the great staple of the country, its wool, is in a great measure to be ascribed to the short and sweet pasturage which these very sandstone districts afford. These lands produce, except during the prevalence of excessive droughts, a nutritious herbage, and form a dry healthy soil, on which sheep thrive particularly well, although it is said that one hundred acres of this description, of average quality, will not support more than five or six head of cattle.

In seasons of drought, the flocks and herds are driven into the interior The year of our visit (1839) was accounted a wet one, and some parts of the sandstone district which produced good crops of grain,* in drier seasons would have been dry to barrenness.

In such a climate it is not surprising that there are hardly any streams that merit the name of rivers. It is necessary to guard against being misled by the inspection of maps of the country, and forming from them the idea that it is well watered. Such an impression would be erroneous, and yet the maps are not inaccurate; streams do at times exist in the places where they are laid down on the maps, but for the greater part of every year no more is to be seen than the beds or courses, in which, during the season of floods, or after long-continued rains, absolute torrents of water flow, but which will within the short space of a month again become a string of deep pools. Were it not for this peculiar provision of nature, the country for the greater part of the year would be without water, and, consequently, uninhabitable.

The principal rivers which are found to the east of the Blue Mountains are, the Hunter, George, Shoalham, and Hawkesbury. None of these streams are navigable further than the tide flows in the estuaries, which sometimes extend twenty or thirty miles inland, for beyond them they are usually no more than twenty inches in depth. Each of these streams has numerous tributaries, which drain a large area of country, and during heavy rains the main branches are suddenly swelled, and cause the floods which have been spoken of. To the west of the mountains, the water-courses are of a very different character. The Darling, for instance, through a course of seven hundred miles, does not receive a single tributary, although it is said to drain an extent of sixty thousand square miles. It possesses the other character which has been mentioned, of being frequently reduced to a mere string of pools. The Darling, Morrumbidgee, and Lachlan, unite about one hundred miles from the ocean, and their joint stream is known by the name of the Murray, which after passing through Lake Alexandria, enters the sea at Encounter Bay. The surface drained by these streams is about two hundred and fifty thousand square miles.

Another remarkable occurrence observed in these western waters, is the disappearance of a river in swampy lands, where, as is supposed, it is swallowed up by the caverns in the limestone rocks. This is the case with the Macquarie, which has its source near Bathurst.

According to all accounts, salt is very generally diffused throughout New South Wales, and even all Australia. It has been reported as being found in masses in the sandstone, but no specimens of it were

* In the diluvial flats along the rivers, the wheat crop is usually about twenty-five bushels to the acre. Forty to forty-five bushels have been obtained, but such crops are very unusual.

obtained by the Expedition. Scarcely a well is dug in the interior which is not brackish; and, according to Major Mitchell, Captain Sturt, Oxley, and others, many of the rivers are quite saline in parts of their course. The northern tributaries of the Hunter and Darling are instances of this.

The lakes are also said to be saline, and in some instances sufficiently strong to afford a large and profitable yield of salt; but being very far in the interior, and without the means of transportation, they are of little value. Along the south coast of Australia, such lakes are described as existing near the sea, and may possibly prove of some value to that portion of New Holland.

Lead and iron have been found in small quantities; the deposits of the former are all trifling. Those of the latter afford too impure an ore, and not in sufficient abundance, to be worked.

The minerals stated to be found in Australia, specimens of which were procured for the Expedition, are, chalcedony, agates, jasper, quartz, augite, and stilbite; feldspar, arragonite, gypsum, chlorite, mica in granite; sulphur and alum, galena and plumbago, magnetic iron, iron pyrites, and basalt.

Fossils appear to be confined to particular localities, but are by no means rare.

Columns of basalt of great regularity are found on the coast of Illawarra, but the articulations are all plane.

The water is much impregnated with alum and iron, and its use is avoided by the inhabitants.

Deserts covered with saline plants are said to be frequently met with.

Mitchell, in his travels in New South Wales, speaks of the different heights of the ranges of mountains in this country, some of them in the southern and some in the eastern portion as being covered with snow, and rising four thousand five hundred feet above the sea. To the Blue Mountain range he ascribes a height of three thousand four hundred feet, composed entirely of sand; beyond this the granite or dividing range occurs, which is only two thousand two hundred feet in height. The Canobolus Mountains, further to the westward, are four thousand four hundred and sixty-one feet high, and of primitive rock; beyond these, the extensive plains of the interior, the valley of Millewa or Murray river, seem again to be entirely composed of sedimentary rocks, similar to the sandstone of the coast.

The climate of Australia may be considered generally as very dry; the irregularity of the rains, and the nature of the soil, all prove that it is so; yet the aridity is not marked, as in other countries, by a general tendency in the plants to produce thorns, although the peculiarity of the vegetation makes the dryness apparent in other ways. From all accounts, New South Wales is subject to as great atmospheric vicissitudes, as the middle United States. For a series of years, droughts will occur, which in turn give place to years of successive floods, and these prevail to an extent that can hardly be credited, were it not that the account has been received from good authority. As a striking instance of it, Oxley, in his exploring journeys into the interior, in 1817, found the country every where overflowed, so as to prevent him from proceeding; while Mitchell, in 1835, in the same districts, was continually in danger of perishing from thirst. The latter states that he found unios (or fresh-water mussels) sticking in the banks of rivers and ponds above the level of the water; and also dead trees and saplings in similar situations.

This alternate change must exert a great influence on the productions of the soil; the rivers ceasing to flow, and their beds becoming as it were dry, with the exception of the pools heretofore spoken of, must likewise have an influence. The prevailing westerly winds sweep with force over the whole country, blighting all they touch. The effect of these hot winds is remarkable, for they will in a few hours entirely destroy the crops, by extracting all the moisture from the grain, even after it is formed, and almost ready for harvest; and the only portion that is left is that which has been sheltered by trees, hedges, or fences. They thus destroy the prospect of the husbandman when his crops are ready for the sickle. It is thought, and I should imagine with reason, that were the Blue Mountains a more lofty range, this would not be the case, as they would have a tendency to continue the supplies to the streams throughout the year, by the condensation of the vapour from the sea.

These hot winds come from the direction of the Blue Mountains, and, what seems remarkable, are not felt on the other side of the mountains, or in their immediate vicinity. Yet the extent between the coast and the mountains is not sufficient to produce these winds, being only forty-five miles; and if they proceed from the interior, they must pass over those mountains, an elevation in some places of three thousand four hundred feet. Their great destructiveness is undoubtedly caused by their capacity for moisture, although few observations have as yet (as far as I was able to obtain information) been made upon them, except in relation to the blight they occasion. It has been found that fields which have a line of woods on the side whence they blow, escape injury. The harvest immediately on the line of the coast does not suffer so much, being exempted in part from their withering influence by the moisture that is imbibed from the sea.

There is a portion of this country that is an exception to the genera. rule of aridity, namely, the district of Illawarra. This forms a belt of from one to ten miles wide, and has the range of the Kangaroo Hills just behind it, of one thousand feet; these are sufficiently high at this distance from the coast to condense the moisture, and also to protect the district from the blighting effects of the blasts from the interior.

One is entirely unprepared for the alleged facts in relation to this country; for instance, Mitchell, in his journey to the south and west, during the four winter months, witnessed no precipitation of moisture except frosts in the mornings, and the thermometer was often below the freezing point. Violent winds occur, which have obtained the name of brick-fielders. They are nothing more than a kind of gust, peculiar to the environs of Sydney, after a sultry day. During one of these gusts little or no rain falls, though the wind frequently approaches a hurricane in force. These winds get their name from bringing the dust from the brick-fields, formerly in the suburbs of Sydney, but which are now almost entirely built over. The temperature during the blow generally falls twenty or twenty-five degrees, in the space of as many minutes; the dust is very great, and the wind so strong, as to cause apprehension lest the houses should be unroofed, or the chimneys thrown down. Our standard barometer was carefully watched during the coming on of two of these gusts, and found to fall 0·200 in., the first time; and the second only 0·020 in.; but the temperature fell each time about ten degrees. They were not, however, true brick-fielders, or such as a resident would so denominate.

Snow has been known to fall in Sydney, but so rarely, that we were told some of the inhabitants were doubtful as to its nature. On the mountains it is not uncommon, and in the winter season is always seen on those in the New England district, which, although three or four degrees to the northward of Sydney, enjoys a much cooler climate.

Major Mitchell often found that the temperature exceeded 100° of Fahrenheit. The heat was, of course, very oppressive, and more so on account of the little shade the native trees afford. The difference of temperature between the day and night is great, but upon this point I was able to get but little information; the meteorological registers that have been kept at Sydney, have omitted the night hours altogether.

I have been favoured since my return with the abstract returns of the meteorological registers during parts of the years 1840 and '41, kept at the South Head of Port Jackson, two hundred and fifty-four feet above the level of the sea. Being kept immediately at the sea-coast, this record does not furnish so satisfactory an account of the climate, as if the place of observation had been farther in the interior

I have also those kept at Port Macquarie, to the north, and at Port Philip, to the south; the following are some of the results from them.

During the winter months, June, July, August, and September, on the eastern coast, they have the winds from the south to the west quarter, blowing strong and cold, not unlike our northeast winds, accompanied by rain. The quantity of rain which fell in the winter of 1840 was 35·25 inches; in 1841, 45·00 inches. The temperature during this season is from 40° to 78°, mean temperature, 66°. During the summer months, October, November, December, January, and February, easterly winds prevail, and the temperature varies from 56° to 90°; the mean temperature being about 78°. On the south coast, as appears from the tables kept at Port Philip, the winds pursue a reversed order; for during the summer months they are found to prevail from the southward and westward, whilst in winter they come from the eastward. The mean standing of the barometer seems to be higher at Port Philip than at Port Jackson and Port Macquarie; its range is within 0·5 in., and at the two latter places it seldom reached 30·000 in. During our stay at Sydney, the range of the barometer was generally higher than this. The temperature of the winter months varies from 35° to 65°: the mean is about 47°; and that of the summer months from 50° to 98°, the mean being 68°.

The above observations relative to the winds on the south and eastern coast, will serve to explain some of the difficulties that have resulted from vessels taking the southern route round New Holland, to proceed west. These, during some seasons of the year, have met constant head winds and storms. It would seem that the most feasible time to make the southern passage to the westward, is during the winter months. The probable cause of this difference is the immense vacuum which is formed on the vast plains in the interior of New Holland during the summer, that is supplied by these southerly winds. But it is not altogether certain that these winds prevail at any distance from the land. They were described to me rather as regular breezes, prevailing during the day, moderating towards night, and succeeded by light land-winds until the following morning. The intermediate months between the summer and winter, or those about the equinoxes, are attended with variable winds and uncertain weather, but from information I received, they do not appear to suffer here from very violent gales during these seasons. Severe gales are, however, experienced at these seasons at New Zealand, in the same latitude, of which I shall speak hereafter.

Our own results for the time we stayed in Australia will be found under their appropriate head. At Sydney I found a great variety of opinions existing about the climate. During our stay, the weather was unfavourable for all astronomical observations, and almost the whole time cloudy or rainy. It was amusing to find many of those to whom I had the pleasure of an introduction, apologizing for the badness of the weather. It brought forcibly to my recollection, the fault that Captain Basil Hall finds with the people of the United States, but was far from being annoying to me. I have but little doubt, that the climate is, generally speaking, a healthy one, and not unlike that of some parts of our own country. The colony is subject to occasional epidemics, and from the best information I could procure, it is thought that the mortality is about one in forty-three; this may be called a very small proportion, when one takes into consideration the great quantity of ardent spirits that is consumed.

The general appearance of the vegetation of New South Wales presents many peculiarities. The character of its productions is totally distinct from those of the other portions of the globe. The gum trees, Norfolk pines, and those of Moreton Bay, attract attention from their scattered appearance, and peculiar foliage. All these have a dark and sombre hue. A remark made by one of our gentlemen is characteristic of the former, "that they were ghosts of trees." The leaves being set edgewise causes this appearance, and in consequence give little or no shade. This peculiar position of the leaf is more conspicuous in the Eucalypti than in other genera, for in them the leaves are all pendant, while the leaves in the other genera are usually upright, rigid, and somewhat as may be seen in the acacias and other tribes. It was observed that both surfaces of the leaves were much alike, having as it were, two upper surfaces. Whether any physiological purpose has been assigned for such an arrangement I have not been informed.

According to observations made by the gentlemen of the Expedition, the great mass of the vegetation of Australia belongs to the natural orders of Myrtaceæ, Leguminosæ, Proteaceæ, Epacrideæ, and Compositæ.

The most common genera are Eucalyptus and Acacia. Many trees of one of the species of the former genus were seen one hundred and twenty feet in height, and with a girth, about six feet from the ground, of eighteen feet. This species is called by the settlers black gum, and is much used for domestic purposes, particularly its bark, with which they cover their huts and houses. The dilated foliaceous acacias are very numerous, and are objects of attraction from their gay and fragrant blossoms. The trees which present the greatest variety of species, are those known as gum trees, viz.: blue gum, gray gum, flooded gum, iron bark, and stringy bark. The leaves of these gum trees have a powerful aromatic flavour, and a taste approaching to camphor. They are used in the colony for a variety of purposes, according to their quality. Thus, the blue gum, (Eucalyptus piperita,) is employed for ship-building; the iron gum, (Eucalyptus resinifera,) for fencing; and the gray gum and black-butted gum, for boards and plank.

The Norfolk Island pine, (Araucaria excelsa,) is used for cabinet-work; the swamp oak, (Casuarina torulosa,) for shingles and cabinet-work, as is the cedar (Cedrela australis,) which grows to a very large size; the turpentine-wood, (Tristania arbicans,) for boat building; the pear tree, (Xylomelum pyriforme,) the apple, (Angophora lanceolata,) the mountain-ash, sallow, sassafras, and several kinds of wood which they called "Curagong," were also observed in use, but the trees were not seen.

The grass tree (Xanthorrhœa hastilis) did not equal our expectations, which were probably too highly raised by the descriptions of those who had gone before us; yet when in flower it must be a conspicuous object, and in all stages of growth suits well for the foreground of a picture.

Among the most singular of the productions of Australia are the wooden pears, as they are called. These have a close external resemblance to the fruit whose name they bear, but are ligneous within. Another of the fruits is a cherry, whose stone is external, and would be similar to our fruit of that name were the kernel in its proper place. The pit adheres firmly to the pulp, which is of the size of a pistol-bullet, but the fruit shrinks when ripe to that of a buck-shot. The pear grows on a low shrub, the cherry on a large bush.

I have before remarked how different the "forest," so called in New South Wales, is from what is understood by the term elsewhere. The want of close growth is not the only remarkable appearance, but the absence of all decayed foliage is also extraordinary. The ground is clear of any fallen leaves, and every thing betokens that perennial verdure is here the order of things. These two features combined, give the forests of Australia the air of a neatly-kept park. Annual plants, (if so they can be called,) abound in the forest, requiring, it is said, more than a single year to bring their seeds to maturity. There were instances we were told of crops of grain remaining three years in the ground. A few plants found in other parts of the world, are, it is well known, only brought into existence after a lapse of years, and others give repeated crops during the same year. That these types, so rare in other countries, should be abundant in Australia, is not remarkable, when it is considered that they are but instances of an almost complete diversity between the natural history of this country and that of other regions.

The timber of the Australian trees is generally of greater specific gravity than water.

The remark, that the leaves of the trees are wood, and their wood iron, is not inappropriate to most of the plants of this country. It is not, however, to be inferred that all the plants are different from those of other countries; so far from this being the case, a considerable admixture of ordinary forms was met with. Among these were a great variety of grasses, some of which were before considered to be peculiar to North America. Many other forms decidedly North American were also met with, a circumstance which, from the difference of geographical position, distance, and climate, was not to be expected; but for these details I must refer to the Botanical Report.

All seem to have been struck with the apparent monotony of the scenery, foliage, and flora, although in reality the latter presents great variety. The general sentiment was, that they were fatigued by it, which is not a little surprising, as the Australian Flora rivals in number of species that of Brazil. This feeling may be accounted for by the overpowering impression that is made by the gum trees, whose foliage is of a dark sombre green. There is also something in the general absence of underbrush; and the trees are so distant from one another that there is no need of roads, so that a carriage may drive any where.

The trees are in general tall in proportion to their diameter, with an umbrella top, and have the appearance of being thinly clad in foliage. No woody vines are to be seen, nor any parasitic plants. In many places a stunted growth of detached shrubs, called in the colony "scrub," exists, which might be termed one of their "forests" in a dwarf shape.

In the Illawarra district a totally distinct state of things exists. Here is to be found all the luxuriance of the tropics—lofty palms, among them the Corypha australis, with tree-ferns of two or more varieties, different species of Ficus, a scandent Piper, and very many vines. The forest of this district is thick, and alive with animal life.

This district is about fifty miles long, and forms a semicircular area about thirty miles in its greatest width. The peculiarity of the situation of this district would tend to show what would have been the probable state of New Holland, or rather its eastern side, if the mountains were sufficiently high to intercept the moisture of the ocean, and prevent the access to it of the dry hot winds from the interior. Illawarra may be termed the granary of New South Wales; here the crops seldom if ever fail, and are very abundant.

I had the pleasure of visiting the seat of the Hon. Mr. M'Leay. It is situated on Elizabeth Bay, beyond or within Wooloomoloo. The house, which displays much taste, is built of sandstone, and is situated beneath a hill, and on a knoll about fifty feet above the water. In front of it is a lawn bounded by a parapet wall, and between this and the water are several acres of land very tastefully laid out as an arboretum and flower-garden. Copses of native trees have been judiciously left on the north and south sides of the grounds, and not only protect them from the injurious winds, but add greatly to their beauty. A splendid specimen of the Acrostichum grande, or Stag horn, from Moreton Bay, was seen suspended from a tree.

ACROSTICHUM GRANDE.
N. S. WALES.

The garden is intersected by many walks, leading to the best points of view. It contains many rare and fine specimens of plants from

England, Mauritius, the Cape of Good Hope, the East Indies, and America. The flower-gardens at and in the neighbourhood of Sydney would do credit to any part of the world.

The Norfolk Island pine, and that from Moreton Bay, (the Araucaria Cunninghamii,) thought by some to be as handsome a tree as the one from Norfolk Island, were also among the plantations. From the flower-garden a walk leads through a lattice-work bower, covered with native Bignonias and passion-flowers, into the kitchen and fruit-garden. These we found well stocked with fruit and vegetables of almost all European kinds, which seem to thrive particularly well here. The kindness, attention, and hospitality, with which we were received and treated, will long be gratefully remembered. To Mr. M'Leay, the Expedition is indebted for much aid, and through him our collections were increased. The English oak thrives well, and many fine specimens of it were seen. From the information I have received, very many of the Australian plants succeed admirably in England.

The soil of Sydney consists of black mould, mixed with a clean white sand. The quantity of sand is such, as in the dry seasons to affect the vegetation. This sand I understood is now exported to England at a great profit, being found a valuable article in the manufacture of plate glass. This soil, however, is made to yield a plentiful supply of fruits and vegetables; and the display exhibited at the Horticultural Exhibition was highly creditable, not only for the perfection to which the productions had been brought, but for their great variety. The exhibition was held in the large market-house in George Street, which was tastefully decorated for the occasion with branches and festoons of flowers. In front of the door was an arch formed of beautiful flowers, with the motto "Advance, Australia!" surmounted by a crown, and the letters V. R. in yellow flowers. Behind this the band was stationed, which, on our entrance, struck up Yankee Doodle. Tickets were sent to the consul for those belonging to the squadron. There were a great many South American plants in pots. A premium was received for Tropæolum pentaphyllum, Maurandya Barclayana, and for two species of Calceolaria. There were likewise Amaryllis belladonna and umbellata, Bouvardia triphylla, Cobæa scandens, and several Passifloras, and a variety of hyacinths, dahlias, tuberoses, &c., all fine.

The grapes exhibited were beautiful, and some of them in very large clusters. Nectarines, peaches, apples, pears, small oranges, shaddocks, pine-apples, chestnuts, and walnuts, were also in abundance.

After viewing the fruit we examined the vegetables, which consisted of potatoes, carrots, turnips, very large pumpkins, cucumbers, cabbages of different kinds and very fine, particularly the curled Savoy and early York, tomatoes, celery, squashes, vegetable marrow, beets, capsicums, and beans.

After the vegetables came specimens of native wines, and a silver cup was given as a premium for the best. The white wine resembled hock in taste; the red, claret. The climate is thought to be favourable to the production of the grape. The first wine made in the colony was by Mr. Blaxland, on his estate at Newington.

The premiums were silver medals. A very handsome gold one was exhibited, which was to be given the next year for the best crop of wheat.

There was a large concourse of visiters, all seemingly much interested in the exhibition, which was open from one o'clock until six. The proceeds of the exhibition of the following day were for the benefit of the poor.

There are a variety of other fruits and vegetables grown near Sydney, which, being out of season, were not exhibited; but to show the great variety this climate produces it is as well to mention them, viz.: English cherries, plums, raspberries, figs, water and musk-melons, filberts, citrons, lemons, strawberries, granadillas, olives, and a species of cherimoyer; and for vegetables, asparagus, cauliflower, lettuce, radishes, spinach, broccoli, artichokes, egg-plant, mustard, and capers.

They have many imported varieties of grapes. Sir John Jamison is now making experiments on his estate to effect their successful growth, and manufacture wine. He has obtained cultivators both from Madeira and the Rhine, to superintend his vineyard and vintage. The reports made yearly to the Agricultural Society, hold out well-founded hopes of success.

Several good nurseries of fruit trees exist in the vicinity of Paramatta, and the Botanical Garden at Sydney also furnishes trees to the cultivators.

The grains grown in the colony are, wheat, rye, barley, Indian corn, and oats. The wheat yields from six to twenty-five bushels to the acre, and some low ground as high as thirty-five bushels. Its weight per bushel is sixty-two pounds. The crops of this grain are subject to great fluctuations, and the most promising appearance may in a single day be entirely destroyed.

Tobacco has been cultivated, and it is thought will succeed; but the frequent frosts render it a very uncertain crop.

Cotton has been attempted, but with little success. The value of pasturage, and its profitable yield in sheep-walks, will long be a bar to the extensive cultivation of any plants that require much labour in their production. Our horticulturist remarks, that cherries do not succeed well, being affected by the dry cutting winds which occur in the blossoming season.

The orange, citron, and lemon trees present a scraggy and yellow appearance, and produce small and insipid fruit, in comparison with that of the tropics. Peaches thrive, and grow in large quantities, and of high flavour. Every farmer has his peach orchard; and the fruit is so plenty that they fatten their pigs on them.

The natives of Australia are fast disappearing. The entire aboriginal population has been estimated as high as two hundred thousand; this estimate is founded on the supposition that the unexplored regions of the country do not differ materially from that part of it which is known, which cannot well be the case. Other estimates, and probably much nearer the truth, are given at from sixty to seventy-five thousand.

The ravages of intoxication and disease, combined with their occasional warfare, will readily account for the rapid disappearance of the native population; and but a few more years will suffice for the now scanty population to become extinct. In 1835, the Surveyor-General, Mitchell, estimated that in about one-seventh of the whole colony, which he had examined, the natives did not exceed six thousand in number; they are in many parts most wretched-looking beings, and incorrigible beggars: the moment they see a stranger, he is fairly tormented to give something; a shilling or a sixpence contents many, and when laid out for rum, or bread, is shared by all present.

The introduction of European arts has caused but little improvement, while the vices which accompany them have been the bane of the native population, which has thus acquired a fondness for ardent spirits and tobacco. The natives usually lead a wandering, vagabond life, hanging about the houses of the settlers where they are well treated, and doing little jobs for a slight recompense in the above articles. Their habitations are mere temporary shelters, formed of boughs and bark piled up against the stump of a fallen tree, rather to shield them from the wind than for a regular habitation; the reason for this may be, that owing to superstitious scruples they never encamp in one spot three nights in succession. At Illawarra, their huts were made by setting two forked sticks upright, on which another was laid horizontally; on the latter, one end of pieces of bark, taken from the nearest gum tree, is laid, while the other end rests upon the ground. A fire is built on the open side, which not only warms them, but keeps off the myriads of musquitoes and other insects. As many as can enter such a hut, take shelter in it, lying upon the soft bark of the ti tree.

NATIVE HUT, NEW SOUTH WALES.

The natives of Australia differ from any other race of men, in features, complexion, habits, and language. Their colour and features assimilate them to the African type; their long, black, silky hair has a resemblance to the Malays; in their language they approximate more nearly to our American Indians; while there is much in their physical traits, manners, and customs, to which no analogy can be traced in any other people.

The natives are of middle height, perhaps a little above it; they are slender in make, with long arms and legs. From their wandering life, irregular habits, and bad food, they are extremely meager; and as their thinness is accompanied by considerable protuberance of the abdomen, it gives to their figure a distorted and singular appearance. The cast of the face is between the African and Malay; the forehead usually narrow and high; the eyes small, black, and deep-set; the nose much depressed at the upper part between the eyes, and widened at the base, which is done in infancy by the mother, the natural shape being of an aquiline form; the cheek-bones are high, the mouth large, and furnished with strong, well-set teeth; the chin frequently retreats; the neck is thin and short. Their colour usually approaches chocolate, a deep umber, or reddish black, varying much in shade; and individuals of pure blood are sometimes as light-coloured as mulattoes. Their most striking distinction is their hair, which is like that of dark-haired Europeans, although more silky. It is fine, disposed to curl, and gives them a totally different appearance from the African, and also from the Malay, and American Indian. Most of them have thick beards and whiskers, and they are more hairy than the whites. The proportions of two of them will be found in the table of comparative proportions at the end of the fifth volume.

They are difficult to manage, taking offence easily when they are ill treated; and if any one attempts to control, thwart, or restrain their wandering habits, they at once resort to the woods, and resume their primitive mode of life, subsisting upon fish, grubs, berries, and occasionally enjoying a feast of kangaroo or opossum-flesh. They eat the larvæ of all kinds of insects with great gusto. Those who reside upon the coast, fish with gigs or spears, which are usually three-pronged; they have no fish-hooks of their own manufacture.

NATIVE OF AUSTRALIA.

When they feel that they have been injured by a white settler, they gratify their revenge by spearing his cattle; and it is said upon good authority, that not a few of the whites, even of the better class, will, when they can do so with impunity, retaliate in the blood of these wretched natives; and it is to be regretted that they are not very scrupulous in distinguishing the guilty from the innocent.

The natives of New South Wales are a proud, high-tempered race: each man is independent of his neighbour, owning no superior, and exacting no deference; they have not in their language any word signifying a chief or superior, nor to command or serve. Each individual is the source of his own comforts, and the artificer of his own household implements and weapons; and but for the love of companionship, he might live with his family apart and isolated from the rest, without sacrificing any advantages whatever. They have an air of haughtiness and insolence arising from this independence, and nothing will induce them to acknowledge any human being as their superior, or to show any marks of respect. In illustration of this, Mr. Watson the missionary is the only white man to whose name they prefix "Mr.," and this he thinks is chiefly owing to the habit acquired when children under his authority. All others, of whatever rank, they address by their Christian or surname. This does not proceed from ignorance on their part, as they are known to understand the distinctions of rank among the whites, and are continually witnessing the subservience and respect exacted among them. They appear to have a consciousness of independence, which causes them, on all occasions, to treat even the highest with equality. On being asked to work, they usually reply, "White fellow work, not black fellow;" and on entering a room, they never remain standing, but immediately seat themselves.

They are not great talkers, but are usually silent and reserved. They are generally well-disposed, but dislike to be much spoken to, particularly in a tone of raillery. An anecdote was mentioned of a gentleman amusing himself with a native, by teasing him, in perfect good-humour, when the man suddenly seized a billet of wood, threw it at him, and then in a great rage rushed for his spear. It was with great difficulty that he could be pacified, and made to know that no insult was intended; he then begged that they would not talk to him in that manner, as he might become wild and ungovernable. They look upon the whites with a mixture of distrust and contempt, and to govern them by threats and violence is found impossible. They are susceptible of being led by kind treatment, but on an injury or insult they immediately take to the bush, and resume their wandering habits. They do not carry on any systematic attacks, and their fears of the whites are so great, that large companies of them have been dispersed by small exploring parties and a few resolute stockmen.

Though they are constantly wandering about, yet they usually confine themselves to a radius of fifty or sixty miles from the place they consider their residence. If they venture beyond this, which they sometimes do with a party of whites, they always betray the greatest fear of falling in with some Myall or stranger blacks, who they say would put them to death immediately.

Their great timidity has caused a false estimate to be put upon their character, by ascribing to it great ferocity; and, as an instance of it, it is mentioned, that if a party of natives be suddenly approached in the interior, who are unacquainted with white men, and taken by surprise, supposing that they are surrounded and doomed to death, they make the most furious onset, and sell their lives as dearly as possible: this arises from the panic with which they are seized, depriving them temporarily of reason.

They have not, properly speaking, any distribution into tribes. In their conflicts, those speaking the same language, and who have fought side by side, are frequently drawn up in battle-array against each other, and a short time after may be again seen acting together Their conflicts, for they do not deserve the name of wars, are conducted after the following manner. The quarrel or misunderstanding generally arises from some trivial affair; when the aggrieved party assembles his neighbours to consult them relative to the course to be pursued. The general opinion having been declared, a messenger is sent to announce their intention to commence hostilities to the opposite party, and to fix a day for the combat. The latter immediately assemble their friends, and make preparations for the approaching contest. The two parties on the day assigned meet, accompanied by the women and children. The first onset is made by the oldest women (hags they might be termed) vituperating the opposite side. Then a warrior advances, and several throws of spears take place. These are parried with much dexterity, for all the natives possess great art and skill in avoiding missiles with their shields. This exchange of missiles continues for some time, and not unfrequently ends without any fatal result. When one of either party is killed, a separation takes place, succeeded by another course of recrimination, after which explanations are made, the affair terminates, and hostility is at an end; the two parties meet amicably, bury the dead, and join in the corrobory dance.

These dances are not only the usual close of their combats, but are frequent in time of peace. They appear almost necessary to stir up their blood; and under the excitement they produce, the whole nature of the people seems to be changed. To a spectator, the effect of one of these exhibitions almost equals that of a tragic melodrame.

A suitable place for the performance is selected in the neighbourhood of their huts. Here a fire is built by the women and boys, while such of the men as are to take a share in the exhibition, usually about twenty in number, disappear to arrange their persons. When these preparations are completed, and the fire burns brightly, the performers are seen advancing in the guise of as many skeletons. This effect is produced by means of pipe-clay, with which they paint broad white lines on their arms and legs, and on the head, while others of less breadth are drawn across the body, to correspond to the ribs. The music consists in beating time on their shields, and singing, and to it the movements of the dancers conform. It must not be supposed that this exhibition is a dance in our sense of the word, nor is it like any thing that we saw in the South Sea islands. It consists of violent and odd movements of the arms, legs, and body, contortions and violent muscular actions, amounting almost to frenzy. The performers appear more like a child's pasteboard supple-Jack than any thing human in their movements.

Drawn by A. T. Agate. Engd by E. G. Dunnell.

CORROBORY (DANCE) N.S.W.

This action continues for a time, and then the skeletons, for so I may term them, for they truly resemble them, suddenly seem to vanish and reappear. The disappearance is effected by merely turning round, for the figures are painted only in front, and their dusky forms are lost by mingling with the dark background. The trees illuminated by the fire, are brought out with some of the figures in bold relief, while others were indistinct and ghost-like. All concurred to give an air of wildness to the strange scene. As the dance proceeds, the excitement increases, and those who a short time before appeared only half alive, become full of animation, and finally were obliged to stop from exhaustion.

These corroborys are the occasion of much intercourse among the tribes, as they frequently make visits to each other for the sole purpose of carrying a new song for the dance. They have several kinds of these dances, which appear to be their only amusement. In their music they do not sound any of the common chords, and the only accompaniment was a kind of bass, as written below, which was in fact only a very deep-toned grunt, sounded, as ho, ho, ho, very deep in the throat. At the end of each dance they finished with a loud whoo, or screech, an octave above the key-note.

The above is thought by Mr. Drayton not to be entirely native music, but the following he has no doubt of; the words are given as he heard them.

The above, as well as those which follow, were obtained from a native who was on his way with the new song to his tribe.

We have seen that the combats, of which mention has been made, are attended with little loss of life; nor are their set battles bloody. In all their contests they seem to act upon certain principles, well established in their code of honour, and firmly adhered to. According to eye-witnesses of their battles, these are conducted with system and regularity. On one occasion, the parties advanced towards each other drawn up in three lines, with the women following in the rear, and when they had arrived within a few rods of each other, they threw their boomerengs or curved sticks. These, as they fell among their opponents, were picked up by the women, and given to the warriors, who hurled them back to their original owners, by whom they were again used. When these weapons were lost or broken, they then had recourse to their spears, which they threw, parried, and returned in like manner. They then closed and fought hand to hand with their clubs, for a considerable time. Their extraordinary quickness of eye, alertness, and agility of movement, protect them from much harm; and their thickness of skull may also be taken into account, for nothing worse than a few bruises and broken limbs resulted. The fight gradually dwindled down to a single combat between two of the most determined warriors, and when one of these was knocked down by a stunning blow, another took his place and continued the fight until one was severely injured. The battle then terminated.

Some of their personal or private quarrels are settled by a sort of duel, or rather a trial of whose head is hardest. The accused or challenged party extends his head, with the crown uppermost, towards his adversary, who strikes a blow with his utmost force with a waddy, which is the weapon they usually carry about them, and with which they punish their wives, who exhibit generally many marks from the use of it. The challenger then presents his head in return, and blows are thus continued alternately, until one or the other is disabled, or both declare themselves satisfied. Those who have witnessed these encounters are quite astonished that every blow does not stun or kill, for each of them would be fatal to a white man; but the great thickness of their skulls enables them to sustain this violence with but little injury.

Their weapons are the spear, club, or nulla-nulla, boomereng, dundumel, and the bundi, of which drawings are given in the tail-piece at the end of the last chapter. Their spears are about ten feet long, and very slender, made of cane or wood tapering to a point, which is barbed. They are light, and one would scarcely be inclined to believe that they could be darted with any force; nor could they without the aid of the wammera, a straight flat stick, three feet in length, terminating in a socket of bone or hide, into which the end of the spear is fixed. The wammera is grasped in the right hand by three fingers, the spear lying between the fore-finger and thumb. Previous to throwing it, a tremulous or vibratory motion is given to it, which is supposed to add to the accuracy of the aim; in projecting the spear, the wammera is retained in the hand, and the use of this simple contrivance adds greatly to the projectile force given to the spear. They are well practised in the use of these weapons.

The nulla-nulla, or uta, is from thirty to thirty-six inches in length, the handle being of a size to be conveniently grasped.

The dundumel is a weapon used by the natives of the interior; it has a curved flat handle thirty inches in length, and terminates in a projection not unlike a hatchet; it is thrown from the hand before coming to close quarters, and usually at a very short distance.

But the most extraordinary weapon is the boomereng. This is a flat stick, three feet long and two inches wide by three quarters of an inch thick, curved or crooked in the centre, forming an obtuse angle. At first sight one would conclude it was a wooden sword, very rudely and clumsily made; indeed one of the early navigators took it for such. It is an implement used both for war and in the chase. In the hands of a native it is a missile efficient for both, and is made to describe some most extraordinary curves and movements.

NATIVE THROWING THE BOOMERENG.

It is grasped at one end in the right hand, and is thrown sickle-wise, either upwards into the air, or downwards so as to strike the ground at some distance from the thrower. In the first case it flies with a rotary motion, as its shape would indicate; after ascending to a great height in the air, it suddenly returns in an elliptical orbit to a spot near its starting-point. The natives in its use are enabled to strike objects which lie behind others with great precision, and to reach those near as if by a back-stroke, by throwing it at a particular angle. The diagram at the end of the chapter, exhibits the curves at the angles of 22°, 45°, and 65°, respectively, which I have obtained in making experiments with it. Some facts which were spoken of in its use, are remarkable. On throwing it downwards on the ground, it rebounds in a straight line, pursuing a ricochet motion until it strikes the object at which it is thrown. Birds and small animals are killed with it, and it is also used in killing ducks. The most singular curve described by it, is when thrown into the air, above the angle of 45°; its flight is always then backwards, and the native who throws it stands with his back, instead of his face, to the object he is desirous of hitting. The diagram also exhibits its fall in case it loses its rotary motion. It is a favourite weapon with the natives, and is frequently seen ingeniously carved.

As a defence, they use a shield made of the thick bark of the gum tree; this they call hiclemara. It is peculiar in shape, and on the coast is three feet long by six or eight inches wide, with a handle in the centre; it is made rounding. Those in the interior are only a three-cornered piece of wood, with a hole on each side, through which the hand is thrust. The size of the latter is smaller, being only two feet long and three or four inches broad. It would seem almost impossible that so small a shield should be sufficient to guard the body of a man; and nothing but their quickness of eye and hand could make it of any value as a protection against the spear or club.

The mode in which the natives climb trees was considered extraordinary by those who witnessed it, although they had been accustomed to the feats of the Polynesians in the ascent of the cocoa nut trees. The Australians mount a tree four or five feet in diameter, both with rapidity and safety. As they climb they cut notches above them, with a stone or metal hatchet, large enough to admit two of their toes, which are inserted in them, and support their weight until other holes are cut.

The natives who reside upon the coast use canoes which are constructed as follows.

A gum tree that has a thick and tough bark is selected; this is girdled, and the bark slit so that by care a piece of it may be stripped from the tree large enough to make the canoe, which is usually about fourteen feet long and seven wide. This piece of bark is charred on the inside, after which it is folded in each end, so as to bring the edges of the two halves of the entire circuit of the bark together; in this position these edges are fastened by cords and wooden rivets. The simple canoe is now complete, is usually about three feet wide in the middle, and will convey half a dozen persons.

They use paddles of different sizes, say from two to five feet in length. In using the shorter kind, a paddle is held in each hand.

A fire is commonly carried upon a layer of gravel in the middle of the boat: a custom which appears to arise either from a natural or superstitious reluctance to be without a fire at any time. In this custom, as will be recollected, they resemble the Fuegians, who, however, far excel them in the art of constructing canoes.

The social system and intercourse of the Australians is regulated by custom alone. As no system of government exists, or any acknowledgment of power to enact laws, they are solely guided by old usage, and can give no account whatever of its origin. The universal reprobation of their associates, which follows a breach of ancient customs, has a strong tendency to preserve a strict observance of them. Many of these customs struck us as remarkable; those that have not been actually seen by the officers of the Expedition, have been described by persons entitled to the fullest credit.

The custom (to use the language of the settlers) "of making young men," is singular. The object of the institution seems to be to imprint forcibly upon the youth the rules and observances by which his after life is to be governed; and so strikingly are they adapted to insure good conduct, that it can hardly be believed that they could have originated among savages, such as the natives of Australia now are.

When the boys reach the age of fourteen, or that of puberty, the elders of the tribe prepare to initiate them into the privileges of manhood. A night or two previous, a dismal cry is heard in the woods, which the boys are told is the Bùlù calling for them. Thereupon all the men of the tribe set off for some secluded spot, previously fixed upon, taking with them the boys or youths to be initiated. No white man is allowed to be present, and the precise nature of the ceremony is therefore unknown; but it is certain that the ceremonies are designed to try their courage, fortitude, and the expertness of the boys in reference to their future employments in the chase and in war. There is probably some difference in these ceremonies among the different tribes. The Wellington station, or those of the interior, for instance, never knock out a front tooth, which is always done on the coast.

From the time the youths are initiated, they are required to yield implicit obedience to their elders. This is the only control that seems to prevail, and is very requisite to preserve order and harmony in their social intercourse, as well as to supply the place of distinctions of rank among them.

The youths are likewise restricted to articles of diet, not being allowed to eat eggs, fish, or any of the finer kinds of opossum or kangaroo. Their fare is consequently of a very poor description, but as they grow older these restrictions are removed, although at what age we have not learnt; but after having passed the middle age, they are entirely at liberty to partake of all. The purpose of this is thought to be not only to accustom them to a simple and hardy way of living, but also that they should provide for the aged, and not be allowed to appropriate all to themselves. Selfishness is therefore no part of their character, and all observers are struck with their custom of dividing any thing they may receive among each other, a disinterestedness that is seldom seen among civilized nations.

To protect the morals of the youths, they are forbidden from the time of their initiation until their marriage to speak to or even to approach a female. They must encamp at a distance from them, and if, perchance, one is seen in the pathway, they are obliged to make a detour in order to avoid her. Mr. Watson stated he had been often put to great inconvenience in travelling through the woods with a young native for his guide, as he could never be induced to approach an encampment where there were any women.

The ceremony of marriage is peculiar. In most cases the parties are betrothed at an early age, and as soon as they arrive at the proper age, the young man claims his gin or wife.

The women are considered as an article of property, and are sold or given away by the parents or relatives without the least regard to their own wishes. As far as our observation went, the women appear to take little care of their children. Polygamy exists, and they will frequently give one of their wives to a friend who may be in want of one; but notwithstanding this laxity they are extremely jealous, and are very prompt to resent any freedom taken with their wives. Their quarrels for the most part are occasioned by the fair sex, and being the cause, they usually are the greatest sufferers; for the waddy is applied to their heads in a most unmerciful style, and few old women are to be seen who do not bear unquestionable marks of the hard usage they have received. The husband who suspects another of seducing his wife, either kills one or both. The affair is taken up by the tribe, if the party belongs to another, who inflict punishment on him in the following manner.

The guilty party is furnished with a shield, and made to stand at a suitable distance, and the whole tribe cast their spears at him; his expertness and activity often enable him to escape any serious injury, but instances do occur in which the party is killed. Such punishments are inflicted with great formality, upon an appointed day, and the whole tribe assemble to witness it. The person most injured has the first throw, and it depends upon the feelings of the tribe respecting the offence committed, whether they endeavour to do injury to the culprit or not; and thus it may be supposed that there is some judgment evinced in this mode of punishment.

The following account of the burial of their dead, was received from the missionary who was an eye-witness to it. He was called out one evening to see a native, who they said was dying. On repairing to the camp, he was too late, for the man was already dead, and notwithstanding the short space of time that had elapsed, the corpse was already wrapped up for burial. The legs had been bent at the knees and hips, and tied to the body, and the head bent downwards towards the legs. In this position the corpse was enveloped in a blanket, and bound round with many ligatures, so as to form a shapeless lump. There were about fifty natives present, seated within a small space in front. The women were raising dismal lamentations and cutting themselves with sharp sticks; while the men were engaged in an earnest consultation as to the place which should be fixed upon for the burial. At length it was determined to be on the banks of the Macquarie, at no great distance from the mission station. On the following day the missionary proceeded to the place, and found that the natives had already cleared the grass from a space about twenty feet in diameter; in the centre of this the grave was marked out, of an oval shape, six feet long by three feet wide. After digging to the depth of about a foot, they left a ledge all around the grave of a few inches in width: the excavation, thus diminished in size, was continued to the depth of five feet, the sides not being exactly perpendicular, but sloping slightly inwards. At the bottom of the grave was laid a bed of leaves, covered with an opossum-skin cloak, and having a stuffed bag of kangaroo-skin for a pillow; on this couch the body was laid, and the implements of hunting and war which the deceased had used during his lifetime were laid beside him. Leaves and branches of bushes were strewed over him, until the grave was filled up to the ledge or shelf above mentioned. Across the grave were laid strong stakes, with the ends resting on this shelf, and on these a layer of stones, which filled the hole to the level of the soil. The excavated earth was then put over the whole, forming a conical heap eight or nine feet high. The trees on each side were marked with irregular incisions, but whether intended as symbols, or merely to identify the place of sepulture, was not understood. All the time this was going on, fires were kept burning around the place, to drive away evil spirits, and the women and children uttered loud lamentations, inflicting at the same time wounds upon themselves. When the grave was completed, all the women and children were ordered away, and the missionary, perceiving that it was expected that he would do the same, retired also. His presumption was that they intended to give utterance to their grief, and that they were ashamed to do it in his presence, or before the women and children.

The day after the burial the natives visited every spot in which they recollected to have seen the deceased, and fumigated it, for the purpose of driving away the evil spirits. They even went into the missionaries' houses, greatly to the annoyance of the ladies.

Their style of mourning consists in bedaubing themselves with pipe-clay; and a more hideous object than an old woman thus tricked out can hardly be conceived. The body and limbs are streaked with it, and the face completely covered as with a mask, in which holes are left for the eyes, nostrils, and mouth. The mask is gradually removed until the last that is seen of it is a small patch on the top of the head.

They have some idea of a future state, although some assert that the whole man dies, and that nothing is left of him; while others are of opinion that his spirit yet lives, either as a wandering ghost or in a state of metamorphosis, animating a bird or other creature of a lower order than man.

Their opinions on religious subjects generally partake of the same unsettled character, which makes it very difficult to obtain any clear idea of them. The great difficulty is the unwillingness of the natives to talk upon the subject, either from superstition or shame; and it is the opinion of the missionaries that no full account of their religious notions will be obtained, until one of the well-informed adults is converted to Christianity, an event which is not soon to be expected. The missionaries have had little or no success; none of the adults have hitherto shown any desire to embrace Christianity; and it is remarked, that there appears to be a want of susceptibility in their character to religious impressions. Some of their ceremonies which partook of that character have been discontinued of late years, and no others have been adopted in their place. They have, however, some indistinct notions of a Deity. The missionaries at Wellington have heard from them of a being whom they call Bai-a-mai, and whom, with his son Burambin, they deem the creator of all things. To this Bai-a-mai they pay a kind of annual worship by dancing and singing a song in his honour. This song, they say, was brought from a distant country by strangers who went about teaching it. This annual worship took place in the month of February, and all who did not join in it were supposed to incur the displeasure of the god.

Bai-a-mai was supposed to live on an island beyond the great sea of the coast, and to eat fish, which, when he required food, came up at his call from the water. Burambin, others say, was brought into existence by Bai-a-mai, when the missionaries first came to Wellington.

Dararwirgal is a brother of Bai-a-mai, and lives in the far west. To him they ascribe the origin of the small-pox, which has made such ravages among them. They say that he was vexed for want of a tomahawk, and therefore sent that disease among them; but they now suppose that he has obtained one, and that the disease will come no more.

Balumbals are angels, said to be white, who live on a mountain to the southwest, at a great distance. Their food is honey, and their employment like that of the missionaries.

Wandong is their evil spirit, whom they have learnt from the whites to call the "Devil." They describe him as a gigantic black man, always prowling about at night, ready to seize and devour any unfortunate wanderer. So great is their horror of this imaginary being, that they never venture from their fires at night, except under the pressure of great necessity, when they always carry a firebrand to intimidate the monster.

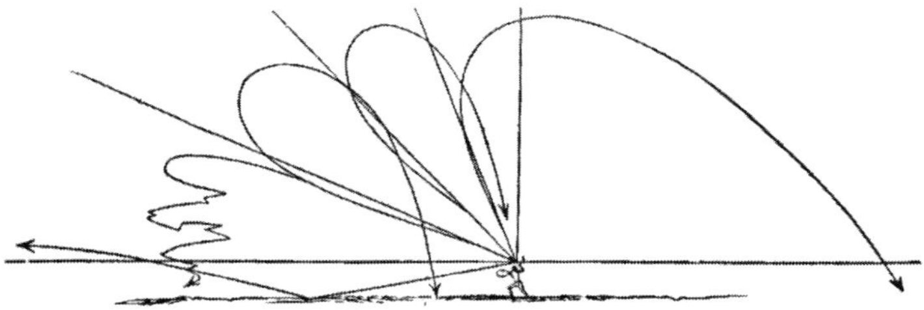

FLIGHT OF THE BOOMERENG.

CHAPTER VII.

CONTENTS.

EXTENT OF SEA-COAST OF AUSTRALIA—PROGRESS OF DISCOVERY—EXTENT OF THE GOVERNMENT OF NEW SOUTH WALES—HISTORY OF NEW SOUTH WALES—AUSTRALIAN LAND COMPANY—LAND FUND—POLICY OF GOVERNOR MACQUARIE—CONVICT-SHIP—VALUE OF CONVICT LABOUR—DISPOSITION OF CONVICTS ON ARRIVAL—ALLOWANCES TO CONVICT SERVANTS—WORKING OF THE ASSIGNMENT SYSTEM—SEPARATION OF FEMALE CONVICTS FROM THEIR CHILDREN—FACTORY AT PARAMATTA—ITS DISCIPLINE—CHILDREN IN THE FACTORY—IRON-GANGS—CAPTAIN MACONOCHIE'S SOCIAL SYSTEM—ITS OPERATION—ITS PARTIAL ADOPTION AT SYDNEY, AND AT NORFOLK ISLAND—TWO DIFFERENT FORMS OF SOCIAL SYSTEM—CONDITION OF CONVICTS AND SOLDIERS COMPARED—TICKET-OF-LEAVE MEN—DISTINCTIONS IN SOCIETY—FREE EMIGRANTS—LABOURING CLASS OF FREE EMIGRANTS—CRIMPS—HIGHER CLASS OF SOCIETY—GOVERNOR, HOW APPOINTED, AND FOR WHAT TERM—HIS POWERS—HIS SUCCESSOR IN CASE OF DEATH OR ABSENCE—LEGISLATIVE COUNCIL—COLONIAL RULES AND REGULATIONS—EXPENDITURE AND MODE OF TAXATION—WORKING OF THE LEGISLATIVE SYSTEM—STATUTE OF NEW SOUTH WALES—DESIRE FOR A NEW CONSTITUTION—BOUNTY ON IMMIGRATION—EMIGRATION—MODE OF PROCURING GRANTS OF LAND—STATE OF CRIME IN NEW SOUTH WALES—JUDGE BURTON'S CHARGE—HIS REPORT ON THE JURY SYSTEM—OPINIONS OF OTHER JUDGES—COURTS OF LAW—EDUCATION—BISHOPRIC OF NEW SOUTH WALES—COLLEGES.

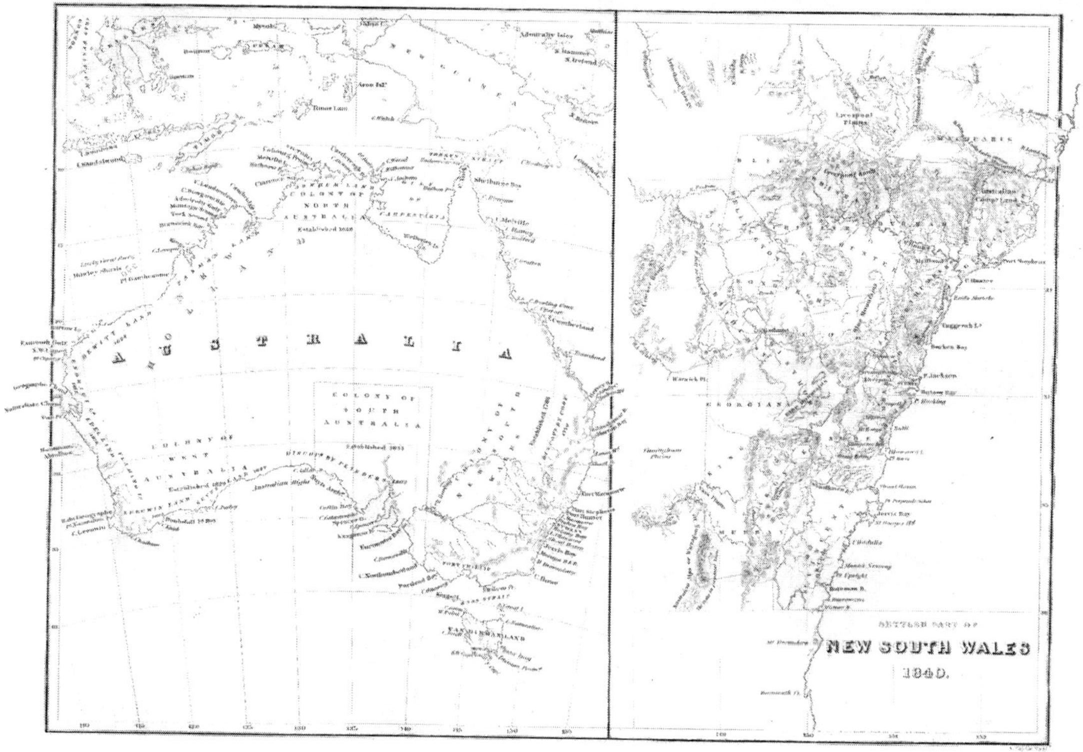
AUSTRALIA
COLONY OF SOUTH AUSTRALIA
COLONY OF NORTH AUSTRALIA
COLONY OF WEST AUSTRALIA
NEW GUINEA
Admiralty Isles
Jervis Bay
Botany Bay
P. Jackson
Port Stephens
SETTLED PART OF
NEW SOUTH WALES
1840.

CHAPTER VII.

HISTORY, GOVERNMENT, AND SOCIAL STATE OF NEW SOUTH WALES.

1839.

New South Wales is known in the United States almost by its name alone.

It happened from the circumstances of our visit, that we were enabled to obtain much information in relation to this rising colony.

Australia, or New Holland, of which New South Wales is as yet the most important part, requires no description of its dimensions and geographical position. It is sufficient to note the fact that it possesses a sea-coast of the vast extent of eight thousand miles. It was first discovered by the Dutch, while engaged in exploring the coast of New Guinea, who saw the portion of it to the south of Endeavour Straits, and gave it a name from that of their own country, in 1606. A few months after this discovery, Louis de Torres made the north-eastern point of Australia.

In 1616, Theodoric Hartog fell in with that part of the western coast which lies within the tropic and latitude 28° S. This he called Endracht's Land, after the name of his vessel.

In 1618, the coast between latitude 11° and 15° S., reaching from the Gulf of Carpentaria to Cape Talbot, was seen by Zeachern.

In 1628, De Witt and Carpenter discovered that part of the western coast known as De Witt's Land, and surveyed the Gulf of Carpentaria.

In 1667, Van Nuyt sailed along the southern coast, from Cape Leeuwin to Spencer's Gulf; and to this part of Australia his name has of late been restored upon the maps.

Tasman, in 1642, discovered Van Diemen's Land, which was long after considered to be connected with the main land of Australia. Finally, between 1766 and 1770, after an interval of a century, during which no researches had been made, and some of the discoveries already mentioned had been forgotten, Cook explored the eastern coast, from Cape Home to Cape York, and called its whole extent New South Wales.

Researches were again suspended until after the establishment of the convict colony, and in 1798 Flinders and Bass discovered the straits which bear the name of the latter, which separate Van Diemen's from the main land, and sailed around that island. To the country adjacent to these straits, the name of Bass's Land was given; and in 1803, Grant explored the coast to the west of it. Flinders, who was for several years engaged in making surveys on the eastern coast, also connected the land discovered by Grant, with that of Van Nuyt, and re-examined the latter. It is to Flinders that we owe our most precise knowledge of the general geographical features of the eastern and southern coast of Australia; and since the close of his labours, Captain P. P. King, of H. B. M. Navy, has been engaged, and other officers are now assiduously employed, in surveying the northern coast. The interior, as has been already stated, has been made the subject of numerous exploring tours by the surveyors of the colony, and other persons employed by the British government.

The territory included under the name of New South Wales is the eastern portion of Australia, and extends from the twenty-third to the thirty-eighth degree of south latitude. The power of its governor, however, extends further, and within his jurisdiction are included the whole eastern coast, from Cape York to Wilson's Promontory, or between latitude 10° 37′ and 39° 2′ S.; with the country inland as far as the meridian of 129° E. Moreton Bay on the northern, and Port Philip on the southern coast, with Norfolk Island, and all others between it and the coast, are also placed under his authority.

The epoch whence the history of the colony dates, is the year 1787, when the eastern coast was chosen by the British government as the site of a penal colony. The convicts hold so prominent a part in the events which have occurred since that period, that their history may be almost considered as that of New South Wales.

Botany Bay, in consequence of extravagant ideas formed of its excellence as a harbour, and the fertility of the country around it, was the portion chosen for the settlement. The first gang of convicts sent out was composed of six hundred male and two hundred and fifty female criminals, who were guarded by a body of troops consisting of two hundred officers and soldiers. About forty of the soldiers were married, and had their wives and children with them. The whole were under the direction of Captain Philips, who held the appointment of governor.

The expedition, in conformity to the instructions, proceeded in the first instance to Botany Bay. This, to their disappointment, was found an unsafe harbour; and the country far from suitable for the intended purpose. On exploring the coast, they unexpectedly found, about seven miles to the northward, a capacious and beautiful harbour, which Cook had reported as fit only for boats; this, which they called Port Jackson, proved to be adapted to the intended purpose, and to it they removed. The people of the colony were landed on the 26th of January, 1788. The Governor delivered an address to his settlers on the 7th of February, strongly recommending marriage to the convicts; and in consequence of this admonition, fourteen marriages took place the succeeding week.

In 1790, one of the severe droughts to which the country is liable occurred, and the colony was reduced to great distress for provisions. All the live-stock, which had been imported for breeding, was killed off for food, and the inhabitants were reduced to an ear of corn per day. From the exhaustion which prevailed, all labour was suspended.

In February, 1792, the first lieutenant-governor arrived. He was also commandant of the New South Wales corps. This corps was specially raised for service in the colony, and was one of the greatest evils under which it suffered for many years.

In December, 1792, Governor Philips returned to England.

In September, 1795, Governor Hunter arrived, and assumed the direction of affairs. His administration lasted until 1802.

During the interregnum between Governors Philips and Hunter, Captain Paterson acted as governor.

In 1794, the first free settlers arrived in the colony.

The officers of the New South Wales corps soon became merchants, and dealt in all that was issued from the public stores. Rum was the great article of traffic; and an act was passed, that on the arrival of any vessel with stores, an issue of spirits from her cargo should be made to each officer in proportion to his rank.

The officers also obtained the manifest of every vessel that arrived, selected what they thought proper from her cargo, and afterwards disposed of it to the soldiers, settlers, and convicts, at a large profit.

They claimed the privilege of importing spirits, which was refused to others, and of selling it to the non-commissioned officers, many of whom held licenses to sell spirits by retail. In this way, many of the officers of the New South Wales corps realized large sums by trade, and counteracted all the endeavours of the governor to effect a reform in the colony.

In September, 1800, Captain King assumed his duties as governor, and during the whole of his administration, which lasted till 1806, provisions continued to be imported into the colony at the expense of the home government, principally from the Cape of Good Hope and Batavia.

The military were gradually acquiring more power, and their officers generally set the laws at defiance, assuming the right of landing spirits from every vessel that arrived. Governor King endeavoured to put a stop to this practice; but the military, who had become powerful and influential, overawed him. In an evil hour, hoping to check their power and influence, he gave licenses not only to the constables, but to the jailer, to sell rum; and the latter, it is said, was allowed to keep a public tap-room opposite the jail door.

In consequence of this state of things, the Governor's power was very much weakened, if not entirely destroyed, and the whole settlement was thrown into confusion. The convicts were under no efficient control, and bands of them, under the name of Bush-rangers,* traversed the country, and entered the houses of the settlers even in the open day, committing the most fearful atrocities. Anarchy and confusion reigned every where.

The Castle Hill convicts now mutinied, but were overcome, and some of them executed.

Captain Bligh, R. N. (who had commanded the Bounty), succeeded Captain King. During his administration, *rum* was the medium of exchange, and the settlers had no other purchasers for their produce but the privileged dealers in that article, who took every advantage of them.

In 1807, two stills for manufacturing spirits were imported by Mr. M'Arthur and Captain Abbot, of the 102d regiment. The Governor seized them, and prohibited distillation in the colony. Much discontent grew out of this prohibition; and other difficulties occurred, which resulted in the arrest of Governor Bligh by the military, and other turbulent persons in the colony, in 1808. The home government now saw the necessity of putting down this lawless spirit, and reinstated Governor Bligh; but although he was also promoted to the rank of admiral, he is said to have died of a broken heart.

Governor Macquarie was his successor. He seems to have endea-

* The Bush-rangers are still very troublesome at times. In addition to the runaway convicts, of which their bands are principally composed, they also include soldiers who have deserted. They occasionally commit great barbarities, and are consequently much dreaded. Few indeed of the lonely settlements are safe from their depredations. In order to suppress them, there is a body of mounted police; but its numbers are too small to put an effectual stop to the evil.

voured to build up Sydney and Paramatta at the expense of the morals of the community, and appears to have discouraged free emigration. The emancipated convicts were admitted by him to the same privileges and immunities as the free settlers; and he treated the whole colony as if it were the gift of the mother country to those of her subjects who had outraged her laws. This policy soon had its effect at home, where it is said that crimes were committed in the hope of being sentenced to transportation; and it is asserted that the emancipated convicts, known as ticket-of-leave men, were much more desirably situated than honest persons of their own rank of life in the mother country. Many of these in consequence wrote to their friends to come out to them, and pointed out the means of doing so at the expense of the crown. This state of things was offensive to the free settlers, who opposed the endeavours of the colonial government to break down the distinctions that naturally exist between the polluted and unpolluted. Their repugnance was increased by the attempt, on the part of the emancipated convicts, to make property the only qualification for acquiring civil and political rights. So strongly was the line of distinction drawn by the free settlers at that time, that it remains unbroken to the present day, and affects even the third and fourth generation. From the countenance shown to the convicts by Governor Macquarie, their minds have become impressed with the idea that the colony is intended for their benefit; and they consider that they have the best right to administer the government, while the free emigrants in their opinion are interfering with their rights, by occupying all places of emolument and trust. Sufficient interest was excited by the complaints of the free emigrants to cause a commission of inquiry to be appointed. Much good resulted from its action, and a settled policy was at length adopted in relation to the treatment of the convicts.

Governor Macquarie was succeeded, in 1821, by Sir Thomas Brisbane, during whose administration all restrictions on the press were removed.

In 1824, by an act of Parliament, a Supreme Court was established, having equity, common law, admiralty, and ecclesiastical jurisdiction. Trial by jury was authorized under certain limitations, and the Legislative Council instituted. This was a great improvement upon the former system, and reduced the power of the Governor, which had before been absolute, while it at the same time gave him the best advisers.

The Legislative Council consisted of the Governor, with three civil officers, and three gentlemen not holding office. The establishment of this body was a step towards a constitutional form of government. It continued in the original form, until, by an act of Parliament in 1828, it was increased to not less than ten or more than fifteen members, all of whom were appointed by the crown. At the same time an Executive Council was created, and in these two bodies the government of the colony is still vested.

During the administration of Sir Thomas Brisbane and Sir Ralph Darling, many improvements in policy took place. Among them were inducements to free settlers, and many officers of the army and navy, with other respectable persons, received grants of land. The acquisition of a population of a better character furnished the means of enforcing the laws, and removed the necessity of employing those who had been guilty of crime, to effect its suppression. A better state of things succeeded. Agriculture, and particularly the raising of herds and flocks, were promoted by the facilities afforded by the government under the assignment system; and the resources of the colony were developed, particularly in the growth of wool, which has now become its great staple. The success of these agricultural efforts, excited in England, particularly among its manufacturers, a more direct interest in the colony, and attracted much attention to it; in consequence of which the Australian Agricultural Company, in which many influential persons in England became interested, was incorporated under a royal charter. The avowed objects of this association were to further the improvement of the cultivation of land in New South Wales, and the rearing of cattle, horses, and fine-woolled sheep. The capital of the Company was a million of pounds sterling, and government agreed to grant in addition a million of acres of land, in any part of the territory that might be selected.

The agent of the Company, Mr. Dawson, commenced operations in 1826, at Port Stephens, to the north of Hunter's river, on a tract selected by him and the surveyor-general; he continued to manage their affairs until 1829, when, in consequence of a misunderstanding between Mr. Dawson and the Company, growing out of the unavoidable difficulties he had to contend with, and the many misrepresentations made by his enemies, he was removed, and Sir Edward Parry, the celebrated polar navigator, appointed in his stead.

Sir Edward Parry continued in the management until 1836, when he retired, and was succeeded by Captain P. P. King, the able surveyor, and who still conducts the affairs of the Company. I regretted much not having been able to accept of the pressing invitation to visit Captain King. It affords me great pleasure to bear testimony to the correctness of his charts and sailing directions, which I have on several occasions been called upon to verify and trust to in navigating the squadron.

The difficulties encountered by Mr. Dawson, were chiefly owing to the excessive droughts that occurred in the years 1827, 1828, and 1829. In these years there was in some districts an entire, and in others a partial failure of the crops, while the pasture-grounds were all dried up.

At other times, continued rains would cause great floods; whole districts of country were overflowed; and along the rivers, not only stacks of grain, but the buildings, were swept away. From this it may be easily understood what difficulties the settlers of this country have to contend with. To these also are to be added the contests between the two parties, as to whether this shall remain a penal colony or become a free one.

It is only to be wondered that it should have continued to flourish, notwithstanding all the impediments it has met with from misrule and anarchy, growing out of a neglect to establish any well-combined system of policy in its early stages. The governors, for want of any positive enactments, were left free to adopt such measures as circumstances might dictate, and having their attention engrossed by the difficulties with which they were continually surrounded, were compelled to neglect the improvement of society, and took no pains to frame prospective regulations for the well-being of the colony.

In 1831, Governor Darling was superseded by Sir Richard Bourke. The country had, during the administration of his two predecessors, improved rapidly by the aid of convict labour. When the latter gentleman came into office, the policy of selling lands, instead of making free grants, was adopted; this was considered an important change for the colony.

The money arising from the sales of land was set apart by the home government, to be applied to the immigration of free settlers; but great complaints have been made that this fund has been diverted from the original object, or that a surplus remained in the government coffers unexpended. Labour is, in consequence, in the greatest demand in all parts of the colony, and the inconveniences of convict labour begin to show themselves. From what I could understand, the assignment system is getting into disrepute, and all the respectable settlers are now turning their attention to the moral condition of the colony. Strong representations have been made to the home government, and an act has been passed, by which New South Wales is no longer to be a penal settlement, and transportation thither is to cease. The only points that are now used for this purpose are Moreton Bay and Norfolk Island.

The principles upon which free grants are made, have been subject to great fluctuation. In Governor Macquarie's time, no grants to a single individual could exceed one hundred and fifty acres. Many of the difficulties and evils that resulted from these free grants, are said to have grown out of his disposition (miscalled humane), which led him to view the convicts as men in misfortune, rather than as the outcasts of society.

His regulations in entering on his duties were explicit, "that a convict should receive a grant of land only, when from good conduct and a disposition to industry he should be found deserving of favour, and of receiving emancipation." In contradiction to the spirit of this regulation, he made grants of land to any emancipated convicts, and even appointed them to office as constables, &c. These emancipations were easily obtained, and transportation became, as has already been stated, rather a reward than a punishment for crime. Instead of bestowing his indulgences only on the deserving, a rule of action which, if strictly carried into execution, would have been productive of good, both to the rising community and the convicts themselves, he dispensed his favours indiscriminately. He committed a double error, when in addition he appointed them to office, thus placing them over their companions in crime. Good could not be expected to result from such a course, and the profusion with which land was distributed among the whole class of emancipated convicts, whether they had been set free before the expiration of their term for good behaviour, or had served their time out, produced positive injury.

During our stay at Sydney, a convict-ship arrived; and being desirous of obtaining a view of her accommodations, and the mode of treating the convicts, I visited her. This vessel was prepared expressly for the purpose. Between decks, a strong grated barricade, well spiked with iron, is built across the ship at the steerage bulkhead. This affords the officers a free view of all that is going on among the prisoners.

Bunks for sleeping are placed on each side all the way to the bow, resembling those in a guard-room. Each of these will accommodate five persons. There is no outlet but through a door in the steerage bulkhead, and this is always guarded by a sentry. Light and air are admitted through the hatches, which are well and strongly grated. The guard is under the command of a sergeant, and is accommodated in the steerage, the whole being under the orders of a surgeon, whose duty it is to superintend and regulate every thing that relates to the prisoners, inspect the ship daily, and administer punishment, even unto death if necessary. The surgeon also has control over the master of the vessel, and his regulations. The master and mates, on receiving a certificate from the surgeon, are allowed a small sum for every convict landed, in addition to their pay.

The criminals have prison fare, and are supplied with wooden-ware for their eating utensils, which are kept in very nice order. The quarter-deck is barricadoed near the main-mast, abaft of which all the arms and accoutrements of the guard and vessel are kept. The master and officers are usually lodged in the poop-cabin. The prisoners are habituated to the discipline of the ship, on board the hulks, before leaving England. The usual, and most effectual, punishment for misbehaviour is to place the culprit in a narrow box on deck, in which he is compelled to stand erect. This punishment is said to be effectual in reducing the most refractory male convicts to order, but it was not found so efficacious in the female convict-ship; for, when put in the box, they would bawl so loudly, and use their tongues so freely, that it was found necessary to increase the punishment by placing a cistern of water on the top of the box. This was turned over upon those who persist in using their tongues, and acted on the occupant as a shower-bath, the cooling effect of which was always and quickly efficacious in quieting them. I was informed that more than two such showers were never required to subdue the most turbulent.

I was struck with the ruddy, healthy, and athletic looks of the young convicts that were arriving, and from their deportment and countenances I should hardly have been inclined to believe that they had been the perpetrators of heinous crimes.

I am not at all surprised that many of the settlers of the colony should be opposed to the change in the assignment system; for when such a fine body of men is seen, the reason is easily understood, as the possession of such strong and hale persons to all intents and purposes as slaves, and at the expense of their maintenance alone, must be very lucrative to those requiring labourers. I am, on the other hand, at a loss to conceive how the assignment system can be looked upon in any other light than as a great evil, which must be abolished if it be designed to make the inhabitants of New South Wales a moral community, and to reform the convicts. It acts most unequally on the parties, and is a barrier to the reformation that the punishment of transportation is intended to effect.

The convicts on arriving are sent to the barracks at Sydney. The government selects from them such mechanics as are required for the public service, and then the numerous applicants for labourers are supplied. Those assigned to private employers, are sent to the interior under the charge of a constable or overseer.

They build their own huts, and the climate being very fine, require but little shelter. The hours of labour are from six to six, and the quantity of labour exacted from them is about two-thirds of what would be required in England. They are treated in all respects as if they were free, and no restraint is imposed, except that they cannot leave their masters, who, when they have no further use for them, return them to the government to be reassigned.

When on ticket of leave, they may reside in any place they choose to select.

The convict's time of probation depends upon the original term of his sentence; but on a commission of crime within the colony, it begins from his last conviction. For refractory conduct, they may be taken to the nearest magistrate, who orders punishment on the oath of the master. The magistrate has also power to send them to the nearest chain-gang employed on public works. Here they are worked in irons, and kept on scanty food for a limited period, after which they may be returned to their masters. If badly treated, the convict may have the affair investigated, but redress comes slowly.

One of the great evils of the system is, that many of the convicts on arriving are assigned to persons in Sydney and other towns, the consequence of which is that they are exposed to the contaminations and temptations that are likely to beset them in those thickly-peopled places, and this too only a few months after their conviction in the mother country. This influence removes all hopes of reform, and they are usually soon found among the criminals of New South Wales.

All persons who are landholders may receive convicts as assigned servants, in the proportion of one to every three hundred and twenty acres, but no one proprietor can have in his employ more than seventy-five convicts.

Written application for labourers is made to the Board of Assignment, and the applicants must bind themselves to keep the assigned convict for at least one month, and to furnish him with food and clothing agreeably to the government regulations, which are as follows, viz.:

The weekly rations consist of twelve pounds of wheat, or nine pounds of seconds flour; or, in lieu thereof, at the discretion of the master, three pounds of maize meal, and nine pounds of wheat, or seven pounds of seconds flour; with seven pounds of beef or mutton, and four pounds of corned pork, two ounces of salt, and two ounces of soap.

The clothing for a year is as follows, viz.: two frocks or jackets, three shirts, of strong linen or cotton, two pairs of trousers, three pairs of shoes, of stout durable leather, one hat or cap, and the use of a good blanket and mattrass belonging to the master.

Custom, however, has extended the above allowances, and the quantity of luxuries added in tobacco, sugar, tea, and grog, makes the amount nearly double. These additions have become absolutely necessary in order to procure work from the convicts, and the free supply of them is the only way in which they can be made to work in the harvest season. I was informed that a settler considered it all-important to have a large stock of these luxuries on hand at the season of pressure; for although the assigned servants do not actually refuse to work, they do so little, that, in order to save his crop, the master must yield them the extra indulgences.

Another evil attendant on the assignment system is the difference in the treatment they receive from those to whom they are assigned. On the arrival of a convict-ship, a large number of persons who have made applications to the Board, are in waiting; they of course know nothing of the character of the convicts, and, as I learned from a good source, no record is kept, or sent with the convicts themselves. The Board is entirely ignorant of their character or crimes, and thus can exercise no discrimination in assigning the convict to the hands of a good or of a hard master. The greatest villains may, therefore, fall into kind hands, while one who is comparatively innocent may suffer much more than he deserves.

The punishment of transportation must continue very unequal until a classification be resorted to. Many convicts, by bad treatment, are confirmed in their vices.

For any misbehaviour, they are, as has been seen, subject to severe castigation upon their master's making oath before a magistrate. This not unfrequently drives the culprit or convict to further crime, and in revenge for these wrongs, he either neglects his master's interest, or has been known to set fire to his harvest when gathered.

The present system appears fitted to entail evil and misery on the colony, and there are few disinterested men who do not view it as calculated to prevent any moral improvement. Murders, robberies, and frauds are brought about by it, for which extreme punishments are of such frequent occurrence that it is a matter of astonishment that a stranger should remark that an execution had taken place. The day before our arrival five criminals had been hung, and more were to suffer in a few days.

These executions take place without causing any unusual excitement. There is little doubt that the convict population contains among its members many of the most abandoned wretches, and I am also aware that the Governor and Council are making every exertion to put a stop to the immorality and vice which so generally prevail; yet I am satisfied that the convicts who are assigned are, in some cases, goaded on to crime by the treatment they receive from their masters, who hold them as slaves, and degrade them to the level of the beast with whom they are forced to labour.

Although Great Britain has a right to assume a proud pre-eminence in her exertions to emancipate the blacks, yet it behooves her to look to her penal settlements, and examine into the tyranny and degradation that a large number of her subjects are suffering there, many of them for slight crimes.

Few except those who have visited this colony can be aware of the extent to which the lash is administered, and oftentimes on the mere pretence of unruly and bad behaviour to their masters, or for the most trivial offences. So many facts of this sort were stated to me by persons in office, and of the highest respectability, that there cannot be a doubt of their correctness. The following extract from a report of the Committee of Transportation in 1835 will show it in its true light.

"In 1835, the number of convicts in the colony of New South Wales was above twenty-eight thousand, and the summary convictions in that year were estimated at twenty-two thousand. In one month in 1833, two hundred and forty-seven convicts were flogged, and nine thousand seven hundred and eighty-four lashes inflicted, which would make for the whole year two thousand nine hundred and sixty-four floggings, and about one hundred and eight thousand lashes. This amount does not embrace one-third of the convicts convicted summarily, but only those sentenced to be flogged, and there yet remain those to be added who were sentenced to other degrees of punishment: male convicts to the iron-gangs and treadmill, and females to the solitary cells of the factory."

The inquiries that I made in relation to the native-born inhabitants, were universally answered by all in favour both of their morals and habits. Judge Burton bears testimony that the free immigrants and native colonists are as exempt from the commission of crime as the inhabitants of any other country.

The defect in the female assignments is equally obvious. They are assigned only to married settlers who are considered respectable. They are accompanied by their children from the mother country, but immediately upon arriving the assignment takes place, and as the party to whom the convict is assigned does not wish to be encumbered with her offspring, they are at once separated. The child is not unfrequently removed from the mother when at the breast, and taken to the factory at Paramatta, where convicts' children are nursed and brought up. The mother is thus severed from her progeny for months, and, perhaps, for ever. The scenes that occur at these separations are often heart-rending, and ought to condemn the whole system. The feelings of the poor creatures may be more readily conceived than described.

Connected with the convict system, are the convict prisons, where the road and iron-gangs are lodged for safe keeping. There is one on the hill at Sydney, which, like most of the buildings at Sydney, bears the name of Governor Macquarie, 1817. In it are shown the guard-room, the working and eating-rooms, and dormitories, all of which are well ventilated. The prisoners sleep in hammocks, swung from parallel bars a few feet above the floor. A whipping-post was shown in an adjoining yard. The older prisoners are kept at work making brooms.

The female convicts who do not conduct themselves well, are sent back to the factory at Paramatta, where they are engaged in prison labour. The practice of keeping them in great numbers there, after they have been sent back, is liable to many objections, and is not calculated to produce reformation. It is very questionable whether their employment in small parties would not have a greater tendency to produce the desired reform.

The factory at Paramatta is situated on the river, about half a mile from the centre of the town, near the place where the steamboats stop. It is a large stone building, enclosed with high walls, and usually contains one thousand inmates. A part of these are those female convicts who have not yet been assigned; and the rest, those who have been remanded for their refractory conduct. Many of the settlers have, from necessity, taken these females into their service, and have been quite glad to get rid of them; for their corrupting influence had often resulted in the total ruin of the male servants who had been assigned in the same family.

It is only lately that good order has been introduced into this establishment, and this is owing to the supervision and care of Sir George Gipps. The accounts of the disorder in it in former times are truly disgusting.

These females are now divided into three classes, according to their ability and behaviour, of which the latter is more especially attended to. The first class is employed in making linen clothing, such as shirts, children's clothes, &c., and do much work for the shop-keepers in Sydney. The second, in making up coarse articles of apparel for government, and shop-keepers; the third, in picking oakum, washing for the establishment, and plaiting straw.

It would be difficult to imagine a more hideous community; and those who visited this establishment could scarcely realize the possibility that Great Britain could have produced such an assemblage of ugly creatures. It is hardly conceivable that the feminine character could ever have existed under faces, in which all traces of gentle expression have long since been annihilated, and where the countenance now indicates only the prevalence of the baser passions.

Some of the rooms were crowded, the inmates sitting on two rows of benches. Upon our entrance they discontinued their occupations, and saluted us with disgusting leering faces, staring at us with a malignity and hatred that were not soon forgotten.

The rooms appeared well ventilated, and scrupulously clean. I wish I could say as much for the prisoners themselves: they were dirty and slovenly in their appearance, and were clothed in a coarse drugget gown, a cap and neckerchief of cotton.

The discipline is very strict, and maintained by a person who at one time was the most unmanageable convict they ever had confined there. She now holds the place of matron, and has the management of the females, under the supervision of Mr. Bell, who is the superintendent of the whole establishment.

Until she was appointed, no sort of order was maintained. We understood that her conduct has been excellent since she filled the place. She is a tall masculine woman, of some intelligence, and has a watchfulness of manner that shows she is an adept at her business. She told us that the punishment for misconduct was solitary confinement, on bread and water, and for smaller misdemeanours, working at the crank of the pumps which supply the establishment with water.

The children are in a room by themselves, and there are about one hundred of them, from the infant to the child of seven or eight years of age. They all looked healthy, were very playful, and appeared to be well taken care of. There are twenty nurses who attend to them. It gave us much pleasure, when the matron entered with us, to see them all come running up to her, demonstrating her kind treatment of them, and the affection they bear her.

To Mr. Bell, the superintendent, we were much indebted for his civility. He appeared well qualified for the management of so extensive and difficult a concern. He explained every thing to us, answering the numerous questions put to him with great pleasure and politeness. This visit was particularly gratifying, as affording a view of one of the most important features of this great penal establishment.

Around Paramatta and Sydney, another such feature is to be seen in the iron-gangs. These consist of the men who have not been assigned, and likewise of all those who are returned by their masters. They are met with upon the roads, working in pairs, chained together. Their dress is peculiar, and they, in consequence, cannot readily escape detection in case of absconding. On the back of the jacket is marked, in conspicuous letters, "chain-gang." They wear a canvass jacket and trousers, and a jockey-cap. They were a rough-looking set, with bad countenances, and, like all other prisoners, stared us broadly in the face. Sentinels or guards constantly accompanied them.

The English are very partial to this mode of treating criminals, and cannot be persuaded that any better course can be devised; yet it is attended with obvious evils.

For a trifling and first offence, a perpetual brand of infamy is set upon a fellow-mortal, his family, and connexions. The natural consequence has been to foster and keep alive a public opinion which tends to the disorganization of society, and to obliterate all that remains of principle in the criminal.

The convict who has just arrived, is regarded by the others as a simpleton and a mere novice; and they undertake to complete his education.

The exploits and crimes performed and committed by these hardened offenders in Australia, New Zealand, and the islands of Polynesia, exhibit a dark picture; and the annoyance thus inflicted upon their inhabitants would not be borne, had they the strength to resist it. Power is the only right that can be urged by Great Britain as a justification of this infliction, and that it would be useless to question.

The majority of convicts are either assigned servants or ticket-of-leave men, and their condition is not unlike that of the slaves in our Southern States. They form a distinct class, and may be considered as the original groundwork of the colony. At present they constitute about a third of the population, but when transportation ceases, their relative numbers will rapidly decrease.

This colony, take it all in all, is in spite of these drawbacks a noble one, and is a new proof of the superiority of the Anglo-Saxon race, and of its enterprise and perseverance in overcoming difficulties.

I understood that Sir George Gipps had determined to adopt Captain Maconochie's system in the management of the road-gangs, and shall therefore proceed to examine it.

Captain Maconochie's argument for the necessity of a change is founded on the admitted fact, that the example of severe suffering on the conviction of crime, has not hitherto been found effective in preventing its recurrence. He maintains that the sole and direct object of secondary punishments should be the reformation of the individual culprit, or at all events his subjugation, and his training to self-command, by the latter of which he may give satisfactory proof that he deserves a restoration to his privileges in society. He does not proscribe punishment, but on the contrary believes it indispensable to induce penitence and submission; he regards it as necessary as a deterring example, and not for a vindictive end.

An entire reform, or a self-control tantamount to it, can, in his opinion, be obtained only by specific punishments for the past, and by a training for the future. To effect this latter he proposes to group prisoners together in associations, made to resemble those of common life as closely as possible, subdividing them into small parties or families, as may be agreed on among themselves, with common interest; that they shall receive wages in the form of marks of commendation, which they may exchange at will for immediate gratifications, but of which a fixed accumulation should be required before receiving freedom. He thus hopes to prepare them for society in society, giving them a field for the exercise and cultivation of social virtues, as well as for the voluntary restraint of vices.

Captain Maconochie deems the union of punishment for the past, with training for the future, as totally incompatible with each other, and, therefore, thinks that the former must in all cases precede the latter, and be effectual of itself. He argues, that success in medical treatment by beginning to administer restoratives before the disease is eradicated, might as well be expected as reform while punishment is undergone; and that it is just as necessary to prepare for society in society, as to train man by a preliminary education to the useful employments of life; that it seems idle to expect that mere theoretical instruction, however strongly enforced by short but severe suffering, should be sufficient to enable persons advanced in life to guide their future conduct, as it would be to hope to teach a trade, or any other practical employment, by abstract rules; and that moral lessons, to be taught profitably, require a field of progressive experimental application just as much as engineering.

On these elementary principles Captain Maconochie founds his plan of convict management, to which he applies the name of "Social System," and trusts for its success to the application of moral force in the place of physical coercion. He considers that hitherto the reform of culprits has not been thought the principal object in regulating their treatment.

The object of deterring from the commission of crime has been the duty of the law for the protection of society, and the association of prisoners has been deemed morally hurtful to them.

The Social System proposes to change this course to one in fact directly opposite to it. In criminal administration, according to his views, society is at present placed in one scale, and the culprit in the other, and it is not surprising that the weight of the former should predominate.

He proposes, that the nature of the punishment should be severe and short; that it should melt into probation, and this again into entire freedom, by changes as gradual as possible; thus taking nature as the guide, and copying what occurs on any severe misfortune befalling us, at first overwhelming grief, then a retrospective one, which afterwards slowly gives place to hope and encouragement.

To carry this out, it would be necessary to have solitary imprisonment, with moral and religious exhortations inculcated during sequestration from external influences, with permission to work, and instructions in its performance, but without the power of exchanging the proceeds for indulgences; next, separate imprisonment, with the power of exchanging marks of good conduct for gratifications, to be prolonged until the accumulation of a certain number of marks over and above all those exchanged for indulgences, should exhibit the acquisition of habits of self-control.

To this second stage should succeed social labour through the day, with separate confinement at night, and at length a complete admission to a society, in which the convicts should choose their associates, and be mutually responsible for the good behaviour of each other.

In passing through such a course of discipline, both of the ends which have been spoken of will be attained. The guilty will be first punished, and afterwards rendered fit for society by reformation and training, and will be thus restored to that state in which he was before he committed the crime, after he has been well tried and found worthy of being re-established in it.

As far as I could understand, Captain Maconochie was not prepared to prescribe the exact manner in which his views were to be carried out, and did not appear to set much value upon the mode, provided his principles were kept in view. He was of opinion that the principal error in modern penal science is the importance attached to physical arrangement in the construction of prisons.

According to him, the less stress that is set upon them the better, for it is not the body alone that is to be kept captive, but the will also; and the more care that is taken to guard the former, the less can the latter be attended to.

The peculiar trait in the Social System is, that after punishment every culprit's lot would be in his own hands. His companions would be of his own choice, and the length of his detention and comfort would depend upon the conduct of himself and his associates.

In seeking the reformation of the culprit, the mutual action of companions on each other would be resorted to, and this would be productive of great advantages. No system could be more just; and its language to the criminal would be,—"Having made you pay the penalty for your crime, I now retain you until you are qualified to meet the requisitions of society on your return to it, that you may not fail as you have before done."

The results of this system could not but be far different from the plans in present use, which have reference only to crime and retribution, and may be considered useless in promoting reformation. I understood that Sir George Gipps had already partially and successfully adopted the Social System with the convicts in government employ, by associating them in bands of ten to twenty, and letting them work on the public roads. Some of them were pointed out to me, and in point of appearance were as orderly and cheerful as any free labourers. I was informed that they do more work than when watched by overseers or soldiers.

I saw, however, many iron-gangs, but was informed that they were composed of individuals who had committed offences in the colony. After the commission of several crimes in the colony, they are again transported, to Norfolk Island, where Captain Maconochie is stationed; and it may well excite surprise if he should succeed in reforming these double-refined villains.

Many circumstances have been told me, by those who are well acquainted with the facts, that such is their detestation of Norfolk Island, and their horror of remaining there, that convicts have drawn lots to commit crimes, and even murder, in order to be sent back to Sydney for trial.*

Captain Maconochie's system is looked upon by many as Utopian, and it has excited no little astonishment that any one should conceive the idea of affording to criminals the refined amusements of society, or that books, music, &c., should be furnished them.

I have given this short sketch of the Social System in order to show

* It is the intention of the government to erect on Norfolk Island a penitentiary, on the plan of that at Sing Sing, in the State of New York. The estimated cost was £200,000.

its general plan. For a more full account of it, I would refer to Captain Maconochie's papers, published at different times. I spent several agreeable hours with him; and am satisfied that with the well-educated description of criminals, and with those who may have friends to return to, it will probably answer; but I am disposed to think that the great objection lies in the feelings of society, and its reluctance to readmit its outcasts on any terms, much less place them on a footing of equality.

There are two forms of social management proposed, one by Captain Maconochie, the other by Lord Howick: the former has been sufficiently explained; the latter includes both punishment and training in the insular penitentiaries, from which release may be complete, or merely through the medium of a ticket-of-leave in the colonies.

The latter form I believe is that which has been adopted, and from what I learn, it seems to be succeeding, although I have not been informed that any public account has yet been given of it. The system is about being adopted in Van Diemen's Land, which is a convincing proof that government has become somewhat satisfied with its efficacy; and it is noticed in one of the late Gazettes, that Captain Maconochie had treated his prisoners, on the Queen's birthday (with the approbation of the government), to a play and punch; which is a proof that some had already reached the probationary state.

The ration of the soldiers in New South Wales consists of one pound of meat, one pound of bread, two-thirds of a pint of rum, and an allowance of five-pence for small stores, consisting of salt, sugar, tea, &c. They receive as pay eight-pence per day, and are obliged to serve twenty years before they can claim their discharge.

The convict gets one pound of bread, one pound and a quarter of meat, and one pint of meal. Indeed, there is very little difference in the condition of a soldier and a convict, and were it not for the name, one would be almost induced to prefer the situation of the latter.

There is a description of convicts, as has been mentioned, known under the title of ticket-of-leave men. These, from good behaviour before the expiration of their term of sentence, are permitted to hire themselves out, upon the employer entering into a stipulation to maintain a strict watch over them. This custom has no doubt been forced upon the community by the want of servants, and the necessity of obtaining them. The action of this part of the system will be shown more clearly by the following anecdote.

One day, passing along George Street with a friend, my attention was called to a fashionable equipage, with a well-dressed man driving i On my asking to whom it belonged, I was informed that the person driving it was the owner, and that although a ticket-of-leave man, he was married to a free woman of handsome fortune, living in one of the finest houses in Sydney; that their house was built on the very spot where he stood under the gallows some years since, although through a reprieve, or some accident, he had not been hung; and that it was at any time within the power of the wife to send him off to the whipping-post, and have him severely flogged. There are many convicts who are now the most wealthy people of New South Wales. I do not intend to be understood that they mix at all in the society of the better class; on the contrary, the convicts and their descendants, even to the third and fourth generation, are excluded from it.

Society here is composed of many distinct circles. All those of the first class are entitled to be received at the Government House, or are invited there. This privilege seems at present to be the touchstone of gentility; and if an inquiry is now made of the standing of any one, it is quite sufficient to say he visits at the Government House.

Any connexion with convicts would at once preclude admission to this circle; and so distinctly has this line been drawn, and so closely is it adhered to, that should an officer, or other person, contract marriage ties with any one of the lower classes, he would forthwith be shut out. This state of things naturally leads to many heartburnings among the rising generation, who have every thing to recommend them but a pure descent; whose behaviour is acknowledged by all to be irreproachable, and who among the community stand deservedly very high, some of them occupying posts of high trust and responsibility among men of business, and not a few of them being at the head of large moneyed institutions.

These differences frequently break out when subscription balls are given, and result in challenges being sent to the managers. One occurred on the giving of the St. Patrick's ball. A Mr. D. was admitted as a subscriber by the committee; he afterwards asked for a ticket for a friend of his, which was refused. Objections were then taken to himself, and he was requested to withdraw his name, and receive back his money. This brought forth a challenge, which was disposed of in a summary manner by the committee handing him over to the police, by which he was obliged to apologize to the committee, and bound over to keep the peace. I cannot but believe that this state of society is destined in a very short time to undergo a great change; and many of the inhabitants seem to be of the same opinion, particularly if they obtain a colonial legislature. This it seems almost indispensable they should have, for the wishes and wants of the rising community are too little known and heeded, at the distance of sixteen thousand miles, to insure good government; and the acts and the varying policy of the mother country are so ill adapted to the state of things here, as to strike the most common observers, and only tend to loosen the ties of affection that bind the colonists to it.

The introduction of free emigration, and the discontinuance of the use of the colony as a penal settlement, must soon produce the necessity of legislative bodies, and the elections will give the wealthy part of the citizens, emancipists and their descendants, a powerful voice in those bodies when constituted, which will finally lead to their amalgamation with the higher classes. I was surprised to find among the emancipists themselves the same distinctions kept up.

The labouring class of free emigrants form another class. They have great difficulties to contend with on their landing. As few of them will consent to serve as domestics in association with ticket-of-leave men or convicts, they find themselves placed in many difficult situations. They are compelled to resort to the public inns kept by these people, who endeavour to take every advantage of them, and cause them to part with what little amount they may have brought with them from the mother country. They soon become destitute, and from disappointment betake themselves to all the vices of the convict class. Some steps have been taken to provide for the emigrants on their first arrival, under the government system; but they have not yet been carried into effect, and it is difficult to enforce them.

There is yet another class, and one, as far as my experience goes, now unknown elsewhere, which sets at defiance both law and regulations. I mean a class known here by the name of "Crimps," who are a pest to the trade of the port, and the destruction of all the sailors who visit it. Their trade or employment may be summed up in a few words: it is to entice or kidnap sailors from their ships, and keep them drunk and concealed in some out-of-the-way place. Whole crews of merchantmen are frequently carried off by these fellows, and they are in consequence at times detained until the master or assignee resorts to the agents of these crimps, who are ready to give them a crew at four or five guineas for each sailor. I was told, a few days after my arrival, that the crimps had determined to get some of the men of the squadron; and they succeeded in enticing away the crew of the tender Flying-Fish and three or four other men belonging to the ships. The vigilance and system of these crimps bid defiance to the laws and police, who although quite aware of the existence of the evil, find it out of their power to put a stop to it. Since my departure, the shipping interests have memorialized the Government and Council, and there is a prospect that this nuisance will be abated.

As respects the higher class of society, it is in all respects the same as is met with in England and America, among well-educated persons. Perhaps as to fashion it is a little more colonial, but not more so than the distance from whence fashions originate would account for. The cordial welcome and hospitality we met with could not be surpassed any where.

The Governor is appointed by letters patent, under the great seal of the United Kingdom; but he acts under the direction of the legislature.

The Legislative Council consists of a number not exceeding fifteen, and not less than ten; the members are appointed by the King, and are all residents within the colony.

The Governor is president of this council, and is entitled to vote as a member upon all questions; when it is equally divided, he has an additional or casting vote. To the Governor and Council is delegated the power to make laws for the peace, welfare, and good government of the colony, not repugnant to any act, charter, &c., which may have been issued, or to the laws of England; and no law or ordinance can be passed, unless first laid before the Council by the Governor.

The Governor is, by statute, invested with the right of property in the services of offenders or convicts who have been transported, and he may assign this right to others. He is captain-general and governor-in-chief, and has full control over all the military and civil authorities. He is empowered, and required, to administer oaths to the Chief Justice, and the members of the Executive Council; to keep the public seal; and is invested with authority to suspend members of the Executive Council, and to supply their place, as well as to appoint temporary members to fill vacancies.

He appoints all justices of the peace, coroners, constables, and other necessary officers.

He has the power to grant pardons, reprieves, &c., and to remit punishments for offences, treason, or wilful murder, only excepted; for which upon extraordinary occasions, he can reprieve until the pleasure of the crown be known. His power to shorten the time of transportation is limited, by the condition that all instruments in writing for that purpose are to be approved by the crown.

With the advice of the Executive Council, he is empowered to divide the territory, and its dependencies, into districts, counties, towns, &c., to fortify and erect forts, and provide for the defence of the country.

All public moneys are issued for the support of the government by warrant from the Governor, but only for purposes particularly pointed out.

He may give titles to crown lands. He has also power to appoint fairs, marts, markets, ports, harbours, bays, and havens.

The person who succeeds, in case of the death or absence of the Governor, is the Lieutenant-Governor, and next to him, the Commander of the Forces.

The Executive Council consists of four persons holding office in the colony

1st. The senior officer of the Forces. 2d. The Bishop of Australia. 3d. The Colonial Secretary of New South Wales; and 4th, the Colonial Treasurer: the two latter for the time being. These are appointed by letters patent, under the great seal. It is a council of advice and restraint, and the matters on which they are to be consulted are especially mentioned in their commission.

The Legislative Council consists of fifteen persons, including the Governor, seven of whom hold offices under the government, and during royal pleasure, viz.:

1st. Chief Justice. 2d. Bishop of Australia. 3d. The Commander of the Forces. 4th. The Colonial Secretary. 5th. The Attorney-General. 6th. The Collector of the Customs. 7th. The Auditor-General; with seven others who do not hold offices, but are nominated by the crown.

As is truly said in the colony, they are governed by the royal prerogative, exercised in the person of the Governor.

The rules for his guidance, and that of all colonial officers, are issued by the Secretary of State, and are to be found in a pamphlet form, under the title of "Colonial Rules and Regulations." Great complaints are made in the colony that these are altogether one-sided. In them it is notified that the appointment or term of the Governor's office is limited to a period of six years, from the time of his assumption of his duties; the crown reserving the power of prolonging that period.

The great complaint in the colony is, that the policy of the government at home is always fluctuating with the change of the incumbent who holds the office of Secretary of State. This happens with every change in political parties in the mother country, and the office is often held by persons who have very little knowledge or experience in colonial affairs, and consequently regulations are from time to time issued, and particular orders for the guidance of the Governor are frequently sent him, which leaves him little or no independence in the performance of his duties.

At the distance at which New South Wales is situated, it may readily be conceived what inconvenience is felt by the Governor and

Council in carrying out what they deem best for the interests of the colony. They have no power or control over the revenue, which in reality is under the supervision and direction of the Lords Commissioners of her Majesty's treasury.

The Governor is not allowed to expend any sum over £200 for any one service, (unless under very urgent circumstances,) without the previous sanction of the home government; and although at liberty to draw that amount, it is on his own responsibility; he must account for it, and show the absolute necessity for its use.

The estimates for the ensuing year are made in June, and forwarded for approval. The expenditure must be limited by this estimate, and no further disbursements applied for on account of that year, unless under circumstances entirely unforeseen.

The estimate, after undergoing the supervision of the colonial legislative body, must also undergo the scrutiny of the commissioners of the treasury officers, before any instructions are given by the Secretary of State.

The estimates for taxation follow the same course, and the Council has no control over the funds arising from the property or droits of the crown.

The Governor, in transmitting his accounts for audit, sends them accompanied by certified copies of all estimates of expenditures to which the accounts relate, and of all ordinances for the imposition of taxes, with copies of the despatches sent him by the Secretary of State, conveying the sentiments of her Majesty's government upon them; and it is required that full detailed statements of the revenue and expenditures of the colony be published in the Colonial Gazette immediately after the accounts are transmitted.

These are some of the regulations, which will tend to show how great is the authority still retained by the crown, or in reality by the ministers, and how little discretionary power the Governor has. He is required personally to superintend or authorize things of such small concern that it almost approaches the ridiculous; for instance, a wheelbarrow cannot be mended without an order in writing attested by his signature. Such an order may be necessary, but one would think that other persons might be authorized and trusted to perform such acts. The colony is treated as though it were a den of rogues, and required the constant supervision of the ministry at home. I was told that no one could conceive the mass of despatches containing instructions that a single year produced, and these are often found conflicting with those that had gone before, and thus require a reference back to the Secretary of State. The practical inconvenience is apparent, and it is not surprising that it should excite the ridicule as well as disgust of all thinking men in the colony, to see the attempt to govern the affairs of this rising state by the royal prerogative, exercised by one of her Majesty's principal secretaries of state, in despatches to a governor, whose recommendations are usually adopted, thus making him, at least in part, his own instructer. When the time necessary to pass these communications, which is at least eight months, is considered, there appears great reason for reform, and it is not surprising that the thinking part of the population are very urgent for it.

The high and confidential officer of the crown, which the Governor really is, is looked upon as the mere agent of the ministry at home.

The community do not feel themselves at all protected by the Legislative Council, although they have, apparently, a voice in its proceedings; as its members are composed, to the extent of one half, of persons who do not hold office. In practice, it is not found that this amounts to a check; for on all government questions the members who hold office will be present, and therefore vote in their full strength; while the members of the people, appointed from the most respectable landholders by the crown, do not take sufficient interest in the proceedings to give that punctual attendance that might be required for the interest of those whom they are intended to represent. But even if all were present, the Governor, with his two votes, would always decide the matter in favour of the government; and as before stated, no new law can be considered in council, unless prepared by the Governor, which must effectually prevent any innovations being brought forward by those who represent the interest of the inhabitants of the colony. The only power they have is a negative one: that of entering their protest, and having it sent home for consideration by her Majesty's ministers; but in such case there is little likelihood of its meeting with favour. The official members, on the other hand, are considered as bound to support the Governor or to lose their seats and offices, notwithstanding their oaths faithfully to advise, to the best of their ability and judgment, the government of the colony.

The principal check on the Governor and his Council is the public press, whose conductors are strenuous advocates for reform and colonial rights, and exhibit much ability.

The statute of New South Wales expired in 1839, when it was renewed for a year, and has subsequently been renewed from year to year until the present time (1840). So evident, however, were the defects in the administration, that a clause was added to the act of renewal, declaring that the statute was deemed in many respects inapplicable to the circumstances of the colony and the wants of the inhabitants.

It was made lawful for the colonial legislature to enact any laws or ordinances, subject to the provisions of the statute, for the better administration of justice, and to define the constitution of the courts of law, equity, and juries. This conclusively proves that great difficulty is experienced in governing these rising colonies, and in giving that attention to their wants that they demand; yet Great Britain still manifests a strong desire to retain her control over these subjects, and does not see the necessity of letting them stand alone, and being allowed to feel that they are able to take care of themselves.*

Petitions have been sent home to Parliament and to the Queen praying for the formation of a new constitution, such as they could place confidence in, and in which the people of the colony might be represented. The model of the constitution that they desire is that of the Canadas, and the expectation is that by the great influx of free emigrants, the day will soon arrive when it will be vouchsafed to them.

From the reports of a committee of the Legislative Council of this colony, it is shown that by offering bounties, immigration may be more economically conducted than by the government system. The report states, that during the year 1838 there arrived in the colony seven thousand one hundred and eighty individuals, (exclusive of convicts,) of whom one thousand six hundred and sixty-two made a claim for bounty. In the latter six months of the year 1839, six thousand arrived. The total arrivals from January, 1837, to the end of September, 1839, were sixteen thousand four hundred: by government ships, eight thousand four hundred and eighty-five; by the aid of bounty, four thousand two hundred and sixty-six; unassisted, three thousand six hundred and forty-nine. The amount of bounty to the four thousand two hundred and sixty-six who were introduced by private enterprise, was sixty-five thousand five hundred and eighty-six pounds,—at the rate of fifteen pounds six shillings each; while the cost of introducing eight thousand seven hundred and twenty-one persons by the government system, was one hundred and sixty-eight thousand seven hundred and seventy-five pounds, showing an increase of cost of thirty-seven thousand six hundred and thirty pounds to the colony by the government system.† This subject engrosses the attention of

* The Legislative Council has agreed to take upon the colony the charge of defraying the expenses of the police and jail out of the revenue. This has also caused much dissatisfaction.

† Many curious developements have taken place relative to the colony of South Australia, which was established upon the principles of self-support, having been carried out; no colony

all, now that the transportation, and consequently the assignment system, is to cease. They are desirous of securing workmen and servants, and every exertion is to be made to that end.

There is now a great influx of all kinds of people into this colony, from the capitalist to the labouring man. The colony offers advantages to all of these, but in a very different proportion. There is no country where provisions and the actual necessaries of life are as high as here, and this particularly affects the poor man, for although he receives high wages his expenses are proportionately great. He will therefore be disappointed, if he calculates upon making great savings. On the other hand, the capitalist may at once enter the market and invest his money profitably, and from all that I could learn, securely. Money, however, according to several intelligent and well-informed persons, commanded more than its value; or, in other words, the rate of interest is too high to be sustained. This was in part attributed to the improvements going on, partly for speculating purposes, but generally as permanent investments, the result of profits in business. Money is in fact the best merchandise to carry to New South Wales.

The poor labouring man, if he be sober and industrious, will soon acquire the means of support for himself and family, but he must carefully avoid the contamination to which he will be subject, and avoid improper associates. There is no place where he will be so much led into temptation as here. For the middle class—those who have a small income and do not work—there is every thing to strive against. Labour is high, and so are the necessaries of life. New South Wales is not a place to economize in. A moderate fortune, unless employed in some lucrative and growing business, will finally involve its owner in difficulties; and if he engage in farming, a few bad seasons (very likely to happen) will completely ruin him. From all the information I could obtain, emigration to New South Wales is attended with risk, unless a person be very prudent and can keep himself within his means. The moment he begins to borrow money, he is sure to get behindhand; for few can stand the payment of an interest of fifteen per cent. The great difficulty with all emigrants seems to be, that as land is very cheap in comparison to what they have been accustomed to, they immediately desire to possess large tracts. This it is necessary to look for, and much time and money

under the British dominions has cost the mother country more, nor has any one been conducted so badly, having cost the government about one million pounds for bounty. Some extraordinary circumstances were related to me of the manner in which the government was defrauded, in spite of their stipulations directly to the contrary.

is spent in wandering about the country in search of what is not very easy of attainment. Another difficulty of the newly-arrived settler consists in getting information concerning the unoccupied territory. No land-office or land-agent is found here for the emigrant to apply to, and he not unfrequently falls into the hands of those who defraud him, or is led astray by the reports of the ignorant or prejudiced, and at last is induced to purchase much more than he requires, and in consequence fails of success. The government lands are disposed of in a different way from what ours are. A certain parish having been surveyed and mapped, is advertised as being open for sale; persons select and make application, and if a less quantity than six hundred and forty acres is desired, he is obliged to state the reasons of his wish to obtain it, and the use to which he purposes to put it: the land is then advertised to be sold on a certain day (of the month), at public auction. If the land offered for sale happen to be in the neighbourhood of some wealthy proprietor, he cannot fail to become informed of it; the section is bid up, and the person may be disappointed in obtaining the allotment selected and advertised by his own desire.

The minimum price must be paid, at any rate: this originally was five shillings an acre; it is now twelve. Ten per cent. must be paid down, and the remainder in one month, or the deposit is forfeited. On payment of the money the title-deed is given, subject to the nominal quit-rent of a peppercorn. Before delivery of the deeds, the law provides that forty shillings shall be paid to the colonial secretary, and five shillings to the register. The crown reserves to itself the right of making roads and bridges, as well as of taking timber, stone, and other materials for making and keeping them in repair; as well as all mines of coal and precious metals. No land within one hundred feet of high-water mark on the sea-coast, harbours, or bays, is to be considered open to purchase, unless for purposes of commerce and navigation.

As respects the discontents arising from what the colonists call the misapplication of the land-fund, her Majesty's ministers have determined that she has a right to alienate the waste lands, and divert the appropriation of the proceeds, and that the doubts raised would, if sustained, be laid aside by a declaratory act of Parliament.

All free persons are admitted as purchasers of land, without any limitation whatever as to quantity.

In order to show that the statement of the extent of crime in the colony, however extraordinary it may appear, is not exaggerated, I will give extracts from the charge of Judge Burton to the jury, at the close of the session of the Supreme Court, in November, 1835, and afterwards a report by him to the colonial secretary, in 1836. Both of these may be classed as official documents of the highest authority.*

Judge Burton remarked, that "It was now his duty to discharge them (the jury) from any further attendance this session, but before he did so, he must make a few observations, which they ought to carry to their homes, and there give them a calm and serious consideration; his own mind was sufficiently impressed with their importance.

"It had been his lot to preside alternately with his brother judges in that court, he might say, for three years. It was a period at which he might himself well pause and inquire what he had been doing, what had been the effect of his labours, and especially, considering the numbers of capital convictions which had taken place before him, and the number of sentences passed, it was fitting that he should ask himself the question, what has been the effect of those sentences in the way of example?

"He felt they were equally interested in the same questions; he would therefore lay before them the views and conclusions at which his own mind had arrived. He had requested a return to be made out by the chief clerk of the court of all the capital convictions that had taken place during the last three years, and he thought when he stated the number of them, they would feel he was fully justified in the course of observations he was about to make.

"In 1833, there had been one hundred and thirty-five capital convictions, on which sixty-five sentences of death had been passed; forty-five of these capital convictions, and fifteen of these sentences of death, had taken place upon his judicial responsibility.

"In 1834, there were one hundred and forty-eight capital convictions, on eighty-three of which sentence of death had been passed; forty-eight of which convictions, and thirty-six of which sentences, had been before himself.

"In 1835, one hundred and sixteen capital convictions, and seventy-one sentences of death; fifty-six of which had taken place before him, and twenty-eight of which sentences he had passed. In addition to which, there are thirty-three prisoners who have been capitally convicted, waiting sentence, whether death might be recorded, or passed upon them. The number of capital convictions was a feature sufficiently striking in the administration of justice in the colony; for it was to be remarked, that capital punishment had been taken away from several offences, ever since the 1st of August, 1833,—such as forgery, cattle-stealing, stealing in a dwelling-house under the value of five pounds

* See Appendix X. for tabular statements of crime in New South Wales.

(these were fruitful sources of capital conviction in former times); so that those which had taken place since that time, were all of crimes of violence: murder, rape, robbery, burglary, maliciously stabbing, shooting, and wounding, and offences of similar character.

"The calendar for the present sessions presents the following facts, and had been furnished him by the crown solicitor:

"'There had been convicted of murder, two; stabbing with intent, &c., shooting at, with intent to kill, cutting and maiming, assault, with intent to do bodily harm, six; manslaughter, two; arson, one; piracy and burglary, eight; house-breaking, ten; highway robbery, seven; receiving, one; forgery, two; larceny on the high seas, one; larceny, four; cattle-stealing, one; piracy only, one; robbery, eight;—total, fifty-four.

"The prisoners in jail on the 18th of November, 1835, untried, were seventy-four, from various causes of delay; they were, however, neither unknown nor unheeded. With respect to the causes of this state of crime, he had formed his own conclusions, and begged them to weigh and examine them, and judge for themselves; he thought the number of capital convictions alone, enough to point his own and their attention to it, as an indication of the state of the country as to crime.

"He did not think it necessary to mention the number of convictions before the Supreme Court, during the same period, for offences not capital. He would, however, briefly refer to them, and to all offences which were tried before the several Courts of Quarter Sessions throughout the colony, in the exercise of their summary jurisdiction, and by juries; the mass of offences which were summarily disposed of by the magistrates; and, added to all those, the numerous undiscovered crimes, which every man who had heard him, and to whom the report of his words should come, would at once admit to have occurred within his own circle of knowledge. There the picture presented to their minds would be one of the most painful reflection. It would appear to one that could look down upon the community, that the main business of us all was the commission of crime, or the punishment of it; as if the whole colony was in motion towards the several courts of justice; and the most painful reflection of all must be that so many capital sentences, and the execution of them, had not had the effect of preventing crime, by way of example.

"In his opinion, one grand cause of such a state of things was the overwhelming defect in the religious principles of the community; a principle, which he considered as the polar star, to guide a man in all his conduct, and without which none other would prevent him from crime.

"But that he might not be said to make so grave a charge upon light foundation, he would instance the crimes of violence, the murders, manslaughters, in drunken revels, the perjuries, the false witnessing, from motives of reward or revenge, which in the proceedings before him had been brought to light. There were some indeed of so atrocious a character which had occurred before him, that he would briefly instance some of them, which the time that had elapsed might have caused to pass away from their memory.

"The case of Mullany and his wife, who were convicted of stealing from the person of Patrick Sherry, by administering to him some deleterious drug, which for a time deprived him of sense, and perhaps only the quantity prevented his losing his life. The case of Armstrong, the overseer, who was acquitted upon a false charge, brought against him by a convict under his superintendence, of shooting him with intent to murder.

"The case of Cowan and his wife, who were acquitted of the murder of a man named Kerr, embodies in itself a picture of those evils with which the colony is visited. A person of the name of Campbell, and the deceased Kerr, lived near Liverpool, and kept an unlicensed still, and a house to which the gangs of prisoners in this neighbourhood resorted for drink, and they were cattle-stealers. On a Sunday evening this house was visited by a constable from Liverpool, who arrived about eight o'clock, and found the parties, as he expressed it, 'beastly drunk,' and the two prisoners of the crown in the same state; this was the last time Kerr was seen alive by any respectable person.

"Information was given the next day, by two of Cowan's servants, to the magistrates of Liverpool, against him, for cattle-stealing, and it was proved that their having done so was known to Peter Montgomery, a convict, employed as overseer at the Liverpool Hospital, in the afternoon of the same day, and that he had visited Cowan afterwards, and understood from expressions made by Cowan, during his intoxication, that he expected Kerr would give evidence against him. Kerr was murdered by some one on that night, and his body was afterwards found at fifty rods' distance, but the blood was traced to within seventeen yards of Cowan's door.

"Campbell had given a statement before the magistrates, which, if he had adhered to on the trial, would have brought home the guilt of that murder to both the prisoners; but he recanted the whole of his previous statement, and they were acquitted.

"It appeared in evidence, that Campbell had been forwarded from Liverpool to Sydney, handcuffed with Cowan, and was confined in the same jail-yard with him. It further appeared, (and it deserves mention as an instance of retributive justice, as well as showing the character of the case,) that another dead man was found in the same place three months before, and upon that occasion a coroner's jury had acquitted the prisoner Cowan, upon the evidence of the man Kerr; and this deposition of Kerr's after his death, was given in the court, on evidence in favour of the same prisoner, when Cowan was subsequently tried, and was the main ground of his acquittal.

"In another case, an old man was acquitted of maliciously shooting at a servant in his employment, and the means taken to procure that acquittal, was a charge of felony set up against the principal witness.

"These, and many other instances still more disgusting, had brought him to the conclusion, that there was an overwhelming defect of religious principle in this colony. There was a great deficiency of religious instruction and instructers.

"He had visited the penal settlement, where he saw them herding together without any chance of improvement. A man who had been brought before him for sentence, observed, in a manner which drew tears from his eyes, and wrung his heart, 'That let a man be what he will, when he comes here, he is soon as bad as the rest; a man's heart is taken from him, and there is given to him the heart of a beast.'

"He felt bound to say, that masters of convicts were not sufficiently attentive to the morals of their men. It had been proved before him, that highly respectable persons near a church in the same town, not only neglected to oblige them to attend the worship, but actually suffered them to spend the Lord's day amidst scenes of drunkenness and debauchery. It had been further proved, that the Lord's day, by some masters, was made a day of labour, some other day being allowed to them as an equivalent. He was sorry to add, that many of the worst crimes which had been brought under his notice, were committed on the Lord's day, and he was led to apprehend that there was a very general disregard and desecration of it.

"He had been induced, by what had been proved before him in that court, gravely to consider the question of convicts working out of irons, and felt convinced that it was one of the most fruitful sources of crime to be found in the colony. He had before him a return, from which it appeared that the number of convicts at this time employed upon the roads, is two thousand two hundred and forty; of whom one thousand one hundred and four are out of irons. And when they (the jury) considered who these men were, and what they had been; that they left their huts in any number, armed or unarmed, as they pleased; from the evidence he possessed respecting the conduct of these road-parties of the colony, it would appear that those establishments were like bee-hives, the inhabitants busily pouring in and out; but with this difference: the one works by day, the other by night; the one goes forth to industry, the other to plunder.

"To the careless or worse than careless conduct of the overseers, he did attribute a vast proportion of the burglaries and robberies that were committed in the country districts. It had been proved in a recent case that a party of these men had committed a robbery, under such circumstances of aggravation, that sentence of death had been passed upon four of them.

"The settlers themselves were to blame for many of the crimes committed by convicts belonging to road-parties. It appears they have frequently employed these men, in their leisure hours, or on a Sunday, paying them for their labours in money, which was spent in drink, and so prepared them for crime; and it also appeared that after using their services in harvest, they remunerated them for their services, by granting passes for several days more than was necessary for them to return to their gangs, during which time the whole country they passed through is laid under contribution by their depredations.

"Another source of crime was the occupation of the waste lands of the colony by unauthorized and improper persons, both bond and free, who, commencing with nothing, or a very small capital, soon after acquire a degree of wealth, which must lead every reasonable man to the conclusion that they do not get it honestly.

"The congregation of large numbers of convict servants in the town of Sydney, to which were to be attributed the vast proportion of the burglaries and robberies committed there, the master allowing the convict servants to wander about when and where they please after his work is done.

"The allowing improper persons to have licensed public houses. It had been proved that a great many robberies had been committed at such places, many of the proprietors of these low houses being not far removed from the class of life in which the prisoners were themselves placed.

"Another cause, which comes home to all, is the almost total want of the superintendence of masters over their assigned servants. It had been proved to him that many of the robberies which had been committed are attributed to this alone; also, that convicts, six or seven in number, armed with muskets, and masked, had committed various robberies on their adjoining neighbours. One of them attempted a robbery in the middle of the day, on a Sunday, on the high-road from Sydney to Paramatta, armed with a musket, another person being in his company; and very many robberies were committed through convict servants being left too much at liberty to roam where they pleased, during the hours of night."

In Judge Burton's report to the colonial secretary, as to whether juries in the colony have answered the ends of justice, he gives a full account of the jury system, its formation, &c., some passages of which I shall also quote, as it will tend to show the manner in which the law is administered in the colony, and the difficulties encountered in the proper punishment of crime.

"In civil cases, such as form the ordinary business of the court, the matters in dispute are so simple as to afford but little field for any undue bias on either side.

"It is only in cases occurring between the government and an individual, or involving some point of political or party feeling, that any trial can be had of the principles of the jurymen, and happily there have been no instances of any such during the time (the last three years) that jury trial has been established.

"In criminal cases, there is a greater and more constant ground for apprehension of improper influences, and undue bias upon the minds of the jurymen. The prisoners for trial before the court, are chiefly of a class transported hither for crimes committed out of the colony; and persons of the same condition, and others very low in respectability and character, and frequently allied to them, are qualified, according to colonial law, to serve as jurymen.

"The qualifications are, a clear income, arising out of lands, houses, or other real estate, of at least thirty pounds per annum, or a clear personal estate of three hundred pounds.

"The disqualifications as they now stand are: 'Every man not a natural-born subject of the king, and every man who hath been or shall be attainted of any treason or felony, or convicted of any crime, (unless he shall have received for such crime a free pardon, or shall be within the benefit and protection of some act of Parliament, having force and effect of a pardon under the great seal,) or, secondly, if any person who, either while serving under any sentence passed upon him in any part of the British dominions, or after the expiration or remission of such sentence, shall have been convicted of any treason, felony, or other infamous offence.' "

Respecting the qualifications arising from property, Judge Burton says, "The possession of such an amount as is specified in the act affords no criterion in the colony, where property is notoriously accumulated by every variety of dishonest means. It may be a test of respectability and trustworthiness in a community differently constituted, but wholly fails in a community like this, lacking honesty, but abounding in property. In consequence of this qualification being requisite, many honest and respectable persons in the community, very proper to serve on juries, are excluded.

"Within this range are included a class of persons in the colony who have been transported hither for offences committed out of the colony. They are qualified to act as jurymen under the Local Act, without any proof being required that they had regained that good repute which they once lost, and the mere circumstance of their having served the period of their several sentences, does not establish that fact.

"There are others who, possessing the qualifications in property, have arrived in the colony as free emigrants, the near relatives of transported persons, under such circumstances as justly to lead to the suspicion of an undue bias existing in any case affecting them, and who have connexions in England, not unlikely to follow them to the colonies, possessing ready means of importing into this country property dishonestly acquired, and who speedily accumulate wealth by that and other dishonest means. There is no provision for guarding the administration of justice against the predominance of such persons upon the jury-list. The effect of the colonial law in practice has been, that juries actually empannelled under it have been frequently formed of very improper persons."

From the data submitted with Judge Burton's report, he says, "It appears that a party accused, inclined to exercise his right of peremptory challenge, might insure a large predominance of convicted persons on the jury, inasmuch as the law allows in cases of felony the peremptory challenge of twenty in number, and if a prisoner has professional assistance in his defence, this right of challenge is fully exercised. In one instance I knew gentlemen of high character and respectability thus peremptorily rejected on the part of the prisoner. I took the liberty of asking some of them afterwards if the prisoner was known to them, and was answered that he was not. The conclusion in my own mind was, that they were challenged on account of their respectability. In another case before me, every person of apparent respectability who was called, was peremptorily challenged on the part of the prisoner, which the crown officer observing, challenged all the others, and the case remained over in default of jurors. In both cases the accused had professional assistance.

"Again, the jurors are placed alphabetically on the list, and are summoned in that order; the relatives of convicted persons, qualified, and bearing the same name, are sure to be on the same panel with them. A party may be well informed beforehand, who will be summoned on his jury. An opportunity thus offers for the exertion of improper influence.

"A large proportion of those who have appeared and served are publicans, as many in some cases as eight out of twenty-nine, three having been convicted persons; in other cases, ten out of thirty-one, five having been also convicted persons; and again, eleven out of thirty-five, four of them convicted persons.

"Respecting the large proportion of this class of persons on the jury panels, and the state of crime, and the causes of it, I addressed a letter to his Excellency the Governor, and I now repeat, that the evils arising from the very great number of licensed houses for the sale of ardent spirits, are not restricted to the stimulus which they give to the commission of crime, and concealment of it which they afford, but I have found a very great proportion out of the panel of jurymen before the Supreme Court (who actually attend), to be holders of licensed public houses, frequently very low in respectability, to whose houses, prosecutors, and parties accused, on bail, and their witnesses, bond and free, resort for the purpose of drinking, during the period of time they are in attendance on court; and a reasonable fear is thus excited for the purity of the administration of justice, which I have had occasion as a judge to see realized.

"Upon reference to the jury-list of 1835, I have found that the number to be summoned from criminal issues before the Supreme Court is nine hundred and fifty-three, of whom two hundred and three are publicans and innkeepers. The proportion of those who actually serve, far exceeds that number; and in June, 1835, no less a number than two hundred and twenty-four licenses were granted for public houses in the town of Sydney alone. Few of them do not possess the necessary qualifications, and many are highly respectable persons; but the proportion which they bear to the whole is small."

The keepers of the low public houses in Sydney, are chiefly persons who have been transported to this colony, or are married to convicts, and many of them are notorious drunkards, obscure persons, fighters, gamblers, receivers of stolen goods, harbourers of thieves, and the most depraved of both sexes; they exist upon the vices of the lower orders, and inasmuch as there are no licensed pawnbrokers in Sydney, they act as such, but not as occurs in other countries, upon occasion of some temporary pressure on the poor, for some necessary of life, but for intoxicating liquor.

There is a great unwillingness on the part of respectable persons to appear and serve on juries, arising from a natural repugnance to association and confinement in the jury-room with disreputable persons. Judge Burton goes on to give many instances of the behaviour of the jury in their room, and their determination to acquit; stating, that he had been informed by a respectable inhabitant of Sydney, on whose veracity he could fully rely, that upon one occasion, when a prisoner was on trial for cattle-stealing, he was defended by one of the practitioners of the court, when, during the progress of the trial, a juryman leaned over him towards the practitioner, calling him by name, and said, "It's all right, we'll acquit him." When the prisoner was called on for his defence, the practitioner advised him to say nothing, and call no witnesses, which course was adopted; and he was acquitted.

It is proper to state that the other judges think that the jury trials have met with the success reasonably to be expected, and that matters will grow daily better as the free emigrants arrive and are qualified. From what I understood from gentlemen of the legal profession, there has some improvement taken place since the year 1836.

The courts still adhere to the use of wigs and gowns, and the opinion seems to be that such appendages cannot be dispensed with without injuring their respectability and solemnity in the eyes of the people.

Under the additional clause, amendments have been made by the Legislative Council in the laws regulating trials, and they have also abolished military juries.

Education in the colony of late years has claimed some portion of the attention of the government, which has made allowances to the different sects of Christians for the maintenance of schools. I was obligingly provided with the school return for the year 1838, by William Lithgow, Esquire. This will be found in Appendix XII.

It appears that the whole number educated is only six thousand and thirty-seven, and that the expense incurred by government is twelve thousand four hundred and twenty-six pounds, or upwards of two pounds per head. The number of children attending schools is to the aggregate of population as about one to twenty, which is the same as in 1836. In the return above mentioned, it will be found that there are seventy-six schools, of different denominations; three colleges, and sixty-seven private schools: showing an increase more than fourfold during the last five years. Several attempts have been made to establish the Irish national school system, or a general system of education, but thus far, without success. The chief opposition to this has been from the Church of England.

Among the colleges, two are under the guidance of the Church of England, viz.: King's College or School, at Paramatta, and Sydney

College. The third, called the Australian College, was established by the Reverend Dr. Lang. Of this institution that gentleman is the principal. The college edifices consist of four large buildings, for the accommodation of the professors and their families, with recitation-rooms in each, besides apartments for the students. The expenses, including board, are about forty pounds per annum. The charge at Sydney College is about ten pounds more. Of the latter college, Sir John Jamison is the president. It is in a great measure under the control of the Bishop and Episcopal laity of Sydney. An examination was witnessed at the latter institution, and was thought very creditable to the students. Medals were awarded and appropriate remarks made by the Bishop.

The system of giving to the clergy an allowance from the government, for their support, is the fertile cause of dissension in this community. Many hard thoughts, and harsh expressions, are occasionally felt and uttered, by one sect against the others, in the contest for the stipend distributed among the several denominations. An act was passed in 1836, which appears as liberal as could be expected. The amount appropriated annually is about twenty thousand pounds, of which about three-fourths go to the Episcopal Church, and the remainder is divided among other sects, Roman Catholics included. Regrets were occasionally heard, (perhaps to flatter us,) that the voluntary system of supporting the clergy had not been introduced. It will be well to remark, before quitting the subject, that in all other matters appertaining to the general good and benefit of the community, there appeared a co-operation highly commendable.

The Australian colony was erected into an Episcopal See in 1836, and Archdeacon Broughton was consecrated as the first Bishop. To his lordship we are indebted for many kind attentions, and the lively interest he took in our proceedings.

The exertions that the colonial government and private individuals appear to be making to afford religious instruction, cannot but bring about, in a few years, a very desirable and necessary reform among the lower classes of this colony, of which the facts previously exhibited in the account of the country fully establish the necessity.

NATIVE WEAPONS AND SHIELD.

CHAPTER VIII.

CONTENTS.

JAUNT TO PARAMATTA—DESCRIPTION OF THAT PLACE—OBSERVATORY—GOVERNMENT-HOUSE, AND FARM—VISIT OF NATURALISTS TO ILLAWARRA—WOOLONGONG—VEGETATION OF ILLAWARRA—LAKE ILLAWARRA—JAUNT OF MESSRS. HALE AND AGATE TO HUNTER'S RIVER—TOWN OF NEWCASTLE—COAL MINES—CONVICT MARRIAGES—CONVICT STOCKADE—NATIVES—JOURNEY TO LAKE MACQUARIE—MR THRELKELD—MACQUARIE LAKE—CONVICT SERVANTS—M'GILL, THE NATIVE—SHEEP-SHEARING—VISIT TO PEUEN BEUEN—BANKS OF HUNTER RIVER—MR. HALE'S JOURNEY TO WELLINGTON VALLEY—MAIL CARRIAGE—POST-OFFICE DEPARTMENT—PENRITH—BLUE MOUNTAINS—MOUNT VICTORIA—VALLEY OF CLWYD—BATHURST—LAKES—BUSH-RANGERS—WELLINGTON—NATIVE CHILDREN—MR. WATSON—SHEEP-SHEARING—PROFITS OF SHEEP-FARMS—FLOODS OF THE MACQUARIE—POPULATION OF THE WEST—NATIVE LANGUAGES—MR. PEALE'S JOURNEY—CLIFTON—CAPTAIN M'ARTHUR—HE INTRODUCES THE FINE-WOOL SHEEP—LIVERPOOL—DAM IN COOK'S RIVER—NEPEAN RIVER—WALLABIES—STRATHARA—BIRDS—DECREASE OF NATIVE ANIMALS—VISIT TO NEWINGTON—SALT-WORKS—PREVALENCE OF SALT IN AUSTRALIA—DECREASE OF LAKES—ANCIENT FLOODS—DESCRIPTION OF FISSURES—EARTHQUAKES—BURNING MOUNTAIN—TRADE OF NEW SOUTH WALES—FISHERIES—BANKS—EXPORTS—DEPOTS FOR GRAIN—COLONY AT SWAN RIVER—SOUTH AUSTRALIA—FÊTE AT FORT MACQUARIE—SURPRISE OF THE COLONISTS AT OUR WANT OF PREPARATION—STATE OF THE PEACOCK—ARRIVAL OF H. B. M. SHIP DRUID—SEASON OF OUR VISIT—FACILITIES FOR OUTFITS AT SYDNEY—KINDNESS AND HOSPITALITY OF CITIZENS—CHRISTMAS DINNER—DEPARTURE OF THE SQUADRON—DESERTERS FOUND ON BOARD.

Drawn by A.T. Agate

J. A. Rolph Sc.

FOREST, ILLAWARRA. N.S.W.

CHAPTER VIII.

NEW SOUTH WALES.

1839.

My own time was so completely occupied during the stay of the squadron at Sydney, as to prevent my making any excursions in the colony, with the exception of a brief visit to Paramatta, by invitation of his Excellency the Governor.

The distance of Paramatta from Sydney is fifteen miles. There are two methods of going to it from Sydney: by a carriage on a good macadamized road, or by a steamer up the Paramatta river. The latter is the most agreeable mode, and the scenery on the banks of the river is fine. The whole distance abounds in positions which would furnish beautiful sites for villas, upon the bays, inlets, and headlands. Should the progress of the colony in wealth and population continue, these sites will doubtless be occupied ere long with handsome residences.

The passage by steamer to Paramatta occupies about two hours; here the river becomes narrow, and a mile higher up the stream the tide ceases to flow, and farther navigation ceases.

The town of Paramatta is situated about a mile from the steamboat landing. Although regularly laid out, it has a straggling air, each house having a large space enclosed as a garden, and the attempt at regularity rather injures its appearance by giving it a look of stiffness. It has, and I suppose deserves, the reputation of being a dull place. It is built principally on a single street, about a mile long, at the head of which, on an eminence, is situated the Government-House, where his Excellency the Governor resides during the summer season.

The Paramatta Observatory, established by Sir Thomas Brisbane, is a small building, with several good instruments by Jones, Ramsden, and Troughton; but I regretted to see the dilapidated state it was in. Mr. Dunlap, the present incumbent, obligingly showed me the instruments, and I passed an agreeable morning with him. He is allowed a small salary, but I understood that no allowance was made for repairs of the building, &c.

Paramatta contains some public buildings and works, among which are the female penitentiary or factory, which has already been described, a stone court-house, barracks, and a fine stone bridge over the Paramatta river. I feel greatly indebted to Sir George and Lady Gipps, with whom I passed two days, for the kindness and attention they showed me.

Previous to Sir George Gipps's going to New South Wales, he had been in Canada, and on his return had paid a short visit to the United States. It afforded me pleasure to find the liberal views and feelings he evinced towards our country. It is needless to say that at the Government-House my time passed agreeably, and that I sensibly felt the exchange of such agreeable society for the routine of my duties on shipboard. Here, also, I had the pleasure of meeting several agreeable people.

The houses of Paramatta are generally no more than two stories high, and are built of sandstone. The town contains several churches. The Government-House is a commodious, unpretending two-story building. The grounds are extensive, but not remarkable for beauty. A farm is attached to the domain, where many government cattle are kept, and there are numerous outbuildings and dairies. The ground had a familiar look to me, for the grass was burnt up, and reminded me of my home at Washington during the heat of summer. It was, therefore, an unfavourable time to see its beauties. I understood that this place was laid out as an experimental farm; but this plan, I should suppose, is now laid aside, for the people of the colony are abundantly able to take care of themselves.

A telegraph is placed in a conspicuous position within a short walk of the house, which communicates with Sydney, and was formerly in constant use. It is now in contemplation to remove it, as it is no more needed, which is a farther proof of the advancement of this colony towards a well-regulated government.

There are also schools at Paramatta under the direction of the resident chaplain.

Several of the gentlemen who were left at Sydney, visited the Illawarra district, which has already been more than once spoken of. They made the passage from Sydney to Woolongong in a steamer. Owing to the steamer not being well-adapted for a sea-voyage, much inconvenience, delay, and disappointment occur on this route, although upon the whole it facilitates the intercommunication between this district and the city. Woolongong, the port at which the steamer stops, is a small thriving town, and will be the principal one of this district. It has no natural harbour, but one is now under construction, at the expense of government, by excavating the solid rock (limestone), for the accommodation of steamers and small vessels: a large number of convicts were at work upon it. The port will never be fully protected until the proposed pier or breakwater is built, for during half the year, the sea makes it dangerous to lie at anchor in the roadstead, notwithstanding the strong moorings which have been laid down. It will also be very difficult to enter the basin in bad weather, until such a breakwater is in existence to protect it. The basin, when completed, will contain about half a dozen vessels. The construction of the breakwater is carried on at the same time as that of the basin, and the stone excavated from the one is used in the construction of the other. Both were to have been finished in 1842.

The district of Illawarra is held by a few persons, who have large grants of land. The roads are constructed and kept in order at the expense of the government. When one of the residents was asked whether the road was a public one, he answered, it was a "government road."

The convict population, including ticket-of-leave holders, in this district bears a proportion to the free as one to three. Of the remaining two-thirds, more than one-half are emancipists and expirées. The proportion of women to men is also about one to three.

For the hospitable reception given them by Mr. Plunket, the Attorney-General of the colony, our gentlemen are under great obligation. He happened to be spending some time at his farm, near Woolongong. This contains about two hundred acres, and is exceedingly pretty. The residence of Mr. Plunket is a neat cottage, built after the manner of the settlers, and is well adapted to the country. It is surrounded by the most luxuriant foliage, nearly all of which has a tropical character, and includes palms, cabbage-trees, and several varieties of tree-ferns, all growing to a great height.

A drive through the woods, accompanied by the ladies of the family, afforded many opportunities of making collections, and getting information.

Some idea may be formed of the advancement of this district, and the rise in the value of property, from the fact that Mr. Plunket sold his farm for fourteen thousand pounds, which, but two years before, he had bought for seven hundred.

Dr. Osborne, R. N., has a farm likewise, near Lake Illawarra, which is now divided by a narrow sand-beach from the sea. This lake is shallow, and is about six miles long, by four miles wide. It contains a great quantity of fish, principally mullet.* Large quantities of shells are to be seen on its banks. These are burnt into lime, which is used both for building and as manure. On the borders of the lake reside several fishermen, and it is a general resort for the natives. Mullet, caught in large quantities, are salted and dried.

Daisy Bank, the seat of Dr. Osborne, is about ten miles from Woolongong. Here also our gentlemen met with that kind hospitality which reigns throughout this country. This part of the district is nearly all brought into cultivation. The mountain scenery is fine, and a few very large trees are conspicuous objects in it. The side of the mountain affords a good field for making botanical collections, as it is not easily accessible to cattle. A large accession was made to our collection of seeds. The woods were alive with birds, among which were the white cockatoo, which collects in flocks, and does infinite mischief to the wheat-fields. They are difficult to approach in consequence of the good look-out kept by the old birds. The small species of the kangaroo, called the wallaby, is found here, as are large black and diamond snakes, lizards, black and white cockatoos, and sand-leeches. The latter is much dreaded, as its bite is venomous, and produces ulcers. It is very troublesome, crawling up and attaching itself to the flesh, where it gloats upon the blood, and not unfrequently bursts from repletion.

This district is level, and was thought to resemble some parts of our own country, after the harvest was gathered in. Silicified wood is very common in Illawarra, and many stumps of it are seen in passing along the road. In some of them the texture of the wood is well preserved; and so natural is their aspect, that at first sight they appear as if they were now standing where they had originally grown. The diameter of some of them is about two and a half feet, and the whole mass is completely petrified. They are quite black, except where bleached by exposure.

The Illawarra district extends from Woolongong to Shoalhaven, and is the most interesting portion of Australia to visit. In this small compass is found some of the most remarkable of the sandstone scenery, and there is also an opportunity of viewing a basaltic formation, which is no where else to be found in the colony.

Kiama is remarkable for the number of deep and wild caverns,

* One of our gentlemen was assured by the fishermen that there were thirteen kinds of fish in Illawarra Lake.

through which the sea forces a passage to the distance of one hundred yards or more, sweeping along at a furious rate; and when the noise of its progress has nearly died away, loud thunderings are heard rushing through its vaults. The Blow-Hole of Kiama Point is already a place of some celebrity, and it merits to be so. A subterranean passage of about twenty feet broad by eighteen high, receives the advancing wave, which passes quietly along for two hundred feet. It then meets a basaltic wall, against which it dashes with a sullen roar, and passes upwards through a narrow opening above, rising at times to a height of one hundred feet, throwing off innumerable jets in all directions, and which fall around in ever-changing forms.

Some of the basaltic scenery about Kiama, will bear comparison with the far-famed Giant's Causeway, and the rocks of Staffa, if it does not surpass them, united as it is with the luxuriant and splendid forests of palms, tree-ferns, and the woody creepers of the tropics.

About Shoalhaven is one of the largest and finest farming and grazing districts in the colony. Its scenery is extremely picturesque, particularly when viewed from the summit of Coolomgata. The broad Shoalhaven river is seen to the southward, flowing through rich meadows and farms, enclosing a delta; while the deep and sinuous bays with which the coast is indented, and which enclose innumerable islets, appear like a crowded cluster of lakes.

To the north, a wide verdant plain extends to a mountain bluff, called Broughton's Head. Through this the Broughton river winds, and beyond it is seen the Illawarra mountain range.

On a wide platform around Woolongong Point, are to be seen at high-water mark, globular concretions, that resemble cannon-balls in appearance. They vary in size, from one inch to four in diameter, and are very compact and tough. They generally contain some foreign body, and in about a third of them, Mr. Dana found a single fossil shell in a beautiful state of preservation. For a full detail of the geological structure of this district, which is exceedingly interesting, I must refer to the Geological Report.

Mr. Hale and Mr. Agate made a jaunt to the Hunter river, and thence to Lake Macquarie, to the establishment of Mr. Threlkeld, the missionary employed among the aborigines.

The passage to Hunter river, a distance of eighty miles to the north of Sydney, is made in a steamer. The boat was small and ill-adapted for the sea.

Leaving Sydney just before dark, they reached Newcastle, at the mouth of the Hunter river, about noon the next day. They, however, had a head wind and much sea to contend against.

Among other accidents, the shipping of a sea caused much fright among the women on board, and threw one poor girl into hysterics. They were all glad to pass within the island of Nobboy, off the mouth of Hunter river, and to get on shore at Newcastle.

The town of Newcastle is a small village of seventy or eighty houses, built on the side of a hill; it contains two taverns and several grog-shops, a jail, convict stockade, hospital, court-house, and a venerable old-looking church. On one of the neighbouring hills is a flagstaff, and on another a windmill. The business of a coal-mine and that of the building of a breakwater for the protection of the harbour, give the place an air of life and animation.

Our travellers put up at Rowell's "Commercial Hotel;" and on proceeding to make inquiries relative to the mode of reaching Mr. Threlkeld's, they were referred to Dr. Brook, the surgeon of the hospital, and a friend of Mr. Threlkeld. He offered them every attention, and advised them to wait for Mr. Threlkeld's conveyance. This delay gave them an opportunity of seeing something of the place, and the natives, as well as to make drawings. The view of the surrounding country from the windmill was extensive, overlooking the town; the Hunter river was seen winding through a well-wooded country, rising occasionally into low hills. At a bend of the river the steamer was seen aground, on her way to Maitland, about twenty-five miles farther up the river. The coast trended to the north, and was visible as far as Port Stephens, about fifty miles distant.

There are two coal-pits, one on the hill, the other in the valley. The former is the older, and has been worked about eight years. Both are the property of the Australian Agricultural Company, and are under the direction of Mr. James Steel. The coal is first seen along the cliffs, forming black horizontal strata, separated by sandstone and clay shale, from twenty feet to forty feet in thickness. They formerly quarried it from the cliff, but the greater part of the coal is now obtained by mining.

From the older coal-pit they have excavated an area of twenty-four acres. The shafts are carried down about one hundred feet, to the fifth or lowermost coal-seam, which is about sixty feet below the level of the sea. The coal is at first taken out in small narrow areas, the passages in which are but four feet high, leaving about as much standing as is removed, the roof above being of fragile shale, and requiring propping every three or four feet. The work is all performed by convicts, who, after digging the coal out, take it in small carriages on railways, which pass to the shaft, where it is raised by steam-power. The lower bed only is considered sufficiently extensive and pure to pay for its exploration, and is about three feet thick. The coal is pure, except a layer of one and a half inches of bluish sandstone. It is bituminous, and burns readily, with abundance of flame, somewhat like kennel coal. It is compact, though less so than the best Pittsburg and Liverpool, and is of fair quality, although sometimes impregnated with clay, which causes it to leave a large quantity of ashes.

Pyrites is occasionally disseminated in masses through it. Coal abounds throughout the valley of the Hunter, appearing at the surface in many places.

The average quantity of coal produced is sixty tons a day, which is piled up near the mouth of the pit, and thence sent to the pier on a railway, where it is shipped to Sydney, Van Diemen's Land, and even to the Cape of Good Hope.

The new shaft in the valley is only sixty feet deep, the difference of the two being in the height of the hill.

Dr. Brook was formerly superintendent of this station, and gave a droll account of the summary manner in which marriages were concluded with the female convicts. If he saw a man who had just come in from the country with a clean shirt on, he was sure he had come for a wife, and the event always justified his surmise. The man usually intimated his wish with a modest sheepish grin. The fair frail candidates for matrimony were paraded for his inspection, and if he found one whose looks pleased him, he put the plain question at once, "Will you have me?" He was seldom answered in the negative, for marriage liberates the lady from the restraint she was under. The banns were then announced by the parson for three Sundays, when the lucky swain returned to claim his bride.

From the known licentious and unruly character of the female convicts, it is not to be supposed that these marriages can be very fruitful of happiness; but as both parties had been felons, they are probably as well matched as could be expected.

The greatest difficulty the superintendent of a station has to contend with, is the management of the female convicts.

Captain Furlong, commandant of the garrison, was kind enough to show the convict stockade; it encloses a prison for the convicts, and a guard-house for the soldiers. The convicts all belong to the iron-gang, composed here, as at Sydney, of those who have been guilty of some crime in the colony. They were kept constantly in irons, and are employed on the public works. They eat and sleep in the same apartments, and their bed is a blanket on the floor; to guard two hundred convicts, there are seventy soldiers stationed here.

At Dr. Brook's they had the pleasure of meeting with Mr. Dawson, the first agent of the Australian Land Company, and the founder of Port Stephens, who is well acquainted with this colony, and has published a popular work in relation to it. He of course possessed much information, and among other opinions seemed to entertain the idea that no free colony can succeed, and that in all cases the first settlers of a new country ought to have the use of slave labour, in order to be successful. He argued that these only had realized fortunes; where they had been left to their own resources they had generally failed, and left it to their successors to reap the advantages of their labour. As evidence of this opinion he contrasted the settlements of New South Wales and Swan River. At the latter establishment it is well known that the first settlers have lost almost every thing, and have struggled with every difficulty, and that they now desire to have the advantages of convict labour. This remark, however, is not true as respects South Australia; and its general accuracy would undoubtedly much depend upon the location.

In their walks they came across a group of several blacks (natives) seated around a small fire; they were pointed out as the remnant of the tribes which about forty years ago wandered in freedom over the plains of the Hunter and around the borders of Lake Macquarie. Their appearance was wretched in the extreme: emaciated limbs, shapeless bodies, immense heads, deep-set glaring eyes, thickly-matted hair, and the whole begrimed with dirt and red paint, gave them an aspect hardly human. The dress (if such it could be called) of the women, was a loose ragged gown, and of the men, a strip of blanket wrapped round the middle, or a pair of tattered pantaloons, which but half performed their office.

Mr. Threlkeld's conveyance did not arrive, and not being able to get another, they determined to walk to Lake Macquarie, and for this purpose they resorted to the natives as guides, and by a great deal of coaxing and promises of bull (grog), their natural repugnance to make an exertion was overcome. An evidence of the pride which characterizes these natives was shown in this interview. One of them, whose *sobriquet* was Big-headed Blackboy, was stretched out before the fire, and no answer could be obtained from him, but a drawling repetition, in grunts of displeasure, of "Bel (not) me want to go." After promises and expostulations enough to overcome all patience, Mr. Hale, tired of his obstinacy and stupidity, touched him slightly with his foot, telling him to get up and listen. He immediately arose, and seizing his spear, which was lying near him, turned his side towards Mr. Hale, and stood looking at him askance, with an expression of demoniac malice, as though he would have run him through with pleasure; but he did not speak a word in reply to all that was said to him.

Friday, 13th December, the morning being chilly, the blacks, who are very susceptible to cold, did not make their appearance till some hours after sunrise. At half-past eight our travellers set out in company with a troop of natives, headed by the two whom they had hired. The first of these was named *Jemmy*, the best-natured and most intelligent of all; the other was Big-headed Blackboy, who had got over his sulks. Jemmy refused to start until he had received a couple of shillings, which he forthwith converted into a loaf of bread and a bottle of grog. When about a mile from the town he asked permission to take a drink; and a cup of bark was produced from a thicket where it had been hidden, whereupon the contents of the bottle as well as the loaf were shared out among the troop. The two guides took no more than an equal portion; for, according to the custom of the natives, all share alike. The cup was made of a piece of the bark of the ti tree, which resembles that of the birch, about a foot square. The ends were folded in and tied together, to form a cavity of trough-like shape. Such cups are called by them taudé. The path or cart-road they followed, passed through a hilly country covered with forests. The gum trees were the most prevalent, and many of them were of great size, growing close together without any underwood.

The gum tree, of which there are many kinds, is peculiar to New Holland. It has an inner bark of about an inch thick, enclosed by an outer one which is quite thin. The latter is shed every year, which gives their trunks and branches a peculiar appearance of many colours, from pure white, through all the shades of yellow, olive, and red, to a deep brown. These colours, showing through the green foliage, produce a very striking effect on a stranger, and the contrast is heightened by an occasional sight of a black and withered trunk, from which the bark had been stripped by the natives to make canoes, or by settlers to roof their houses.

Ten miles brought them to Lake Macquarie, but on the opposite side to Mr. Threlkeld's house, and they found themselves disappointed in finding a canoe, which they were assured would be met with at a settler's on the banks of the lake. They were thus obliged to walk ten miles further. The guides were here again taken with sullenness, and refused to proceed. They were proof against all promises and abuse, and kept replying, "Me marry (very) tired, bel (not) me want to go." Through the kindness of Mr. Warren, the settler referred to, this obstacle was overcome, by his offering to send his son as guide, with a horse to carry the portmanteau. This offer was thankfully accepted.

After proceeding a few miles they came upon a little encampment of natives, crouching around fires in front of their huts, which were as rude as possible, made of a few pieces of bark laid against a stump and covered with bushes; they barely sufficed as a screen to keep off the wind. One of the women was quite good-looking, with large black eyes, white teeth, and small features. She was better dressed, too, than any of the others, and the pretty half-caste child that was clinging to her skirts, made it sufficiently evident in what manner her finery had been obtained. As a part of the lake was said to be fordable, it was determined to take advantage of it, in order to shorten the route. One mounted the horse to pass over. Whilst they were proceeding quietly along, the horse suddenly reared and plunged, relieving himself of his rider and load, which were thrown into water two feet deep, without any further injury than a good ducking, and the disparagement of the wardrobe. It was found that the horse had trodden upon a stingray, which fully accounted for his sudden gambols. It was sunset when they arrived at Mr. Threlkeld's station, which at first sight appeared like a comfortable farm-house, such as is often seen in our western country. Mr. Threlkeld was found busy attending to his cattle, and gave them a warm and friendly reception, which made them at once feel at home.

As Mr. Threlkeld has occupied a conspicuous place in this colony, it may be well to give a short sketch of his labours in the missionary field, in order to show the progress he has made, and the difficulties he has had to contend with. I do this more readily from the feeling that great injustice has been done him, and that he has suffered much contumely and persecution from those who were too prone to listen to the scandalous reports of interested individuals.

Mr. Threlkeld left England in 1814, as a missionary to the Society Islands; he resided with Mr. Williams, at Raiatea, until 1824, when the death of his wife determined him to pay a visit to England. About this time the inspectors of missionaries, Messrs. Tyerman and Bennet, arrived at the islands, and he took passage with them to Sydney. On their arrival at Sydney, these gentlemen, supposing that a favourable opportunity offered to establish a mission among the Australian aborigines, requested him to take charge of it, which he consented to do. Moreton Bay was at first proposed as the location, but it was afterwards changed to Lake Macquarie, the latter place being a favourite resort of the natives. Ten thousand acres were granted by government to the Missionary Society, in trust for the natives. The establishment was accordingly begun on this lake, on the opposite side to that now occupied by Mr. Threlkeld, who at once planned his station on the only footing by which he thought a reasonable chance of success would be insured, that of a farming establishment, extensive enough to give employment to the natives, and induce them to settle. Their number, as is usually the case, had been greatly overrated; he soon, however, collected about fifty around him, and began to employ them in felling trees, turning up the ground, and building; at the same time labouring with them himself, in order to obtain such a knowledge of their character, language, habits, &c., as might enable him to become useful on the great subjects of his mission.

The expense of forming such an establishment was far greater than had been anticipated, but was indispensable in a country like New South Wales, where all the necessaries of life, at the commencement of a settlement, have to be purchased. Added to this are the droughts to which they are subject, and the expenses of transportation.

In consequence of the demands made upon them, the directors of the Society became alarmed, and after reproving him severely for his extravagance, finally dishonoured one of his drafts, and refused to pay it until compelled by a lawsuit. This, of course, broke his connexion with the Society, as Mr. Threlkeld was naturally indignant at the undeserved disgrace to which they had subjected him.

The directors offered to pay his passage to England, but this he refused, having determined to carry on the work by his own unassisted efforts.

That he might be independent of any funds of the Society, and to prevent its being said that he had derived any profit from them, he removed in 1828 to the opposite side of the lake, a position far less advantageous.

After struggling for two years to conduct the mission and maintain his large family, he received a stipend of one hundred and fifty pounds from the government, with the assignment of four convicts. With this assistance he has been able to provide for his family, and devote himself to the instruction of the aborigines; but he has found his means inadequate to keep a number employed about his station, in such a manner as to overcome their natural tendency to a wandering life.

The consequence was, that the blacks, from the attraction held out to them of indulging in drunkenness and other vices, left his neighbourhood to frequent the towns, where they had been rapidly diminishing in number.

Mr. Threlkeld did not find the natives deficient in intelligence; but he has not been able to overcome their aversion to a fixed residence

In proof of this, they abandoned comfortable and substantial huts, which he built for them, after a few days' residence, on the plea that they were infested with vermin.

Frequently, they would all quit him to attend some meeting of their tribe, for war, hunting, or some religious ceremony, and stay away for months.

He laboured in vain against these disadvantages, and it is not difficult to perceive how impossible, under such circumstances, it would be to meet with success in teaching and converting a set of savages, so wedded to their usages.

Mr. Threlkeld's labours have, however, been turned to some advantage. He has published a grammar, and translated several of the books of the New Testament. His influence has been productive of a better tone of feeling between the blacks and the settlers than prevails elsewhere, and has prevented those outrages which have occurred in other parts of the country. He has been able to render essential service as an interpreter, both to the natives and government, in the courts.

A circumstance occurred about two years ago, which was the means of setting Mr. Threlkeld's whole conduct in its true light before the public.

The Rev. Dr. Lang, a minister of considerable notoriety in New South Wales, established a newspaper, which was in the habit of holding up and assailing all the abuses in the colony. Among others, he attacked Mr. Threlkeld, accusing him of malversation, unfaithfulness, and incapacity in his trust, and in a style of gross abuse, seconded by vulgar doggerel, gave grounds to the belief that he was actuated by any other than a proper zeal in the cause of missions. After great forbearance, Mr. Threlkeld wrote him a letter of remonstrance, which was at once published in the newspaper, accompanied with insulting comments. Mr. Threlkeld then instituted an action for libel, and obtained a verdict in his favour, which, although the damages were only nominal, is an uncommon thing in New South Wales, when a libel case is submitted to a jury. In the progress of the trial, the merits and sacrifices of the missionary were made apparent, and the faithfulness and diligence with which he had laboured, under so many disadvantages, became well known, for ever silencing the aspersions of his enemies. He had, in consequence, the satisfaction, not long since, of receiving a letter from the directors of the London Missionary Society, expressing their regret that they should have been led into such unjust suspicions and misplaced severity towards him.

Macquarie Lake communicates with the sea by a narrow inlet. Its shape is irregular, having several long narrow bays extending into the land, and from this cause it is in reality much more extensive than it appears. The soil around is sterile, and its principal ingredient is sandstone. The lake is surrounded by the sombre green of the gum trees, and the landscape is uninviting.

Many ant-hills were passed, each appearing to contain a numerous colony of different species of ants. They are of various colours, red, black, gray, and yellow, and of all sizes, from that of minute animalculæ, to that of a wasp. Most of them were said to give poisonous bites, and those of the largest kinds had visible stings. Most of the snakes, small as well as large, are venomous to a high degree.

Mr. Threlkeld, like many others in the colony, had convict servants assigned for the use of the station. It is thought almost impossible for a settler to manage his affairs without them, and it is somewhat curious to see a clergyman associated and in daily intercourse with thieves and abandoned felons. There is scarcely a person in comfortable circumstances, who has not derived much of his fortune from their exertions, although not without suffering very much from the constant vexations attendant on such aid. Mr. Threlkeld had hired a family of emigrants as intermediate assistants, but he was doubtful if he had benefited himself by it.

The difference between the two kinds of servants is great. The convict, on the one hand, is obliged to do the work his master appoints, and in the exact manner he directs; but the master suffers from his vices and dishonesty; and on the other hand, the emigrant is under all his English prejudices: self-willed, and conscious of his superiority over the other servants, he will not be driven, and is hardly to be coaxed into adopting the necessary alterations which the difference of soil and climate requires. Both try, in no small degree, the temper of a settler in New South Wales.

At Mr. Threlkeld's, Mr. Hale saw M'Gill, who was reputed to be one of the most intelligent natives; and his portrait was taken by Mr. Agate. His physiognomy was much more agreeable than that of the other blacks, being less strongly marked with the peculiarities of his race. He was about the middle size, of a dark chocolate colour, with fine glossy black hair and whiskers, a good forehead, eyes not deeply set, a nose that might be described as aquiline, although depressed and broad at the base. It was very evident that M'Gill was accustomed to teach his native language, for when he was asked the name of any thing, he pronounced the word very distinctly, syllable by syllable, so that it was impossible to mistake it. Though acquainted with the doctrines of Christianity, and all the comforts and advantages of civilization, it was impossible for him to overcome his attachment to the customs of his people, and he is always a prominent leader in the corrobories and other assemblies.

M'GILL.

Mr. Threlkeld has a son, who is also engaged in missionary labours near Darling river, about three hundred miles in the interior, and who understands the language. A boy was sent down by the son for the father to take charge of. There was no difference perceived between him and the natives of the Hunter river.

NEW HOLLAND BOY.

Inquiries for their implements of the chase and warfare, caused M'Gill, King Ben, and Shingleman, to set to work to furbish up their arms, including spears, shields, boomerengs, clubs, &c. The natives are seldom seen without arms, for they have not only to fear attacks from other tribes, but assaults from their own. This not unfrequently happens; and it is not long since the brother of King Ben was speared while asleep, for some private grudge, by Dismal; and it is said that Big-headed Blackboy, who has already been introduced to the reader, has committed several murders, and not long since burnt his mother nearly to death, in revenge for the loss of his brother, who died whilst under her care. This was not because he had any suspicions of unfair conduct, but simply from one of the unaccountable customs or superstitions of these people, which holds the nearest relative of a person accountable for his death, if it takes place under his care.

From the destructive influence of their own vices, and those of the community, these blacks are rapidly dying off. As an instance of this, Mr. Threlkeld mentioned that a tribe which occasionally visited the lake, and consisted at the time of his arrival of sixty, is now reduced, after a lapse of fifteen years, to twenty, only five of whom are females.

During our travellers' stay, two natives of some note arrived: King Ben and King Shingleman. The natives had no distinctions of rank among themselves, but when a native had performed any great service for one of the settlers, he was rewarded by giving him a large oval brass plate, with his royal title inscribed thereon. At first the natives were greatly pleased and proud of this mark of distinction, but as is the case every where, when the novelty was over, and these honorary medals became common, they began to hold them in disrepute, and now prefer the hard silver.

Sheep-shearing is performed in the neighbourhood of Lake Macquarie by men who make it their business. This operation was witnessed by some of our party, and was thought to be performed in a slovenly manner. It generally takes place in November and December.

Some others of our gentlemen paid a visit to Peuen Beuen, the seat of Mr. Stevens, near the head waters of the Hunter river. The route was by steamboat to Newcastle and thence to Maitland. The river at Newcastle is about one-third of a mile across, and the distance to Maitland, by water, about thirty miles, although it is only about twenty miles by land. The tide reaches Maitland, where the water is found to be brackish.

The banks of the river are extended flats. This is one of the principal agricultural districts of the colony, the soil enjoying the advantages of being naturally irrigated; but on the other hand, the crops are liable to destruction from heavy floods. These floods frequently occur, when there has been no sign of bad weather on the coast; but storms of rain occur seventy or eighty miles in the interior, which raise the streams thirty or forty feet, doing great damage.

On the way up the Hunter, a steamboat was seen building. The best ship-timber is said to be the flooded gum tree. The steamboat stopped at Green Hill, and they rode to Maitland, about three miles. Maitland is a widely-scattered village, with many neat dwellings, stores, and shops, &c., built of brick and other materials, and much better than could have been anticipated. Near Harper's Hill, a place noted for the fossils which have been found there, a chain-gang was seen at work on the road, with their attendant guard. They were generally young and hearty-looking men.

Some natives were passed who were quite naked, but they did not attempt to approach. There are no wild tribes in this vicinity. These poor creatures are becoming rapidly exterminated by the whites, who are not over-scrupulous as to the means. The natives have now and then committed a murder, but in general they are more sinned against than sinning. It is remarkable that they do not complain of their lands being taken from them, but confine their lamentations to the destruction of the kangaroos by the whites; and they think it very hard that they should be punished for killing the white man's kangaroo, (a sheep or a bullock.)

Mr. Hale made a journey to the Wellington Valley, about two hundred and thirty miles to the northwest of Sydney, and on the frontiers of the colony. It was first occupied, seventeen years ago, as a military post, when several small brick buildings were erected, and some of the land, which is considered the most fertile in the colony, brought into cultivation. It was afterwards converted into a penal station, for a description of convicts called "Specials," or such as were superior in education and social rank.

In 1832, it was granted by government to the Church Missionary Society, in trust for the aborigines, with an annuity of five hundred pounds, in part as the support of a mission establishment on the grant; and ever since, there have been two ministers of the Society resident at the place, employed in endeavouring to convert and civilize the natives.

The only conveyance is the mails, unless a vehicle is purchased, the outlay for which would be about four hundred dollars. The mail was taken in preference to this mode, both as avoiding cost and as less liable to the dangers of journeying alone. On account of the numerous Bush-rangers and runaway convicts, travelling in New South Wales is not considered safe.

The mail leaves Sydney once a week for Wellington Valley. There is some difficulty in procuring a seat, and the fare is thirty-two dollars and fifty cents; a very exorbitant charge considering the mode of conveyance, which was a two-wheeled vehicle, with seats for five persons. It had no top, and was in all respects a very uncomfortable conveyance. Formerly more commodious coaches were employed; but the government, finding that the contractors, in their anxiety to obtain passengers, were accustomed to delay the mail, ordered that none but two-wheeled vehicles should be used. The party left Sydney about 5 P.M. Three miles from town is an inn at which the mail-carts from all parts of the country meet, so as to enter the city in company at 8 A.M. For every minute of delay after this hour, the penalty of a shilling is exacted.

The post-office department is now under excellent regulations; the number of miles of mail route travelled in the colony is nearly three hundred thousand, and the gross revenue amounts to eight thousand three hundred and ninety pounds, being two thousand pounds more than the expenditure. The rate of postage is high, especially on ship-letters. The post was established in 1828, and at the end of the first year only eight post-offices were opened. In 1839, there were forty, showing the great increase of population and business.

The route towards Wellington Valley lay through Paramatta; and about 11 P.M. Penrith, thirty-six miles from Sydney, and on the Nepean, was reached. The mail left Penrith at four o'clock in the morning, and crossed the river on a raft. The Nepean, on its course towards the sea, assumes the name of Hawkesbury, and becomes the largest stream in the eastern part of the colony. At Penrith it is about one hundred and fifty yards wide, and forms the eastern boundary of the Emu Plains,—an interval of level ground, five or six miles broad, between the river and the Blue Mountains.

These mountains are the dividing range between the lands of the coast and the interior, and were, for many years after the establishment of the colony, considered as impassable, although many unsuccessful attempts to cross them were made previous to the administration of Governor Macquarie. During his administration, he sent out many expeditions by land and sea, and in 1814, a passage was effected, and the plains of Bathurst were discovered.

On reaching this part of the country, one is no longer surprised that these mountains were considered impassable. The barrier consists of a broad belt of mountainous country, about fifty miles in width, and varying in height from one thousand to three thousand five hundred feet, according to Mitchell. The route which was followed through them was about eighty miles in length, and for the whole distance there were not more than five or six miles of level, and those are chiefly due to the planning of the engineers. The road is constantly ascending or descending, and on every side, as far as the view extends, is a succession of mountain ridges, their summits rising in detached peaks, and their declivities terminating in narrow and deep gorges. Their sides are sometimes clothed with a scanty growth of dark evergreens, but in very many places presented only bare and rugged masses of brown sandstone rock. The whole scene for the first forty miles, is wild, dismal, and monotonous beyond description. In the latter part of the route through the mountains, the scenery begins to improve, and finally becomes very striking, the sandstone being succeeded by trap and granite. The descent of Mount Victoria is celebrated for its beauty throughout the colony. This road was laid out by Major Mitchell, the Surveyor-General of the colony, and by him the mountain was named. The descent of this mountain is more than a mile in length, and in some parts is inclined at an angle of five degrees. The road is cut in the solid rock, it is hard, smooth, and accurately graduated, and notwithstanding its great angle of declivity, heavily laden teams ascend with less difficulty than would be supposed. At the foot, the road is carried along a high embankment or viaduct, which has been thrown across a deep chasm, and the river flowing on either side is fine. On the left is a wide deep gorge, encircled by high and naked precipices topped with the sombre hue of the gum trees; on the right, an open valley, with a rivulet winding through it, sloping gently towards the northeast, gives a totally different current to the feelings. Governor Macquarie has named this the Vale of Clwyd, after a similar scene in Wales.

A little beyond this descent is the Weatherboard Inn, the land about which is, according to Major Mitchell, the only spot among the mountains fit for cultivation. He mentions, in order to show the difficulties the surveyors had to encounter, that one of them, a Mr. Dixon, penetrated the valley of the Grose, which, until then, had not been visited, where he was lost for four days, having been bewildered by the intricate character of the valleys; and when he finally emerged from them, he, in his official letter, "thanked God he had found his way out of them."

Shortly after leaving the inn, two small rivulets are passed, pursuing opposite directions. One of them falls into Cox's river, a branch of the Hawkesbury; the other, the Fish river, discharges into the Macquarie. Not far distant is Mount Lambie, the last and highest eminence of the range, from whose summit the lighthouse of Port Jackson is visible, at a distance of sixty miles. The road passes within a few yards of this place, and it was here that Major Mitchell encamped when he was employed in laying down his plans for the construction of it across the whole range. This road will compare advantageously with almost any work of the kind in any country; and this and other public improvements are frequently adduced as the benefits conferred upon the colony by convict labour. There can be but little doubt that the colonial government has many facilities to carry forward improvements, but I very much question, if all things were taken into the calculation, that it would be found to result in so great a difference as is generally supposed.

After leaving the mountains, the road leads for several miles through an undulating country, covered with an open forest of stunted gum trees, and then comes in sight of the plains of Bathurst. These are of moderate extent, being little more than the valley through which the river Macquarie finds a channel. In the month of December there was no flowing stream, and the river, which at some seasons is a broad and powerful current, consisted merely of a string of pools.

The appearance of the town of Bathurst disappoints. It consists of a few hundred houses, scattered in detached groups over the plain. The absence of trees and cultivation serves to increase the want of interest in the landscape. The town-plot was first laid out on the eastern side of the river, but after several houses had been erected, it was removed to the opposite bank, a circumstance which accounts for the dispersed appearance of the village. Most of the wealthy inhabitants have their dwellings two or three miles removed from the town, among the low hills in the neighbourhood; from which circumstance, the importance of the place and the extent of the settlement is not at first apparent.

The low bottom-land in which Bathurst stands is believed from various indications to have been at no distant period a lake. At the time of its discovery it was little better than a marsh, and the Macquarie was flowing in a deep and strong current nearly on a level with its banks, and was navigable for large boats. The plain was covered with long prairie grass, which led to the belief that it was of inexhaustible fertility; but the general opinion of the intelligent residents is, that for the last twenty years the country west of the Blue Mountains has been gradually drying up. Lakes which, when first discovered, were extensive sheets of water, deep enough to float a seventy-four, are now inconsiderable ponds; swamps have been converted into dry pasture-lands; and there is hardly a river which now continues running throughout the year. It is remarkable, that in these lakes and ponds, which have become dry, there are found the stumps of large trees, showing conclusively that these places must have been dry at some former period, and that they had continued so for a long time, giving rise to the opinion that the country must be subject to long periodical alterations of climate.

On the morning of the 16th, Mr. Hale started in the mail-cart for Wellington. For the first twenty miles the road was a mere cart-track, through a piece of hilly country called "the Rocks," which is a repetition of the Blue Mountains on a smaller scale. Beyond, there is a succession of valleys, bounded by ranges of low hills, and covered with open woods, like a continuous orchard. This kind of country continues to Wellington Valley, and for the distance of a hundred miles beyond, when it gradually subsides into a level plain, in which many exploring parties have continued their progress for weeks, without meeting any elevation deserving the name of mountain. These plains stretch away towards the interior of the continent, but of their extent in that direction nothing certain is yet known.

Twenty miles from Bathurst brought them to an inn kept by a man named Luck, which had been, about six weeks before, the scene of a tragical incident. During the absence of the landlord, a party of Bush-rangers entered the house at night, and began to plunder. Although they had taken the precaution to disguise themselves with masks of black crape, the landlady recognised one of them, and was so imprudent as to threaten him with the consequences of his crime, whereupon the robber without hesitation drew his pistol, and shot her dead on the spot. What will add to the illustration of the state of society here, is the fact that the murdered woman, though living with Luck as his wife, was not married to him. The laxity of morals which prevails throughout the interior among the lower orders, can hardly be exaggerated.

The next public house to Luck's was a low tavern, in which it was not unusual for stockmen, sheep-shearers, wagon-drivers, &c., to meet and spend a week in drunkenness and debauchery, dissipating, not unfrequently, the earnings of a year, amounting to twenty or thirty pounds. Another inn was kept by the brother of the proprietor, he having committed some crime for which he had been transported to Norfolk Island. The last of the public houses was kept by a native of the colony, and was the best met with. This was another instance of the good character and general deportment, and temperate habits of this class, who in spite of their unhappy parentage, evil example, and inauspicious connexions, offer a remarkable example of the improvement which education, when aided by a change of condition, may effect in a single generation.

The stopping-places for the next two days were the huts of stockmen, and dwellings of settlers, all of which resembled each other in their construction. The sides were made of slabs of wood placed upright in the earth, and were sometimes fastened to a frame; the roof was composed of strips of the bark of the gum tree. In the better sort of houses there were chimneys of brick, and glazed windows; but these were comparatively few; and in the others an elevated hearth of clay, in a recess of the hut, supplied the former, the smoke escaping through the roof. A cupboard, a camp bedstead, a rude table, with a few stools, supplied the want of furniture. In houses of this description, were living gentlemen of education and refined habits, who were submitting to a few years of hardship and banishment from social life, in hopes of realizing rapid fortunes.

SETTLERS COTTAGE N.S. WALES.

On the 18th, Wellington Valley was reached. It is a beautiful plain, about four miles square, bounded by low hills, and watered in seasons of freshet by the Bell river, which winds through it, and falls into the Macquarie about two miles below the station. During the season of Mr. Hale's visit the channel was dry.

The buildings at Wellington consist of a dozen small brick houses, erected formerly as barracks for soldiers, and having undergone some slight alteration and repair, they are now inhabited by the missiona ries and a police magistrate. The former are three in number, two clergymen and an agriculturist. They have under instruction forty men, women, and children, but the wandering and capricious habits of these aborigines render it impossible to keep the adults with them. Mr. Watson, the eldest of the missionaries, has now with him fifteen children, whom he does not allow to leave his house, and is endeavouring to teach them the habits of Europeans, and the English language. He considers them as equal to white children in docility and intelligence, and several of them had made as much proficiency in the various branches of education, as could be expected at their age. They could read and write with facility, and solve questions in elementary arithmetic. They had a natural aptitude for music, and they joined with much harmony in singing common English tunes.

Mr. Hale was greatly indebted to the chief missionary, Mr Watson, for his hospitality and the aid he furnished in his researches into the language, manners, and customs of the natives.

While at Wellington, he passed a few days at the station of W. O. Raymond, Esq., one of the magistrates of the colony, who is owner of a large stock of cattle and sheep. His house is situated on the Macquarie, and here an opportunity presented itself of seeing the operation of washing and shearing the sheep. This took place at the time of their visit, and was, on account of the lateness of the season, about a month later than usual.

The sheep were plunged and held in a tub of hot water, until their fleeces were thoroughly soaked; they were then taken out and made to swim about in one of the deep pools of the Macquarie, for half an hour; after this they were held under the spout of a pump, where they were rubbed, combed, and rinsed, until their wool was considered sufficiently clean.

The sheep are shorn when dry, and the fleeces assorted according to their fineness, in lots, which are afterwards packed in bales of from two to three hundred pounds: these are then compressed by a lever-press.

The average weight of a fleece is about two and a half pounds. Mr. Raymond calculates the cost of transportation to Sydney at about two pence per pound, and the average price of the wool there is eighteen pence per pound. The freight to England is one and a half pence; and there it has to compete with fine wools from other countries. As to the question whether this can be done profitably, there is a considerable difference of opinion between well-informed persons in the colony. According to some, it can be afforded even at a much lower rate, but in this estimate the labour of those who are employed as shepherds is no doubt calculated as being that of convicts, and it may be questionable whether, when this source of labour fails, the price will be a remunerating one.

The flocks of sheep kept near Wellington are pastured beyond the legal limits, which is a meridian line, in the neighbourhood of that place. Beyond this line the government refuses to make any grants of land; but any respectable inhabitant, on the payment of ten pounds, may obtain a license to pasture his flocks beyond this artificial boundary.

Each flock consists of from five hundred to a thousand sheep, and is under the care of a single shepherd. There are usually two flocks to each station, where a servant is employed as hut-keeper. The cost of these when convicts, is no more than their food and clothing, which is, however, rendered greater than would at first seem probable, by the necessity of bringing even flour from Sydney.

The land and labour may, however, be put down at an expense merely nominal, for the increase of the flocks at present more than counterbalances this item; but this advantage will cease when the assigned convicts are withdrawn from the colony; the wages of a hired servant will then amount to from seventeen to twenty pounds a year, exclusive of his clothing and food.

The cost of a sheep varies much in different parts of the colony; the average price is from three shillings to one pound, so that the outlay for the smallest flock would be from seventy-five to five hundred pounds. Comparing this with the price of wool, (eighteen pence per pound,) an estimate may be formed of the probable profits.

The climate seems peculiarly well adapted to a fine-woolled sheep, and it is calculated that the flocks double themselves in three or four years. In 1807, the quantity of wool exported was not more than two hundred and forty-five pounds, in 1838 and 1839 it exceeded five millions of pounds. With these facts, the rapid accumulation of fortunes in New South Wales will no longer be a mystery.

It is said that the owners of stock have already pushed their stations one hundred and twenty miles beyond the boundary, and the only impediment to their farther extension seems to be the scarcity of water, of which the more remote country is almost destitute.

The country about Wellington becomes almost impassable during heavy rains, for the waters are then so much swelled as to put a stop to travelling. Mr. Hale was detained a week from this cause; and at Wellington, the Macquarie, which was before only a string of pools, became a large river, flowing with a rapid current; yet at a distance of twenty miles farther down, it had ceased to flow, thus exhibiting the phenomenon of a large stream losing itself. This remarkable circumstance is usually ascribed to the many dry pools it has to fill on its route, each of which must be overflowing before there can be any farther current; but this is hardly sufficient to account for the almost sudden disappearance of a body of water sixty feet wide and two feet deep, flowing at the rate of three or four miles per hour. It would seem more probable that water may make its way into some of the vast caverns that are known to exist in this limestone region.

The population beyond the Blue Mountains amounts to ten thousand, and it is supposed that there is little room for its farther increase, as all the stations capable of supporting flocks are now occupied, and as there is little or no chance for the extension of husbandry. Wellington Valley, although it was considered when first discovered, as fitted to be the granary of the district, has disappointed all such expectations; and out of seven harvests which have occurred since the missionaries commenced operations in it, six have wholly or partially failed.

According to Mr. Hale, the number of languages in Australia has been greatly exaggerated, and so far from every tribe having, as has been asserted, a separate language, it appears that within the colony, or from Port Macquarie on the north to Port Philip on the south, and extending one hundred miles beyond Wellington to the west, comprising one-tenth of the whole continent, only six, or at most, eight dialects are spoken, and that these are so similar in words and grammatical construction as to place their identity of origin beyond a doubt. From some vocabularies of the language spoken at Swan river, it appears that this similarity of words extends over the entire breadth of the continent. On the other hand, at Port Essington and Melville Island, on the northern coast, though the distance is not so great, the dialect is represented as quite different, notwithstanding the physical characteristics, habits, and customs, are said to be similar to those of the other aborigines. It is not believed, however, that the difference is as great as has been represented, and farther researches, it is thought, will prove the accounts of it to have been exaggerated. The language differs radically from that of the Malay tribes, being highly artificial in its construction, abounding in consonanted sounds, and remarkable for the number and variety of its grammatical inflexions. The verbal modifications are as numerous and comprehensive as in the American languages, but the manner of inflecting is different: the root or radical verb (which is usually a monosyllable) is placed first, and to this the various inflexions or modifying syllables are attached, until they protract the word to an extraordinary length. Thus, in the word Būmarœ, I strike, (Bū or Būm being the root.) Then comes bumal-guain, I have struck; bumal-gurani, I struck yesterday; bumal girri, I shall strike; bumalugidyillinga, I strike myself; bumallanna, we two strike each other; bumalalinga, I strike again; bumalmamblina, I permit to strike again; bumabumara, I continue striking; bumalngarriawagirri, I shall strike to-morrow; and finally, bumalbumalālimambilngarriawagirri, I shall permit to continue striking again to-morrow. Those who are desirous of farther information, relative to this language, are referred to the results of the Philological department.

Mr. Peale made a journey into the interior, in the direction of Argyle, passing through Liverpool, and visiting Camden, Clifton, and Strathara. The last two were the country-seats of gentlemen. Clifton is the residence of James M'Arthur, Esq., who possesses a large estate in its neighbourhood. Mr. M'Arthur, father of the present owner, was the first who introduced sheep into this country. The facts connected with this transaction, as related to me at Sydney, are as follows: Captain M'Arthur, about the year 1797, had procured three rams and five ewes from Captain Kent, R. N., who brought them from the Cape of Good Hope. They were of Spanish blood, and had been sent out by the Dutch government to that colony. Captain M'Arthur soon found by experience, that his ideas as to the fitness of the country for the support of this animal, had not been too extravagant.

In 1803, he visited England, and there made a statement, which was communicated to the government, a copy of which will be found in Appendix XXIII.

In consequence of this statement, Captain M'Arthur's plans were investigated by a committee of the Privy Council—at whose meetings he was present—and were recommended to be adopted. Some sheep were supplied from the flock of George III., and with them he embarked shortly for New South Wales, on board the "Argo," which vessel was so named by himself in reference to the freight she bore.

The government having granted him a large tract of land, in what was termed in the colony the Cow Pastures, he, in gratitude for the assistance he had received, named it Camden, after the distinguished nobleman who had befriended him, and who was then presiding over the Colonial Department. This is now a princely estate, with a magnificent mansion and grounds. The land attached to it contains thirty thousand acres on the Upham river. About the lawns of this mansion, magnolias and other trees of North America flourish by the side of the Acacia pendula, &c., and plants indigenous to the Australian mountains. In the garden are found figs, peaches, pears, plums, and small fruits in the greatest profusion and of the finest quality, besides mulberries, grapes, pine-apples, oranges, &c., growing in the open air The grounds are in beautiful order, and their reputation deservedly great in the colony.

Liverpool is a small town in this neighbourhood, fifteen miles from Paramatta, to the westward. The government has here a large hospital under the direction of Dr. Hill, to whose kindness and attention Mr. Peale was much indebted. This institution is open to the disabled and sick of all nations; is a large building, and admirably kept.

We are sorry that as much cannot be said for the "Wheelwright Arms," at Liverpool, and other hotels in Campbelltown: a larger supply of spiders, flies, and bed-bugs is seldom seen, than that with which the bed-rooms swarm.

In the neighbourhood of Liverpool, a dam is in progress at the head of the tide-water of Cook's river, which empties into Botany Bay. This is a noble work, and is intended for the purpose of giving Sydney a supply of water, of which it is much in want. The work is performed entirely at the expense of government, and the water is led for a long distance by tunnel.

While at Clifton, Mr. Peale made an excursion along the meandering course of the Nepean river. He was much surprised at the productions of the soil, although these were apparently every where deficient of moisture; and also at the singular notes of the birds, particularly the quaint and varied jargon of the Dacelo gigantea, called in the country, the "Laughing Jackass." This is an instance of the ridiculous misapplication of names in this country; for, besides belonging to a different class of the animal kingdom, its notes have little or no resemblance to the braying of an ass, and it feeds upon a very different kind of food, viz.: lizards and serpents. The bird is common in this part of the country, but peculiar to New South Wales.

There are many native magpies, which have somewhat the appearance of a crow. This bird frequents the neighbourhood of houses, and its loud and crow-like note is the matin-call of the country residents. In spite of its hoarse croak, it was spoken of by some as a fine singing bird.

The wallaby, the smallest species of kangaroo, is common here, as well as numerous opossums. On reaching the Wallondilly river, the party stopped to hunt the Ornithorhynchus, which once abounded there, and succeeded in obtaining specimens, although with much difficulty. Proceeding on, they reached Strathara, the seat of Achlan M'Alister, Esquire, to whose kind attentions the whole squadron are much indebted.

His property contains about sixteen thousand acres of the most fertile land in the colony. The soil is composed of decomposed trap rock, and has the appearance of a rich chocolate-coloured mould, which retains the moisture well. The timber which grows upon it is closer and heavier than in the sandstone districts: it is principally Eucalyptus of several species. The grass is thinly spread over the ground, and the cattle and sheep require a great range. One sheep to an acre is the allotment, and even in this proportion they suffer in dry seasons. There are no running streams of water on this estate; but, as has been remarked, the pools are numerous, a most happy circumstance for the country, for from these alone can the cattle be supplied. Argyle is the only place where springs were seen in this part of the country, and they are scanty. The crops of wheat were unusually good, but they were the first that had been collected for three years.

The variety of birds seen here, and the brilliancy of their plumage, are characteristic of Australia. All the birds are remarkable for the closeness of their plumage, and the neatness of their form; many of the species are peculiar to Australia, and are more nearly allied to those of the western part of the Indian Archipelago than of any other region. Even this analogy is extremely limited. Many of the Australian species are said to be confined to peculiar districts, which they only leave on emergency, from want of food, &c.

Mr. Coxen, near Peuen Beuen, informed our gentlemen that several birds had made their appearance around his dwelling that season, that were not known within a hundred miles of his place before. From the little that is known of the ornithology of the rest of New Holland, it seems that the same general character prevails throughout the whole continent, and there are grounds for believing that there is a complete diversity in the species from those of New Guinea. As an instance of this, it may be stated that none of the paradise-birds, so common in the islands to the north, have been found in Australia; and what appears to add strength to this opinion, is the fact that the land birds of Norfolk Island are all known to be peculiar.

The number of parrots that are seen is very great. They usually occupy the tops of trees, and are remarkable for the rapidity of their flight, particularly a green species, little larger than a humming-bird, with which the trees occasionally swarm. Other birds, hardly known to the ornithologist, are also numerous; but Mr. Gould, who is eminent in that department of natural history, is now engaged in making collections, and will probably, ere long, give a full account of the habits and economy of the Australian birds.

The Australian wood-pecker is the famous bill-bird whose note is always hailed with joy by the traveller in these arid regions, as a sign of the vicinity of water. The sound resembles the click of a stone-hammer, and the effect of the united notes of several, is similar to the frog concerts of our springs. According to Mr. Coxen, each bird utters a single note.

It was remarked that the native animals of Australia are fast disappearing. The kangaroo, once so numerous, is now seldom seen; but the native dog still commits ravages among the sheep.* Some of the animals which have become rare are preserved in the Sydney Museum; among these are the woombat (Cheropus), and the Ornithorhynchus, in relation to which so many questions have been raised. Snakes of many kinds still abound, even in the immediate vicinity of Sydney, whose bite is said to be fatal, and which is of course much dreaded. The stories that are related of such poisonous bites, and the dread of them that animals show, make those who wander through the paths extremely cautious, particularly as their small size and grassy colour render them difficult to be seen.

Among the distinguished gentlemen of the colony, to whose hospitality our naturalists were indebted, is John Blaxland, Esq., who resides at Newington, on the river, near Paramatta. The ladies of his family are in possession of a handsome hortus siccus of native plants, collected and prepared by themselves.

A part of this gentleman's estate consists of extensive salt-works, formed by drawing the tide-water from the river into ponds. In these it is evaporated as much as possible by the heat of the sun, and is afterwards boiled. The quantity of salt made at these works during the preceding year (1838) was one thousand tons. About seventy assigned servants (convicts) are employed in the manufacture.

The water of the ocean is far from being the only source of this necessary of life in Australia. Salt springs are abundant, and almost all the wells, particularly those of the sandstone region, are said to afford only brackish water. The small streamlets, and in dry seasons even the rivers, are found to be salt; and there is hardly a traveller or navigator, but has given an account of his disappointment in finding salt water, when every indication gave the promise of fresh.

Major Mitchell attributes the occasional saltness of the Darling river, to salt springs, or to its passing through beds of rock salt. This river, as has been stated, has no tributary for more than six hundred milès, and has at times little or no current; and it is where the stream has no sensible motion, that the saltness is most marked. The salt

* The natives had never attempted to domesticate the dog, and all of the species found, when the country was colonized, were wild.

appears to cover but a small area at any one place, and it has been observed that within short distances of each other, fresh and salt rivulets may be seen, pursuing the same direction, and each retaining its character throughout its whole course.

The lakes in the eastern section of Australia are also nearly all either salt or brackish. Lake George, situated beyond Goulburn, near the source of the Yass river, which empties into the Morrumbidgee, is the largest of these lakes. It is at present only five or six miles in length, by about four in width, although according to unquestionable authority, it was, within twelve or fourteen years, sixteen miles long by twelve wide. Lake Bathurst, which is not far distant from Lake George, has also undergone a similar diminution. In the latter lake there are to be seen stumps of trees, which prove, that although within a few years a considerable lake, and at present decreasing in its extent, it had at a former, and that at no remote epoch, been a marsh, if not actually dry land. Should its present diminution continue, which must take place if the seasons of drought are not interrupted, it will in a few years be again dry land.*

The facts observed at these lakes prove in the most conclusive manner the very great irregularity in the climate of New South Wales. It would appear from them, that, however great the floods now occasionally experienced are considered, those that have occurred must have exceeded them, and filled the basins of these lakes, to such a depth, that within the fifty years that they have been known, the excess of evaporation has not been sufficient to restore them to their pristine state.

In conformity with the condition of these lakes, many places now dry are pointed out, where, within the memory of the settlers, lakes or ponds existed; and near the course of streams, grass is to be seen, attached to the trunks of trees thirty feet above the present level of the water, which must have been lodged there by very great floods.

The great and important changes that floods of such extent and destructive force must produce on the face of the country, may be imagined, and particularly when like New South Wales it is principally composed of soft sandstone. To such causes may be ascribed the numerous coves of the harbours and bays, and the deep ravines which often break the monotony of the table-land. In relation to the bays

* In the basins of the salt lakes of the interior, plants which grow on the shores of the ocean are found in abundance; as for instance the Salsola. These lakes even exceed in saltness the waters of the ocean; those brought by Major Mitchell, and analyzed, contained one hundred and thirteen grains of dry salt in three ounces of water; the specific gravity of the water was from 1·0386 to 1·0553.

and coves, Major Mitchell remarks, that they generally have a direction either from north-northeast to south-southwest, or from west-northwest to east-southeast. Our geologist observed a coincidence of the fissures of the sandstone rock with the same points of the compass. This double and intersecting direction of the fissures, gives to portions of the rock which are bare, the appearance of an artificial pavement of enormous blocks. This appearance is well marked, and can be readily observed in the variegated layers of the sandstone cliffs near the Heads of Port Jackson.

Earthquakes are occasionally felt in New South Wales. The recorded accounts of these are necessarily imperfect; they, however, show that within the last fifty years, six are known to have occurred, viz., on the 22d of January, 1785, the 17th January, 1801, the 7th May, 1804, the 24th September, 1806, the 28th November, 1823, and the 2d August, 1837. That of 28th November, 1823, was also felt at Laurie's Town, Van Diemen's Land.

As far as could be learnt, these earthquakes did no material damage. It may, however, be inferred from the nature of the country, that violent commotions have taken place in former times. Major Mitchell has stated, and the fact was confirmed by the personal examination of our geologist, Mr. Dana, that an alteration in the relative level of the sea is abundantly evident on the cliffs of the coast.

The Burning Mountain of Wingen is something analogous to a crater, which it was not in the power of any of our parties to reach. According to Major Mitchell, it appears to be the same kind of phenomenon as that described by Professor Buckland and Mr. De la Beche, caused by the action of rain-water on iron pyrites, which sets fire to the bituminous shale. The combustion of Wingen extends over an area of about two miles in extent, and occurs near the summit of a group of hills, forming part of a low chain which divides the valley of Kingdom Ponds from that of Page's river. Blue smoke ascends from rents and cracks; the breadth of the widest of which measures about three feet. A red heat appears at the depth of about four fathoms, and no marks of any extensive change appear on the surface near these burning fissures, although the growth of large trees in old cracks on the opposite slope where ignition had ceased, shows that this fire had continued for a very considerable time. The height of this crater is about fifteen hundred feet above the level of the sea.

The trade of Australia is greatly on the increase. Nearly all of it centres in Sydney; and this will account for the rapid rise of that city, which not only has the finest port, but the most central position as respects the east coast. By a reference to the official documents in the Appendix, from XIV. to XX., the rapid increase of the commerce of Sydney will be seen; but they give only an imperfect idea of the life and animation that this port exhibits, or of the bustle attendant on the receiving of produce and forwarding of supplies to the interior, on the arrival of emigrants. The warehouses, and all works connected with this trade, are of a durable description. The number of vessels that entered Port Jackson in 1826, was sixty-two, and their tonnage seventeen thousand one hundred and seventy-eight tons; in 1840 they had increased to seven hundred and nine, and the tonnage to one hundred and seventy-eight thousand nine hundred and fifty-eight tons. The value of imports in 1826 was sixty thousand pounds; in 1840, it had increased to three millions fourteen thousand one hundred and eighty-nine pounds. That of exports, in 1826, was one hundred and six thousand six hundred pounds; in 1840, they amounted to one million three hundred and ninety-nine thousand six hundred and ninety-two pounds.

It will also be seen that in the return of vessels built and registered in 1822, there were but three, of only one hundred and sixty-three tons; in 1840, one hundred and eleven vessels, the amount of whose tonnage was thirteen thousand three hundred and forty-nine tons.

But the most remarkable increase is in the exportation of wool, which in 1822 was only one hundred and seventy-two thousand eight hundred and eighty pounds, while in 1840 it amounted to eight millions six hundred and ten thousand seven hundred and seventy-five pounds. This is not the only instance, for a greater or less increase has occurred in all the productions of the colony. The quantity of timber exported forms also no inconsiderable item. The returns are referred to as showing it, in Appendix XXII.

The fisheries begin to claim attention, and in particular the whale-fishery, as it requires comparatively a small capital, and the returns are quickly realized. The operations of this fishery are conducted with great success, a ready market being found at Sydney, and the great saving by arriving on and returning from their cruising-grounds without loss of time, adds greatly to their gains. They are, however, not as adroit in the pursuit of these treasures as our own countrymen; their vessels are sailed at greater expense, and the officers and crews less enterprising. The value of this trade and its yearly increase is exhibited in the official returns, by which it appears that in 1830, fifty-nine thousand four hundred and seventy-one pounds were derived from it, while in 1840 it amounted to two hundred and twenty-four thousand one hundred and forty-four pounds. I heard many complaints that our whalers were in the habit of taking whales and obtained much of their oil in the bays on the western coast of New Holland; and the remark was made, that if the colonists were not brought into collision with the Yankees, they would succeed well enough. This, I suppose, may be considered as complimentary to the energy and skill of this enterprising portion of our citizens. The whales are reported to be fast leaving their old haunts, in consequence of being disturbed in their calving season, and the places where they used to abound are now only the resort of a few. It is, therefore, supposed they are abandoning the waters of New Holland for other seas, where they are less disturbed.

There is now a large export of salted provisions from the colony, which are well prepared, and there is a considerable trade with the Mauritius and the Cape of Good Hope; the former supplying sugar, &c., in return for the cattle and produce of the dairy; while from the latter wine is imported in considerable quantities, and other spirits to a large amount, as has been already noticed.

The declared value of exports of British and Irish produce and manufactures to New South Wales, for the year ending January 5th, 1840, was one million four hundred and forty thousand four hundred and forty pounds; and of foreign and colonial produce, two hundred and eighty-nine thousand and seventy-two pounds. In return for which the colony sent back, in 1839, six millions eight hundred and ninety-four thousand eight hundred and sixty-eight pounds of wool, besides twenty-three thousand barrels of oil.

The number of vessels employed with passengers was four hundred and sixteen, measuring forty-eight thousand nine hundred and eleven tons.

The export of wool during 1840 was nearly one million pounds of wool more; the increase in the number of sheep in the colony during late years has been very great. In 1796, eight years after the colony was established, they numbered one thousand five hundred and thirty-one; in 1805, six thousand seven hundred and fifty-seven; in 1821, one hundred and twenty thousand; and in 1839, three millions.

There will also be found in Appendix XIV., an official abstract, showing the assets, liabilities, capital, and profits of the banks of the colony, by which it appears that the rate of interest derived from investment amounts to eleven and a half per cent. for the year 1841. Business is almost done entirely on credit, and large discounts are made by the banks for the accommodation of merchants.

The seal-fishery on this coast was formerly very successful; but in consequence of the immense numbers that were destroyed without regard to season, they have been almost exterminated, or driven to new haunts. The seal-fishery, which has heretofore yielded so large an amount of wealth, will soon be at an end in this quarter of the world.

It frequently happens that owing to the failure of the harvests, the colony is under great distress for provisions. Government is erecting deposits for grain, in order to obviate this difficulty; one of which is on an island in the harbour of Port Jackson. Heretofore, on a failure of the crops, large quantities of flour have been imported from France and America, and many cargoes of wheat and rice from India.

The commerce with the United States is very limited, and confined as yet to a single house, in Salem, Massachusetts, which has a few vessels employed in bringing out flour, tobacco, furniture, ice, &c. In exchange, wool and hides have been taken to the United States. This trade has hitherto been profitable; but the uncertainty of crops, and consequent fluctuation in the market, would, with competition, render it of no great advantage.

The Library, and Sydney Museum are creditable institutions, particularly the latter, which contains a large and interesting collection of native productions. It has only been established a few years. There is a reading-room attached to the Library, in which are to be found all the pamphlets and periodicals published in Sydney, and many of those of Europe. Soon after our arrival, we received an invitation to visit these institutions at all times; a privilege which afforded us much pleasure and instruction, and for which we are greatly indebted to the committee and the librarian.

Great Britain has three other colonies in Australia, at Swan river, on the west coast, South Australia on the south, and North Australia on the north coast. The former is considered, in the colony of New South Wales, as a bad speculation, and it is alleged that it needs the aid of convicts or slaves for its advancement. It is believed that all the first settlers, if not completely ruined, have been struggling with difficulties, and its growth, even should it continue, will be slow and precarious.

Orders were received from the home department to raise the minimum price of land; but it being left optional with the Governor, he had declined doing it, under the plea that if individuals would sell land for two shillings and sixpence, it was idle to raise that of government to twelve shillings, particularly as the price allowed to individuals for surrendering their lands was but one shilling and sixpence. This step, of raising the minimum price of crown lands, I was informed, had given much dissatisfaction, and was generally believed to have originated in the desire to force colonization to South Australia, which has lately increased in an extraordinary degree; in 1836, its population was only two hundred, and in 1839, eight thousand two hundred and fifty, principally owing to the encouragement held out by the funds derived from the sale of lands. Adelaide and Port Lincoln are the two principal points, and the latter, according to the accounts of those interested, "has every advantage under the sun!"

The journey over land from Yass to Adelaide, it is said, can be performed with cattle and sheep in sixty days. The schemers here are as eager and enthusiastic for improvements as with us, and among them much stress is laid upon the difficulties of inland transportation, to overcome which, as the country is too thinly settled for railroads, the introduction of camels from Africa has been proposed; and it is believed that the country is well adapted to them.

The population of New South Wales, by the census of 1841, including 26,967 convicts, was 130,856. A very minute table of the census of the above year will be found in Appendix XIII.

I was not able to obtain any accurate statistical returns of the three other colonies; both North and South Australia are rapidly increasing, particularly the latter: fifteen thousand is believed to be the total amount of their population and that of West Australia.

To return again to the squadron and our occupations. By the 18th December, I had finished my observatory duties, and feeling deeply sensible of the great kindness and attention we had received from not only the Governor, military, and civil officers, but from all the society, I gave a lunch at Fort Macquarie previous to delivering it up, and had the honour of entertaining those who had received us so warmly. Although the weather was unpropitious, many of our friends presented themselves; the affair passed off in great good-fellowship, and we had the satisfaction of seeing our guests retire apparently gratified. Owing to the weather, the number of ladies was not so great as we had hoped, but there were enough to add dancing to the other amusements of the occasion. Many patriotic toasts were exchanged, and an effect seldom witnessed produced on the company by the picture drawn by the Polish Count Strezleski, (well known in our country,) of the reception his destitute countrymen had met with on our shores, and the liberality of our government in providing for them. Those who heard his statement will not soon forget the thrill it produced.

During our stay at Port Jackson, our vessels were much visited by all classes; and a great many inquiries made respecting our accommodations, &c. All seemed disappointed at not being able to see the same complete outfits in our vessels as they had seen described in the published accounts of those of the English expedition commanded by

Captain James Ross. They inquired, whether we had compartments in our ships to prevent us from sinking? How we intended to keep ourselves warm? What kind of antiscorbutic we were to use? and where were our great ice-saws? To all of these questions I was obliged to answer, to their great apparent surprise, that we had none, and to agree with them that we were unwise to attempt such service in ordinary cruising vessels; but we had been ordered to go, and that was enough! and go we should. This want of preparation certainly did not add to the character for wisdom of our government, with this community; but they saw us all cheerful, young, and healthy, and gave us the character, that I found our countrymen generally bear, of recklessness of life and limb. The tender Flying-Fish excited their astonishment more than the ships, from her smallness and peculiar rig; and, altogether, as a gentleman told me, most of our visiters considered us doomed to be frozen to death. I did not anticipate such a fate, although I must confess I felt the chances were much against us, in case we were compelled to winter within the Antarctic. From every calculation, we could not stow quite twelve months' provision, even upon short allowance; our fuel was inadequate to last us more than seven months, and the means of protecting ourselves in the ships for winter quarters, were any thing but sufficient. My mind naturally suffered a great deal of anxiety on all these points, and I felt myself not a little depressed by it, particularly when I considered the state of the Peacock. The carpenter of that ship, shortly after our arrival at Sydney, had reported to her commander, Captain Hudson, that the whole of her upper-works were rotten, and required a survey. The vessel was quietly examined into without holding one, and her state was found even worse than represented. I had many long consultations with Captain Hudson, and found it was impossible to put upon her the necessary repairs, without her giving up the southern cruise. We made up our minds that it was absolutely necessary for the credit of the Expedition and the country for her to perform it; for we were well satisfied that improper imputations and motives, would be ascribed to us, if she did not, and was detained undergoing repairs, in a state of inactivity, during the season for operations in the high southern latitudes. The necessity I felt of subjecting so many lives in so unworthy a ship, caused me great anxiety during the whole cruise. The official papers forwarded to the Secretary of the Navy, upon this subject, will be found in Appendix XXI.

All the vessels underwent the necessary repairs of calking, &c., and the Flying-Fish was furnished with two new masts of the Kaurie pine of New Zealand, some feet shorter and larger in diameter than her former ones.

A few days before our departure, the British frigate Druid, Lord John Russel, commander, arrived from England, with Captain Hobson, R. N., the new Governor (under the name of consul) for New Zealand. He was accompanied by a large retinue, and also had all kinds of facilities for his permanent and comfortable establishment there, among which was a house in frame.

The season of our visit to Sydney, was that of their summer, (December,) and it was somewhat difficult for us to realize the luxuriance of vegetation about us. We could hardly become familiar with windows and doors entirely open at Christmas time. Although it was properly the out-of-town season, we found much gaiety existing, and we have great pleasure in acknowledging the attentions and civilities extended to us during the whole of our stay.

The facilities for outfits here are such as are not to be found elsewhere in the Pacific. The mechanics are good, but as artisans are scarce their wages are exorbitant, and the employer is, for the most part, compelled to put up with their demands. From our experience, we inferred they are not to be depended on, and require to be well watched to obtain the requisite quantity of labour from them. Their rations of grog were always a stipulation made by them, and had to be complied with.

During our stay here, our men behaved well. They all received leave in their turn to visit the shore, and I felt gratified in not having a single case reported to me of bad behaviour on shore.

As our departure drew near, one and all of us felt and expressed regret at leaving such kind friends. In very many places and families, we had found ourselves at home, and were always received with that kindness that showed us we were welcome. The seasons, with many other things, may be reversed, yet the hospitality of old England is found here as warm and fresh as ever it was in the parent land. It would be impossible to mention all those to whom we feel indebted for various kindnesses and attentions, or even to cite those from whom the Expedition received many accessions to its collections. Notwithstanding I have mentioned many things that have struck us as requiring great reform, yet the whole impression left on my mind is, that it is a glorious colony, which the mother country, and the whole Anglo-Saxon race, may well be proud of, and that it ought to claim much more attention than it apparently does, from the home government.

After writing our farewell letters, we took our Christmas dinner with many of our friends, and on the morning of the 26th December, at six o'clock—the very day that had been set apart for my departure, before sailing from the United States—we weighed our anchors and stood down the bay. The day was fine, the breeze light and contrary, and we did not get to sea till the afternoon. When we were about passing the Heads, our worthy consul and some others of our countrymen took their leave, and by way of dispelling the gloom that was naturally felt at parting, and to show the good wishes entertained for their welfare, we gave them at parting several hearty cheers, and then bore away on our course.

It falling calm, the Vincennes and tender were obliged to anchor between the Heads. The Peacock and Porpoise succeeded in getting outside, and when the tide made, we weighed and stood after them. On getting to sea, although every search had been previously made by the master-at-arms, I learned that there were two strangers on board, who had contrived to evade his watchfulness, and on beating to quarters, and mustering the crew, they were among the forthcoming. Their appearance was any thing but convict-like; but I felt after all the attentions heaped upon us, it was seemingly but an ungrateful return, to appear to have committed an infraction of their laws, and this after I had received intimation that an attempt would be made through us, to effect desertion among the troops. From their appearance and carriage I thought they showed the drill of soldiers, and at once told them and the assembled crew, that they were mistaken if they expected to be harboured as such, and that on my return from the south, I should send them back to Sydney to be delivered over. I then entered them on the rolls for provisions only, until I ascertained whether they were entitled to receive compensation; and after telling the men they must look forward to a hard and dangerous cruise, and saying a few words relative to what was expected of them by the country and myself, I enjoined upon them the necessity of economy in their food and clothing, in aiding me in my endeavours to promote their health and comfort. We then piped down, and set about preparing the ship for the Antarctic cruise, the events of which will be detailed in the following chapters.

DAISY BANK.

CHAPTER IX.

CONTENTS.

PRELIMINARY REMARKS—PROCEEDINGS OF THE SQUADRON TO THE SECOND OF JANUARY—SEPARATION OF THE FLYING-FISH AND PEACOCK—FIRST ICEBERG SEEN—FALL IN WITH THE ICY BARRIER—PORPOISE LOST SIGHT OF—PROCEEDINGS OF THE PEACOCK FROM THE THIRD OF JANUARY—HER VISIT TO MACQUARIE'S ISLAND—FIRST ICEBERG SEEN BY HER—SHE FALLS IN WITH THE ICY BARRIER—PROCEEDINGS OF THE PORPOISE—SEA-ELEPHANTS SEEN AND TAKEN—LAND SEEN FROM ALL THE VESSELS—PROCEEDINGS OF THE VINCENNES FROM THE SIXTEENTH OF JANUARY—SHE ENTERS A DEEP BAY IN THE BARRIER—PEACOCK AND PORPOISE SEEN—PEACOCK SPOKEN—LAND DISTINCTLY SEEN FROM THE VINCENNES AND PEACOCK—FIGHT BETWEEN A WHALE AND KILLER, SEEN FROM THE PEACOCK—PROCEEDINGS OF THE PEACOCK—SOUNDINGS OBTAINED BY HER—HER PERILOUS SITUATION AND PROVIDENTIAL ESCAPE—HER CRIPPLED CONDITION—CAPTAIN HUDSON RESOLVES TO RETURN—HIS ADMIRABLE CONDUCT, AND THAT OF HIS OFFICERS AND CREW.

CHAPTER IX.

ANTARCTIC CRUISE.

1840.

The subjects of which I am about to treat in the following chapters are exclusively nautical. I shall therefore adopt in treating them more of the form of a log-book, and follow the daily order of their occurrence with more strictness than I have hitherto considered necessary. This will be done in order to illustrate more fully the nature of the remote regions we traversed, and for the purpose of giving a more exact relation of the incidents of this part of our cruise,—incidents that I cannot but hope have made this part of our labours particularly interesting to all of our countrymen who possess a feeling of national pride.

The credit of these discoveries has been claimed on the part of one foreign nation, and their extent, nay, actual existence, called into question by another; both having rival expeditions abroad, one at the same time, the other the year succeeding.

Each of these nations, with what intent I shall not stop to inquire, has seemed disposed to rob us of the honour by underrating the importance of their own researches, and would restrict the Antarctic land to the small parts they respectively saw. However willing I might be in a private capacity to avoid contesting their statements, and let truth make its own way, I feel it due to the honour of our flag to make a proper assertion of the priority of the claim of the American Expedition, and of the greater extent of its discoveries and researches.

That land does exist within the Antarctic Circle is now confirmed by the united testimony of both French and English navigators. D'Urville, the celebrated French navigator, within a few days after land was seen by the three vessels of our squadron, reports that his boats landed on a small point of rocks, at the place (as I suppose) which appeared accessible to us in Piner's Bay, whence the Vincennes was driven by a violent gale; this he called Clarie Land, and testifies to his belief of the existence of a vast tract of land, where our view of it has left no doubt of its existence. Ross, on the other hand, penetrated to the latitude of 79° S. in the succeeding year, coasted for some distance along a lofty country connected with our Antarctic Continent, and establishes beyond all cavil the correctness of our assertion, that we have discovered, not a range of detached islands, but a vast Antarctic Continent. How far Captain Ross was guided in his search by our previous discoveries, will best appear by reference to the chart, with a full account of the proceedings of the squadron, which I sent to him, and which I have inserted in Appendix XXIV. and Atlas. Although I have never received any acknowledgment of their receipt from him personally, yet I have heard of their having reached his hands a few months prior to his Antarctic cruise. Of this, however, I do not complain, and feel only the justifiable desire to maintain the truth in relation to a claim that is indisputable. The following narrative must, I feel satisfied, leave no doubt in any unprejudiced mind of the correctness of the assertion that we have discovered a vast continent; but I would ask in advance, who was there prior to 1840, either in this country or in Europe, that had the least idea that any large body of land existed to the south of New Holland? and who is there that now doubts the fact, whether he admits it to be a vast continent, or contends that it is only a collection of islands?

Examine all the maps and charts published up to that time, and upon them will any traces of such land be found? They will not, and for the very best of reasons—none was known or even suspected to exist. We ourselves anticipated no such discovery; the indications of it were received with doubt and hesitation; I myself did not venture to record in my private journal the certainty of land, until three days after those best acquainted with its appearance in these high latitudes were assured of the fact; and finally, to remove all possibility of doubt, and to prove conclusively that there was no deception in the case, views of the same land were taken from the vessels in three different positions, with the bearings of its peaks and promontories, by whose intersection their position is nearly as well established as the peaks of any of the islands we surveyed from the sea.

All doubt in relation to the reality of our discovery gradually wore away, and towards the close of the cruise of the Vincennes along the icy barrier, the mountains of the Antarctic Continent became familiar and of daily appearance, insomuch that the log-book, which is guardedly silent as to the time and date of its being first observed, now speaks throughout of "the land."

After leaving Sydney we had, until the 31st December,* fine weather and favourable winds. We took advantage of these, and all sail was crowded on the vessels of the squadron. At the above date we had reached the latitude of 43° S.

Under such circumstances, the usual order of sailing, in a line abreast, was easily maintained, and the communications between the vessels were frequent. On the 31st of December, I issued the sailing instructions for the cruise, which will be found in Appendix XXV.

During this favourable weather, all hands were employed in tightening the ports, in order to secure the interior of the vessels as much as possible from the cold and wet, which were to be apprehended in the region to which we were bound. For this purpose, after calking all the openings, the seams were covered with tarred canvass, over which strips of sheet-lead were nailed. The sailors exhibited great interest in these preparations, and studiously sought to make every thing snug; all useless articles were stowed away in the hold, for we were in truth full to overflowing, and places at other times sacred were now crowded.

It was fortunate that the weather for the first few days was so favourable; for so full was every place, that we had been compelled to stow bread in the launch and cutter, and this in bulk; for the quantity was so much beyond that which had been carried on any former occasion, that a sufficient number of bags were not to be had, and in the hurry of its reception on board, time had not been found to provide them. Every ounce of bread thus exposed was looked to with solicitude, for there was a chance that all of it might be needed.

Among other preparations, rough casings of boards were built around all the hatches, having doors furnished with weights and pulleys, in order to insure that they should not be left open. Having thus provided for the exclusion of cold air, I contented myself with preparations for keeping the interior of the vessel at a temperature no higher than 50°. I deemed this preferable to a higher temperature, in order to prevent the injurious effects which might be produced by passing suddenly from below to the deck. I conceived it far more important to keep the air dry than warm, particularly as a lower temperature would have the effect of inducing the men to take exercise for the purpose of exciting their animal heat.

* During the 29th, 30th, and 31st December, the sea was very phosphorescent; temperature 56°.

Aware that warm and dry clothing was an object of the first importance, inspections of the men's feet and dress were held morning and evening, in which the wearing of a suitable number of garments was insisted upon, as well as the greatest personal cleanliness. With the same views, the drying-stoves were particularly attended to; and that every part under deck might be effectually and quickly freed of moisture, additional stoves had been procured at Sydney. Thermometers were hung up in proper places, and frequently consulted, in order by following their indications to secure an equable temperature, and at the time to ascertain when the use of stoves might be dispensed with, in whole or in part. The latter was an important consideration, for we were under the necessity of husbanding our stock of fuel, by expending it only when absolutely necessary.

We also took advantage of the fine weather to bend all our best sails, and to shift our top-gallant masts.

The 1st January was one of those days, which are termed, both at sea and on shore, a weather-breeder. The sea was smooth and placid, but the sky was in places lowering, and had a wintry cast, to which we had long been strangers; the temperature shortly began to fall, the breeze to increase, and the weather to become misty. In a few hours we were sailing rapidly through the water, with a rising sea, and by midnight it was reported that the tender Flying-Fish was barely visible. I shortened sail, but it was difficult to stop our way; and on the morning of the 2d of January, the fog was dense, and the Peacock and Porpoise only were in sight; we hove-to, and the Peacock and Porpoise were ordered to stand east and west, in order to intercept the tender, but they returned without success; we also fired guns in hopes of being heard. In the afternoon, I deemed it useless to wait any longer for her, and that I must take the chance of falling in with her at Macquarie Island, our first appointed place of rendezvous,—a visit to which I had flattered myself might have been avoided, but which it became necessary now to make. We accordingly proceeded on our course for that island, with all sail set. This separation of the tender took place in the latitude of 48° S., and she was not again seen until our return. The officers and crew were not slow in assigning to the Flying-Fish a similar fate with her unfortunate mate, the Sea-Gull. Men-of-war's men are prone to prognosticate evil, and on this occasion they were not wanting in various surmises. Woful accounts were soon afloat of the distress the schooner was in when last seen,—and this in quite a moderate sea.

The barometer now began to assume a lower range, and the temperature to fall below 50°. On the 3d, the fog continuing very hick, the Peacock got beyond hearing of our horns, bells, drums, and guns, and was parted with. This, however, I did not now regret so much, as it was of little consequence whether we sought one or two vessels at our rendezvous, although it might cause a longer detention there.

The wind was now (5th January) veering rapidly to the northwest, with some thunder and lightning, and we in consequence expected the wind to haul to the southwest, but to my surprise, it went back to the northeast, with thick rainy weather. This return of the wind to its old quarter followed a fall of the barometer to 29·60 in., and in a few hours afterwards to 29·30 in., while the weather continued moderate; a large number of albatrosses, Port Egmont hens, and petrels, were seen.

For the last few days we were unable to get any observations, but on the 6th we were favoured with a sight of the sun, and found ourselves in the latitude of 53° 30′ S., and longitude 157° 35′ E. Our variation had increased to fifteen and a half degrees easterly. This being a fine day, we completed our calking, and the more effectual securing of the ship. At midnight we were about fifty miles from Macquarie Island.

The morning of the 7th was misty, with squally weather. A heavy sea rising, and a strong gale setting in, we lost sight of the Porpoise for a few hours. Being unable to see beyond an eighth of a mile, it was thought imprudent to run, for fear of passing the island, and we hove-to to await its moderating. It cleared at noon, and we obtained an observation, by which we found ourselves in latitude 54° 20′ S., and longitude 160° 47′ E. I found that we had been carried to the eastward upwards of twenty miles in less than eighteen hours; this, with the wind hauling to the southwest, brought us to leeward of the island, and the sea and wind increasing, I saw it was useless to attempt to reach it without great loss of time. I therefore bore off to the southward for our second rendezvous, Emerald Island, or its supposed locality.

On the morning of the 8th, the wind, which continued from the same quarter, with heavy cumulous clouds, began to moderate, and we were enabled to make more sail. By our observations, we found a current setting to the southeast, of one mile an hour. Our longitude was 162° 13′ E., latitude 55° 38′ S. The barometer stood at 30·00 in.; the temperature had fallen to 38°; and this change, on account of the rawness of the air, was much felt by the crew.

During the 9th we passed the site of Emerald Isle, situate, as has been stated, in latitude 57° 15′ S., and longitude 162° 30′ E., but saw nothing of it, nor any indications of land, which I therefore infer does not exist in the locality where it is laid down. We again experienced the southeast current of twenty miles a day. Our variation had increased to twenty-two degrees easterly. Making our course with all sail set, the Porpoise in company, we passed to-day some pieces of kelp. The temperature continued at 38°. Numerous flocks of gray petrels around us.

The 10th we encountered the first iceberg, and the temperature of the water fell to 32°. We passed close to it, and found it a mile long, and one hundred and eighty feet in height. We had now reached the latitude of 61° 08′ S., and longitude 162° 32′ E. The current to-day set in the same direction as before, about half a mile per hour. The second iceberg seen was thirty miles, and the third about fifty-five miles south of the first. These ice-islands were apparently much worn by the sea into cavities, exhibiting fissures as though they were ready to be rent asunder, and showed an apparent stratification, much inclined to the horizon. The weather now became misty, and we had occasionally a little snow. I congratulated myself that we had but few on the sick-list, and all were in high spirits at the novelty of the cruise. We continued to meet icebergs of different heights, some of which, though inclined to the horizon, had a plane upper surface.

11th. The fair wind from the northwest, (accompanied with a light mist, rendering objects on the horizon indistinct,) still enabled us to pursue our course southerly. Icebergs became so numerous as to compel us occasionally to change our course. They continued of the same character, with caverns worn in their perpendicular sides, and with flat tops, but the latter were now on a line with the horizon. Towards 6 P. M., we began to perceive smaller pieces of ice, some of which were not more than an eighth of a mile in length, floating as it were in small patches. As the icebergs increased in number, the sea became smoother, and there was no apparent motion. Between 8 and 9 P. M., a low point of ice was perceived ahead, and in a short time we passed within it. There was now a large bay before us. As the vessels moved rapidly, at 10½ P. M. we had reached its extreme limits, and found our further progress entirely stopped by a compact barrier of ice, enclosing large square icebergs. The barrier consisted of masses closely packed, and of every variety of shape and size. We hove-to until full daylight. The night was beautiful, and every thing seemed sunk in sleep, except the sound of the distant and low rustling of the ice, that now and then met the ear. We had now reached the latitude of 64° 11′ S., longitude 164° 30′ E., and found our variation twenty-two degrees easterly. One and all felt disappointed, for we had flattered ourselves that the way was open for further progress to the southward, and had imbibed the impression (from the extraordinary weather we had had at Sydney, and the reports of icebergs having been seen farther to the northward than usual, by all the vessels arriving) that the season would be an open one. What surprised me most was a change in the colour of the water to an olive-green, and some faint appearances resembling distant land; but as it was twilight, and I did not believe the thing credible, I put no faith in these indications, although some of the officers were confident they were not occasioned by icebergs. The barometer stood at 29·200 in.; the temperature of the air 33°, water 32°. We lay-to until four o'clock. As it grew light, on the 12th, a fog set in so thick that we lost sight of the Porpoise, and could not hear any answer to our signals. I therefore determined to work along the barrier to the westward.

We were all day beating in a thick fog, with the barrier of ice close to us, and occasionally in tacking brought it under our bow; at other times we were almost in contact with icebergs. During the whole day we could not see at any time further than a quarter of a mile, and seldom more than the ship's length. The fog, or rather thick mist, was forming in ice on our rigging. From the novelty of our situation, and the excitement produced by it, we did not think of the danger.

I shall now leave the Vincennes and Porpoise pursuing their course to the westward with a head wind, and bring the Peacock up to the barrier.

Previously to parting company on the 3rd of January, the crew of that ship had also been engaged in building hurricane-houses, calking, and chintzing, to secure them from the wet and cold. After parting company, Captain Hudson immediately steered for the first rendezvous, Macquarie Island, and was more fortunate than we were in reaching it, although the Peacock had experienced the same kind of weather that we had, and currents setting to the eastward.

MACQUARIE ISLAND.

On approaching the island, they discovered large patches of kelp, and saw numerous procellaria and albatrosses about the ship. On the 10th of January they made the island, and observed a reef of rocks extending three quarters of a mile off its south end. Passing within a short distance of it, they did not observe any of the signals of the squadron flying as they had anticipated. They, notwithstanding, stood in, lowered a boat, and despatched several officers to put up the signal, make experiments, and collect specimens. The boat approached an indentation on the west side, too open to be called a bay, and found that the surf was running high, and beating with great violence against the rocks, which, together with the kelp, rendered it dangerous to attempt landing. They made for several other places which looked favourable at a distance, but on approaching them, they were found even less accessible. The boat then returned to the first place to make another attempt, which was attended with great difficulty. The boat's anchor was dropped, and she was backed in with great caution to the edge of the rollers; the surf was very high, and rolled in with a noise like thunder, breaking furiously upon the rocks, so as to make the boat fairly tremble, and threatening every moment to overwhelm her; once or twice she was prevented from getting broadside-to by hauling out towards her anchor. At length, after a dozen fruitless attempts, and awaiting a favourable opportunity, Mr. Eld and a quarter-master succeeded in getting ashore, but not without being immersed up to their breasts. It was found impossible to land any instruments; and the quarter-master was despatched to erect the necessary signals, while Mr. Eld proceeded to visit the penguin-rookery not far distant. On approaching the island, it had appeared to be covered with white spots: these excited conjecture; but after landing, the exhalations rendered it not long doubtful that it was birdlime.

Mr. Eld, in his journal, gives the following account of his visit: "Although I had heard so often of the great quantity of birds on the uninhabited islands, I was not prepared to see them in such myriads as here. The whole sides of the rugged hills were literally covered with them. Having passed a deep fissure in the rocks, I ascended a crag that led to what I thought was their principal roost, and at every step my astonishment increased. Such a din of squeaking, squalling, and gabbling, I never before heard or dreamed could be made by any of the feathered tribe. It was impossible to hear one's self speak. It appeared as if every one was vying with his neighbour to make the greatest possible noise. I soon found my presence particularly displeased them, for they snapped at me in all directions, catching hold of my trousers, shaking and pinching my flesh so violently as to make me flinch and stand upon the defensive. As we wanted a number of specimens, I commenced kicking them down the precipice, and knocked on the head those which had the temerity to attack me. After having collected a number, and a few eggs, I laid them aside, whilst I ascended higher on the hill. I had not left them more than eighteen feet, before two albatrosses came down, and commenced picking at the dead birds I had just killed, but not being able to make any impression upon them, deliberately picked up two of the eggs with their beaks, and in spite of my efforts to prevent it, flew away with them. The eggs were about the size of a goose's; the original colour seemed to have been white, but they were so dirty that it was difficult to say with certainty. They were no doubt the eggs of the penguin, as I took them out of their nest, which was only a small place scratched in the earth, just big enough to hold one or two eggs, with little or no grass, sticks, or any thing else to form a nest of. I afterwards picked up a number of these eggs, and another was found, of the size of a hen's egg, white, with a slight tinge of green. On mounting the hill still higher, which was very steep, and composed of volcanic rock, loose stones, and a little soil mixed with birdlime, I found that there were more of these birds than I anticipated. The nests were within two feet of each other, with one or two young ones in each; one of the old ones watching and sitting on the nest, whilst the young were trying ineffectually to nestle themselves under the small wings of the old ones. The appearance of the young was not unlike that of goslings, being covered with a dark thick down.

"These penguins are the Eudyptes chrysocome; they are from sixteen to twenty inches in height, with white breast and nearly black back, the rest being of a dark dove-colour, with the exception of the head, which is adorned on each side with four or five yellow feathers, three or four inches long, looking like graceful plumes. The birds stand erect in rows, which gives them the appearance of Liliputian soldiers. The sight was novel and beautiful, and had it not been for the gabble,—enough to deafen me,—I could have stayed much longer. It was now time to return to the boat, when it occurred to me that live birds would be preferable to the dead; so throwing the latter down, I seized one old and a couple of young ones, and with three or four eggs in my cap, made the best of my way to the boat. It was now found impossible to hand them on board, and not willing to surrender my prize, a lead-line was thrown me from the boat, but did not come near enough, and in my attempts to get it, I was overtaken by a sea, and was thrown violently against the rocks among the kelp, and just made out to crawl on hands and knees beyond the reach of the returning sea, somewhat bruised, wet, and benumbed with the cold."

At this juncture, the quarter-master returned with a large species of penguin over his shoulders, but without the crown of feathers on his head. He described a similar rookery, and also saw some green paroquets with a small red spot on the head, and an oblong slaty or purple spot at the root of the bill, and with straight beaks. Mr. Eld was too much exhausted to return with him to get specimens, and the hour being late, it was necessary to return to the boat, which had been waiting for some time for them. The quarter-master succeeded in getting his penguins to the boat, but Mr. Eld's began floundering about, and although their legs were tied, managed to get into the water, where they were at home, and were soon out of reach. The tying of the legs did not seem any impediment to their exertions in the water, and thus several interesting specimens of natural history were lost, the trouble that it cost making them doubly valuable. With great difficulty Mr. Eld reached the boat; for, having again missed his foothold, he fell among the kelp, but by the timely aid of those on board he was rescued. After an hour's tug at their oars, they reached the ship in safety. During their absence the ship sounded with a line of three hundred fathoms, two and a half miles from the shore; but no bottom was found. The temperature of the water at the surface was 43°, and at three hundred fathoms deep 39°. The current was tried, but none found.

The south end of Macquarie Island lies in latitude 54° 44′ S., and longitude 159° 49′ E. The island is high and much broken; it is apparently covered with verdure, although a long tufted rank grass was the only plant seen by those who landed.

The highest peak on the island is from twelve to fifteen hundred feet high, and as far as our observations extended, it had neither tree nor shrub on it. At 6 P. M. the ship filled away, and at eight was abreast of the Bishop and Clerk. Macquarie Island affords no inducement for a visit, and as far as our examination went, has no suitable place for landing with a boat. The only thing I had to regret was not being able to make it a magnetic station.

On the 11th and 12th nothing particular occurred on board the Peacock. All sail was set, and running to the southward on the 13th, in latitude 61° 30′ S., longitude 161° 05′ E., the first ice-islands were seen. The dip was observed with Lloyd's and Dolland's needles, which made it 86° 53′.

There was no occasion on the night of the 13th to light the binnacle-lamps, as newspaper print could be read with ease at midnight. On the 14th, while still making much progress to the south, and passing occasionally icebergs and brash ice, the water appeared somewhat discoloured. Robinson's, Lloyd's, and Dolland's needles, gave, the same day, in the cabin, 86° 37′ for the dip, and in the ward-room, 86° 46′. Albatrosses, Cape pigeons, and other birds about.

On the 15th, they passed many ice-islands. The weather was thick and snow fell at intervals; the wind continued from the westward. Many whales were seen; albatrosses, petrels, and Cape pigeons were frequent about the ship. At 4 P. M., the mist raised a little, and to their surprise they saw a perfect barrier of ice, extending to the southwest, with several large icebergs enclosed within it. Shortly after, they discovered a sail, which proved to be the Porpoise.

The Vincennes and Porpoise were left in our narrative near the icy barrier, separated by the fogs and mists that prevailed at times. The Porpoise, on the 13th, in latitude 65° 08′ S., longitude 163° E., discovered several sea-elephants on the ice, and sent a boat to capture them, but without success. The current was tried, and found to set west one-fifth of a mile per hour. Some time afterwards, seeing some sea-elephants near the edge of the ice, a boat was sent, and succeeded in capturing a female. From the numerous sea-elephants, and the discoloration of the water and ice, they were strongly impressed with the idea of land being in the vicinity, but on sounding with one hundred fathoms, no bottom was found; Lieutenant-Commandant Ringgold felt convinced, from the above circumstances, and the report that penguins were heard, that land was near, and thought he could discern to the southeast something like distant mountains. A nearer approach was impossible, as they were then in actual contact with the icy barrier.

On the 14th, at 3 P. M., the water being still discoloured, tried soundings, but found no bottom.

Two sea elephants were seen lying motionless on the ice. On being shot at, the animal would raise its head and look around for an instant, and then resume its former posture. Boats were lowered, when they were captured and brought on board: they proved to be the Phoca proboscidea. Dr. Holmes examined their stomachs, and found nothing but well-digested food. Their dimensions were as follows:

Total length	10 feet, 9 inches.
Length of posterior flipper	1 " 9 "
Breadth	2 " 4 "
Circumference of largest part of body . . .	6 " 3 "

This was a young female. The other was taken afterwards; he measured—

In length	8 feet, 6 inches.
Greatest circumference behind anterior flipper .	5 " 0 "
Length of flippers	1 " 5 "
Breadth "	1 " 5 "

On the 15th the Peacock and Porpoise were in company: the specimens of sea-elephants were put on board the Peacock; and, after having had communication with each other, the vessels again separated, standing on opposite tacks.

On the 16th the three vessels were in longitude 157° 46′ E., and all within a short distance of each other. The water was much discoloured, and many albatrosses, Cape pigeons, and petrels were seen about the ships. On board the Vincennes, we sounded with two hundred and thirty fathoms, and found no bottom; the water had the appearance of an olive-green colour, as if but forty and fifty fathoms deep. At the surface, its temperature was 32°, at the depth sounded, 31°. I should have tried for a deeper cast, but the line was seen to be stranded, when we were obliged to stop; we fortunately saved our apparatus, with Six's thermometers.

On this day (16th January) appearances believed at the time to be land were visible from all the three vessels, and the comparison of the three observations, when taken in connexion with the more positive proofs of its existence afterwards obtained, has left no doubt that the appearance was not deceptive. From this day, therefore, we date the discovery which is claimed for the squadron.

On board the Peacock, it appears that Passed Midshipmen Eld and Reynolds both saw the land from the masthead, and reported it to Captain Hudson: he was well satisfied on examination that the appearance was totally distinct from that of ice-islands, and a majority of the officers and men were also satisfied that if land could exist, that was it.

I mention particularly the names of these two gentlemen, because they have stated the same fact under oath, before the court-martial, after our return.

On board the Porpoise, Lieutenant-Commandant Ringgold states, that "he went aloft in the afternoon, the weather being clear and fine, the horizon good, and clouds lofty; that he saw over the field-ice an object, large, dark, and rounding, resembling a mountain in the distance; the icebergs were all light and brilliant, and in great contrast." He goes on to say, in his report, "I watched for an hour to see if the sun in his decline would change the colour of the object: it remained the same, with a white cloud above, similar to that hovering over high land. At sunset the appearance remained the same. I took the bearings accurately, intending to examine it closely as soon as we got a breeze. I am thoroughly of opinion it is an island surrounded by immense fields of ice. The Peacock in sight to the southward and eastward over the ice; the sun set at a few minutes before ten; soon after, a light air from the southward, with a fog-bank arising, which quickly shut out the field-ice."

In Passed Midshipman Eld's journal, he asserts that he had been several times to the masthead during the day, to view the barrier; that it was not only a barrier of ice, but one of terra firma. Passed

Midshipman Reynolds and himself exclaimed, with one accord, that it was land. Not trusting to the naked eye, they descended for spy-glasses, which confirmed, beyond a doubt, their first impressions. The mountains could be distinctly seen, over the field-ice and bergs, stretching to the southwest as far as any thing could be discerned. Two peaks, in particular, were very distinct, (which I have named after those two officers,) rising in a conical form; and others, the lower parts of which were quite as distinct, but whose summits were lost in light fleecy clouds. Few clouds were to be seen in any other direction, for the weather was remarkably clear. The sun shone brightly on ridge after ridge, whose sides were partially bare; these connected the eminences I have just spoken of, which must be from one to two thousand feet high. Mr. Eld further states, that on reporting the discovery to Captain Hudson, the latter replied that there was no doubt of it, and that he believed that most of the icebergs then in sight were aground. At this time they were close in with the barrier, and could approach no nearer. On this day, the Peacock got a cast of the deep-sea lead, with Six's thermometer attached, to the depth of eight hundred and fifty fathoms, only a short distance from the barrier: the temperature of the surface was 31°, and at the depth sounded, $31\frac{1}{2}$°; current one-fourth of a mile, north-by-east.

The log-book of the Porpoise has also this notice in it: "From six to eight, calm and pleasant,—took in studding-sails; at seven set main-topgallant-studding-sail; discovered what we took to be an island, bearing south-by-east,—a great deal of field-ice in sight; noticed penguins around the brig. (Signed) J. H. North." Dr. Holmes, on the same evening, noted in his journal, a marked appearance of land.

On board the Vincennes there was on the same day much excitement among the crew. All eagerly watched the flight of birds, together with the whales and penguins, and spoke of the proximity of land, which, from the appearance of never-failing signs, could scarcely be doubted. The following is a sketch which I made of what I myself saw, and have called Ringgold's Knoll on the chart, and which at the same time will show the field-ice* as it appeared.

* The field-ice is composed of a vast number of pieces, varying in size, and separated from one another, the long swell keeping the outer ones always in motion. The smallest pieces were about six feet in diameter, while the largest sometimes exceeded five or six hundred feet. Their depth below the surface varies still more, and some appear to be soft, whilst others were hard and compact. The depth of these does not probably in any case exceed twenty feet. Most of them, and particularly the larger ones, had a covering of about eighteen inches of snow. The whole at a distance appeared like a vast level field, broken up as it were by the plough, and presenting shapeless angular masses of every possible figure, while here and there a table-topped iceberg was enclosed.

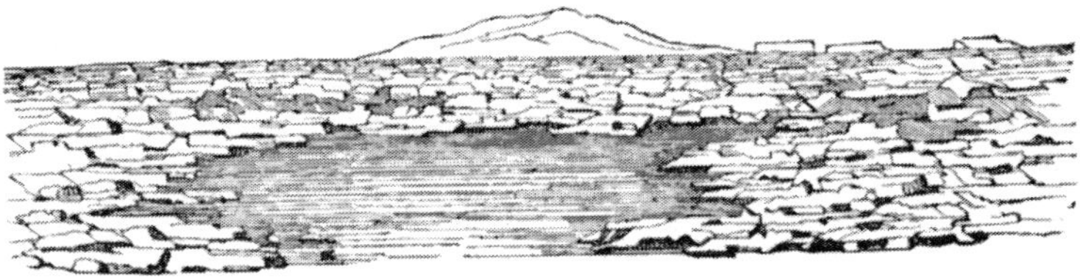

SKETCH OF LAND AND FIELD-ICE.

This night we were beating with frequent tacks, in order to gain as much southing as possible. Previous to its becoming broad daylight, the fog rendered every thing obscure, even at a short distance from the ship. I knew that we were in close proximity to icebergs and field-ice, but, from the report of the look-out at sunset, believed that there was an opening or large bay leading to the southward. The ship had rapid way on her, and was much tossed about, when in an instant all was perfectly still and quiet; the transition was so sudden that many were awakened by it from sound sleep, and all well knew, from the short experience we had had, that the cessation of the sound and motion usual at sea, was a proof that we had run within a line of ice, —an occurrence from which the feeling of great danger is inseparable. The watch was called by the officer of the deck, to be in readiness to execute such orders as might be necessary for the safety of the ship. Many of those from below were seen hurrying up the hatches, and those on deck straining their eyes to discover the barrier in time to avoid accident. The ship still moving rapidly along, some faint hope remained that the bay might prove a deep one, and enable me to satisfy my sanguine hopes and belief relative to the land.

The feeling is awful and the uncertainty most trying thus to enter within the icy barrier blindfolded as it were by an impenetrable fog, and the thought constantly recurring that both ship and crew were in imminent danger; yet I was satisfied that nothing could be gained but by pursuing this course. On we kept, until it was reported to me, by attentive listeners, that they heard the low and distant rustling of the ice: suddenly a dozen voices proclaimed the barrier to be in sight, just ahead. The ship, which a moment before seemed as if unpeopled, from the stillness of all on board, was instantly alive with the bustle of performing the evolutions necessary to bring her to the wind, which was unfavourable to a return on the same track by which we had entered. After a quarter of an hour, the ice was again made ahead, and the full danger of our situation was realized. The ship was certainly embayed; and although the extent of sea-room to which we were limited, was rendered invisible by the dark and murky weather, yet that we were closely circumscribed was evident from having made the ice so soon on either tack, and from the audible rustling around us. It required several hours to extricate the ship from this bay.

Few are able to estimate the feelings that such an occasion causes to a commander, who has the responsibility of the safety of ship and crew operating as a heavy weight upon his heart, and producing a feeling as if on the verge of some overwhelming calamity. All tends to satisfy him that nothing could guide him in safety through, or shield from destruction those who have been entrusted to his charge, but the hand of an all-wise Providence.

17th. In the morning we discovered a ship apparently within a mile of us, to which we made signal and fired a gun, but she was shortly after lost sight of. We also saw the brig to the eastward, close to the barrier of ice. In the afternoon we spoke the Peacock: she had not seen us in the morning; and I should be disposed to believe that the cause of her image appearing so close to us in the morning was produced by refraction above a low fog-bank; but the usual accompaniment of such phenomena, a difference of temperature below and aloft, did not exist.

I now desired Captain Hudson to make the best use of his time in exploring, as to attempt to keep company would only impede our progress, and, without adding to our safety, might prevent the opportunity of examining the barrier for an opening. I was also satisfied that the separation would be a strong incentive to exertion, by exciting rivalry among the officers and crews of the different vessels. This day at noon we were in latitude 66° 20′ S., longitude 156° 02′ E. Many petrels, albatrosses, a few whales, and a seal, were seen from the ship; and the water was quite green.

18th. The weather this day was variable, with light westerly winds; the temperature of air and water 32°. Occasional squalls of snow and mist occurred, but it was at times clear. The water was still olive-green; and the other vessels occasionally in sight, beating to windward.

On the morning of the 19th, we found ourselves in a deep bay, and discovered the Peacock standing to the southwest. Until eight o'clock, A. M., we had a moderate breeze. The water was of a darker olive-green, and had a muddy appearance. Land was now certainly visible from the Vincennes, both to the south-southeast and southwest, in the former direction most distinctly. Both appeared high. It was between eight and nine in the morning when I was fully satisfied that it was certainly land, and my own opinion was confirmed by that of some of the oldest and most experienced seamen on board. The officer of the morning watch, Lieutenant Alden, sent twice, and called my attention to it. We were at this time in longitude 154° 30′ E., latitude 66° 20′ S.; the day was fine, and at times quite clear, with light winds. After divine service, I still saw the outline of the land, unchanged in form but not so distinct as in the morning. By noon, I found we were sagging on to the barrier; the boats were lowered in consequence, and the ship towed off. The report from aloft, was, "A continued barrier of ice around the bay, and no opening to be seen, having the western point of it bearing to the northward of west of us." I stood to the westward to pass around it, fully assured that the Peacock would explore all the outline of the bay.

The Peacock, at 3^h 30^m, according to Captain Hudson's journal, having got into the drift-ice, with a barrier still ahead to the west, tacked to the southeast to work up for an immense mass, which had every appearance of land, and which was believed to be such by all on board. It was seen far beyond and towering above an ice-island that was from one hundred and fifty to two hundred feet in height. It bore from them about southwest,* and had the appearance of being three thousand feet in height, forming a sort of amphitheatre, looking gray and dark, and divided into two distinct ridges or elevations throughout its entire extent, the whole being covered with snow. As there was no probability of getting nearer to it in this quarter, they stood out of the bay, which was about twenty miles deep, to proceed to the westward, hoping to get an opportunity to approach the object more closely on the other side.

We had a beautiful and unusual sight presented to us this night: the sun and moon both appeared above the horizon at the same time, and each throwing its light abroad. The latter was nearly full. The former illuminated the icebergs and distant continent with his deep golden rays; while the latter, in the opposite horizon, tinged with silvery light the clouds in its immediate neighbourhood. There now being no doubt in any mind of the discovery of land, it gave an exciting interest to the cruise, that appeared to set aside all thought of fatigue, and to make every one willing to encounter any difficulty to effect a landing.

20th. This day, on board the Peacock they witnessed a sea-fight between a whale and one of its many enemies. The sea was quite smooth, and offered the best possible view of the whole combat. First, at a distance from the ship, a whale was seen floundering in a most

* Sketches of this land will be seen in the Atlas on the Chart of Antarctic Continent.

extraordinary way, lashing the smooth sea into a perfect foam, and endeavouring apparently to extricate himself from some annoyance. As he approached the ship, the struggle continuing and becoming more violent, it was perceived that a fish, apparently about twenty feet long, held him by the jaw, his contortions, spouting, and throes all betokening the agony of the huge monster. The whale now threw himself at full length from the water with open mouth, his pursuer still hanging to the jaw, the blood issuing from the wound and dyeing the sea to a distance around; but all his flounderings were of no avail; his pertinacious enemy still maintained his hold, and was evidently getting the advantage of him. Much alarm seemed to be felt by the many other whales around. These "killers," as they are called, are of a brownish colour on the back, and white on the belly, with a long dorsal fin. . Such was the turbulence with which they passed, that a good view could not be had of them to make out more nearly the description. These fish attack a whale in the same way as dogs bait a bull, and worry him to death. They are armed with strong sharp teeth, and generally seize the whale by the lower jaw. It is said that the only part of them they eat is the tongue. The whalers give some marvellous accounts of these killers and of their immense strength; among them, that they have been known to drag a whale away from several boats which were towing it to the ship.

There was a great quantity of animalcula in the water, and some large squids (Medusæ) and quantities of shrimp were frequently seen about the icebergs; these are no doubt the attractions which bring whales to frequent these seas.

The last two days we had very many beautiful snow-white petrels about. The character of the ice had now become entirely changed. The tabular-formed icebergs prevailed, and there was comparatively little field-ice. Some of the bergs were of magnificent dimensions, one-third of a mile in length, and from one hundred and fifty to two hundred feet in height, with sides perfectly smooth, as though they had been chiselled. Others, again, exhibited lofty arches of many-coloured tints, leading into deep caverns, open to the swell of the sea, which rushing in, produced loud and distant thunderings. The flight of birds passing in and out of these caverns, recalled the recollection of ruined abbeys, castles, and caves, while here and there a bold projecting bluff, crowned with pinnacles and turrets, resembled some Gothic keep. A little farther onwards would be seen a vast fissure, as if some powerful force had rent in twain these mighty masses. Every noise on board, even our own voices, reverberated from the massive and pure white walls. These tabular bergs are like masses of beautiful alabaster: a verbal description of them can do little to convey the reality to the imagination of one who has not been among them. If an immense city of ruined alabaster palaces can be imagined, of every variety of shape and tint, and composed of huge piles of buildings grouped together, with long lanes or streets winding irregularly through them, some faint idea may be formed of the grandeur and beauty of the spectacle. The time and circumstances under which we were viewing them, threading our way through these vast bergs, we knew not to what end, left an impression upon me of these icy and desolate regions that can never be forgotten.

22d. It was now, during fine weather, one continued day; but we had occasional snow-squalls that produced an obscurity that was tantalizing. The bergs were so vast and inaccessible, that there was no possibility of landing upon them.

The Peacock and Porpoise were in sight of each other this day. A large number of whales, albatrosses, petrels, penguins, &c., were seen around, and a flock of ducks was also reported as having been seen from the Vincennes, as well as several seals. The effect of sunrise, at a little after 2 A. M., on the 23d, was glorious.

As the events which occurred on board the Peacock during the next few days are particularly interesting, I shall proceed to narrate them in detail, leaving the Vincennes and Porpoise to pursue their route along their dangerous and novel pathway, and would particularly refer the reader to the actual condition of the Peacock, a statement of which has been heretofore given, that it may be borne in mind that our vessels had no planking, extra fastening, or other preparations for these icy regions, beyond those of the vessels of war in our service.

The Peacock stood into the bay which the Vincennes had found closed the day before, and saw the same appearance of high land in the distance. The water was much discoloured, and of a dark dirty green. They hove-to, for the double purpose of getting a cast of the lead, and of lowering the boats to carry the instruments to a small iceberg, on which it was possible to land, for the purpose of making magnetic observations. A line of one thousand four hundred fathoms was prepared to sound, and to the lead was attached the cylinder with Six's thermometer. The wind being fresh, several leads at different distances were attached to the line. They were not aware that the lead-line had touched bottom, until they began to haul in, when it was found that the lead bent on at five hundred fathoms was filled with blue and slate-coloured mud. Attached to the lead also was a piece of stone, and a fresh bruise on it, as though the lead had struck heavily on rock. The remainder of the line had evidently lain on the bottom, as the copper cylinder was covered with mud, and the water inside of it was quite muddy. They then beat up a short distance to windward, and again sounded, when, with the line hanging vertically, bottom was reached at three hundred and twenty fathoms; the matter brought up was slate-coloured mud. The temperature of the water at the surface was 32°, and at the above depth $27\frac{1}{2}$°, being a decrease of $4\frac{1}{2}$°.

The boats now returned, and on approaching the ship the persons in them were much startled by hearing the crew cheer ship in consequence of finding soundings. This was a natural burst of joy, on obtaining this unquestionable proof that what they saw was indeed the land; a circumstance that, while it left no doubt, if any had existed, in the mind of any one on board the Peacock, that what they had previously seen was truly terra firma, furnished a proof that cannot be gainsaid, even by those disposed to dispute the evidence of sight, unsupported by so decisive a fact. Mr. Eld and Mr. Stuart, in the boats, succeeded in getting observations, and the mean dip by the needles was 86° 16′.

Mr. Eld's boat succeeded in taking a king-penguin of enormous size, viz.: from tip of tail to the bill, forty-five inches; across the flippers, thirty-seven inches; and the circumference of the body, thirty-three inches. He was taken after a truly sailor-like fashion, by knocking him down. The bird remained quite unmoved on their approach, or rather showed a disposition to come forward to greet them. A blow with the boat-hook, however, stunned him, and before his recovery he was well secured. He showed, on coming to himself, much resentment at the treatment he had received, not only by fighting, but by an inordinate noise. He was in due time preserved as a specimen, and now graces the collection at Washington. In his craw were found thirty-two pebbles, from the size of a pea to that of a hazel-nut.

24th. Bergs and field-ice were in various directions around. They had light baffling winds, clear and pleasant weather, with a smooth sea. The water was of a dark green colour. Standing into the bay for the purpose of approaching the land, they at 5 A. M. passed through drift-ice into an open space, and when they had again approached the field, hove-to for the purpose of sounding. Here bottom was found at the depth of eight hundred fathoms; and the matter brought up was similar to that obtained the day before. The distance between the points where these two soundings were obtained, was but short.

At 8^h 30^m A. M., while attempting to box off the ship from some ice under the bow, she made a stern-board, which brought the stern so forcibly in contact with another mass of ice, that it seemed from the shock, as if it were entirely stove in; the rudder was so much canted from its position, as to carry away the starboard wheel-rope, and to wrench the neck of the rudder itself in such a manner as to render it unserviceable, or even worse than useless. In hopes of lessening the difficulty, relieving-tackles were applied to the tiller, but without effect, for it was discovered that the rudder had been so far twisted as to make a considerable angle with the keel, and every exertion to move it proved ineffectual.

All hands were now called, and every officer and man was speedily at his station. The ship was found to be rapidly entering the ice, and every effort to direct her course by the management of the sails proved fruitless. In this helpless condition scarcely a moment passed without a new shock in some quarter or other from the ice, and every blow threatened instant destruction. The hope was not yet abandoned, that some temporary expedient might be found to bring the rudder again into use, until they should be extricated from this perilous situation. A stage was, therefore, rigged over the stern, for the purpose of examining into its state, but it was found to be so much injured that it was impossible to remedy its defects while in its place, and preparations were forthwith made for unshipping it. In the mean time the position of the vessel was every instant growing worse, surrounded as she was by masses of floe-ice, and driving further and further into it, towards an immense wall-sided iceberg. All attempts to get the vessel on the other tack failed, in consequence of her being so closely encompassed, and it was therefore thought expedient to attempt to bring her head round, by hanging her to an iceberg by the ice-anchors, and thus complete what had been partially effected by the sails. The anchor was attached, but just at the moment the hawser was passed on board, the ship took a start so suddenly astern, that the rope was literally dragged out of the men's hands before they could get a turn around the bits.

The ship now drove stern foremost into the midst of the huge masses of ice, striking the rudder a second time. This blow gave it the finishing stroke, by nearly wringing off the head, breaking two of the pintles, and the upper and lower brace.

The wind now began to freshen, and the floe-ice to set upon the ship. The sails were furled, and spars rigged up and down the ship's sides as fenders. Attempts were again made to plant the ice-anchors, for which purpose the boats were lowered; but the confined space, and the force with which the pieces of ice ground against each other was so great, that the boats proved nearly as unmanageable as the ship. After much exertion, however, the ice-anchors were planted, and the hawser hauled taut. Here they for a time enjoyed comparative security, as the vessel hung by the anchors, which were planted in a large floe. The ice continued to close in rapidly upon them, grinding, crushing, and carrying away the fenders; and the wind, that had changed to seaward, rose with appearances that foreboded bad weather.

At 10^h 30^m this security was at an end; for the anchors, in spite of the exertions of the officers and men who were near them, broke loose, and the ship was again at the mercy of huge floating masses. A rapid stern-board was the consequence; and a contact with an ice-island, vast, perpendicular, and as high as the mastheads, appeared inevitable.

Every possible preparation was made to meet the expected shock. There was no noise or confusion, and the self-possession and admirable conduct of the commander inspired courage and confidence in all. Preparations were made to cockbill the yards, and spars were got out.

While these preparations were going forward, the imminence of the danger lessened for a while: the anchors again held, and there was a hope that they might bring the vessel up before she struck. This hope, however, endured but for a moment; for the anchors, with the whole body of ice to which they were attached, came in, and the ship going astern, struck quartering upon a piece of ice which lay between her and the great ice-island. This afforded the last hope of preventing her from coming in contact with it; and this hope failed also; for, grinding along the ice, she went nearly stern foremost, and struck with her larboard quarter upon the ice-island with a tremendous crash.

The first effect of this blow was to carry away the spanker-boom, the larboard stern-davit, and to crush the stern-boat. The starboard stern-davit was the next to receive the shock, and as this is connected with the spar-deck bulwarks, the whole of them were started; the knee, a rotten one, which bound the davit to the taffrail, was broken off, and with it all the stanchions to the plank-sheer, as far as the gangway.

Severe as was this shock, it happened fortunately that it was followed by as great a rebound. This gave the vessel a cant to starboard, and by the timely aid of the jib and other sails, carried her clear of the ice-island, and forced her into a small opening. While doing this, and before the vessel had moved half her length, an impending mass of ice and snow fell in her wake. Had this fallen only a few seconds earlier, it must have crushed the vessel to atoms.

It was also fortunate that the place where she struck the ice-island was near its southern end, so that there was but a short distance to be passed before she was entirely clear of it. This gave more room for the drifting ice, and permitted the vessel to be worked by her sails.

The relief from this pressing danger, however gratifying, gave no assurance of ultimate safety. The weather had an unusually stormy appearance; and the destruction of the vessel seemed almost inevitable, with the loss of every life on board. They had the melancholy alternative in prospect of being frozen to death one after the other, or perishing in a body by the dissolving of the iceberg on which they should take refuge, should the vessel sink.

When the dinner hour arrived the vessel was again fast in the ice, and nothing could for a time be done: it was therefore piped as usual. This served to divert the minds of the men from the dangers around them.

When the meal was over, the former manœuvring was resorted to, the yards being kept swinging to and fro, in order to keep the ship's head in the required direction. She was labouring in the swell, with ice grinding and thumping against her on all sides; every moment something either fore or aft was carried away—chains, bolts, bob-stays, bowsprit, shrouds; even the anchors were lifted, coming down with a surge that carried away the eyebolts and lashings, and left them to hang by the stoppers. The cut-water also was injured, and every timber seemed to groan.

Similar dangers attended those in the boats. Passed Midshipman Eld was sent to plant the ice-anchors: there was no room for the use of oars; the grinding and grating of the ice, as it rose and fell with the swell, rendered great precaution necessary to prevent the boat from being swamped or crushed; and when it is stated that two hours of hard exertion were required to plant the ice-anchors, some idea of the difficulty attending this service will be had. But this was not all; the difficulty of returning was equally great, and no possible way of effecting it seemed to suggest itself. The sides of the icebergs could not be ascended, and to approach the berg on the side next the ship was certain destruction to the boat and crew, for the ice and water were foaming like a cauldron; and to abandon the former was equally out of the question. At last a chance offered, although almost a hopeless one, by passing between two of these bergs, that appeared on the other side of a small clear space. The boat was upon a small piece of ice, from which, by great exertions, she was launched; a few pulls at the oars brought them to the passage; the bergs were closing fast, and agitated by the swell; no time, therefore, was to be lost: the danger was already great, and in a few seconds it would be impossible to pass. They entered; their oars caught, and they got but half-way

PEACOCK IN CONTACT WITH ICE-BERG.

through when the icebergs closed in upon them, and pressed the gunwales together, so as almost to crush the boat; the water entered her, and she was near sinking, when the berg stopped, retreated, and by another hard shove they went through, and were soon alongside the ship.

Every exertion was now made to work the ship and avoid heavy thumps from the ice. The mode resorted to, to get the ship about, was a novel one, namely, by urging her lee bow against a piece of ice, which had the same effect as giving her a lee helm; but this was found rather too expensive a mode of effecting the object, and on the pumps showing an increase of water, it was discontinued. The ice had been rapidly accumulating around the ship, contracting still more narrowly the space or area in which they were, and rendering their situation more hazardous.

At 4 P. M., they clewed up the topsails, the ship being fast in the ice, with the wind directly in from the seaward. The ice-anchors were now again run out, in hopes of relieving her from some of the strain. A short time afterwards the ice clearing from the stern enabled them to unship the rudder, which was taken on board in two pieces: it was immediately placed on the quarter-deck, and all the carpenters employed on it.

It soon began to snow violently, and no clear sea could be seen from the ship in any direction. It becoming obscure, the chance was that they would have to take up their last abode there. About six o'clock the weather cleared a little, and the wind freshened; they parted the hawser attached to the ice-anchor, and made sail again for the clear sea, which could now be seen from the masthead. Towards 8 P. M., as if to blast the little hope that the continuance of clear weather inspired, the ship took a wrong cant, and was forced into a small opening leading farther into the ice to leeward, and towards the massive walls of the berg. Great exertions were made, and fortunately, by the aid of the ice-anchors and sails, they succeeded in getting her round, and her head again pointed towards the clear sea; but they were shortly afterwards wedged in between two large masses of ice. At midnight the sea was observed to rise, although the wind had not increased, causing much motion among the ice; and the stormy appearance of the sky continued, and gave promise of a gale. The only hope left was to force the ship through, and every means were employed to effect this object. The ice they had now to contend with was of larger dimensions, and the increased sea rendered it doubly dangerous. Some of the shocks against it were so heavy as to excite fears that the ship's bow would be driven in, and on one occasion three of the chronometers were thrown out of their beds of sawdust upon their sides. They continued to make but little head way, and the grinding and thumping on the ship was most painful The hope of extricating her lessened every moment; for the quantity of ice between them and the sea was increasing, and the ship evidently moved with it to leeward. Few situations could be more trying, but the emergency was met by Captain Hudson with a coolness, perseverance, and presence of mind, which secured the admiration of all who were present, and inspired full confidence and a firm reliance in his ability to overcome every difficulty that lay within the power of human means.

The annexed sketch of the bay will exhibit the situation of the ship more accurately; it is situated in latitude 65° 55′ 20″ S., longitude 151° 18′ 45″ E.

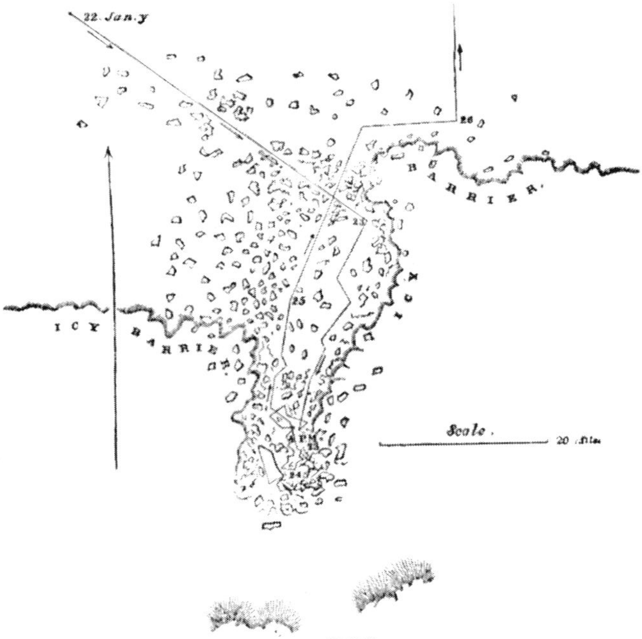

PEACOCK BAY.

In the afternoon of the 25th, the sea continued to increase, and the ship frequently struck against the masses of ice, while every foot they forged ahead carried them seemingly into a more precarious situation. At about 3 A. M., they found that the gripe had been beaten off, and they were now bruising up the stem and grinding away the bows.

There appeared no other course but to drive her out, which was deemed the only chance of saving the ship and crew. All the canvass that would draw was therefore set to force her through; and the wind favouring them, they had by four o'clock succeeded in passing the thick and solid ice, and shortly afterwards found themselves in clear water, without a rudder, the gripe gone, and, as was afterwards found, the stem ground down to within an inch and a half of the wood-ends.

The carpenters were still employed on the rudder, and had succeeded in removing the broken pieces of the pintles from the second and third braces on the stern-post; the upper and lower pintles were broken, leaving only two to hang the rudder by. The weather seemed now to favour them, and about ten o'clock they had finished the rudder, which had been repaired in the best possible manner. Great credit is due to Mr. Dibble, the carpenter, (who left his sick bed on the occasion,) for his exertions, attention, and perseverance. He and the carpenter's crew worked twenty-four hours without intermission. The ship was now hove-to, for it was apprehended that her rolling would render the task of shipping the rudder troublesome. By meridian they were again in a situation to make sail to extricate themselves from a bay some thirty miles in extent, which, with the exception of the small opening by which they had entered, was apparently closed by the barrier.

Shortly afterwards, the wind becoming fair, they made all sail for the outlet. The weather proved fine, and the winds moderate. At midnight they found the only opening left, which was not more than a quarter of a mile wide; they succeeded in passing through this, by 2 A. M., in a snow-storm, and felt grateful to God for their providential escape.

Captain Hudson now came to the conclusion of returning north. "After," as he says, "thoroughly turning over in my own mind the state of the ship,—with the head of the rudder gone, hanging by two braces, and in such a state that we could hardly hope to make it answer its purposes again, in encountering the boisterous weather we should have to pass through before reaching the first port,—the ship considerably strained; her starboard spar-deck bulwarks gone as far forward as the gangway; the gripe off, and the stern mutilated;—fully satisfied from this state of things that she was perfectly useless for cruising among icebergs, and the accompanying dangers, in thick foggy weather, to which, in these latitudes, we should be more or less subject, and where rapid evolutions were often necessary, in which the rudder must perform its part; and that the ship would require exten sive repairs before being employed in surveying operations; and feeling that the season was rapidly coming round when our services would be required in that duty, I held a council of the ward-room officers, and required their opinions as to making any further attempts to cruise in these latitudes.

"There was but one opinion as to the necessity of the ship's returning north, with the exception of Mr. Emmons and Mr. Baldwin, who thought the rudder might stand, provided we did not get near the ice or fall in with icebergs. This of course would be to effect little or nothing, and result only in a loss of time. I accordingly put the ship's head north, determined to proceed at once to Sydney, to effect the necessary repairs, so as to be ready at the earliest possible day to join the squadron."

Such were the dangers and difficulties from which the Peacock, by the admirable conduct of her officers and crew, directed by the consummate seamanship of her commander, was enabled at this time to escape. There still, however, remained thousands of miles of a stormy ocean to be encountered, with a ship so crippled as to be hardly capable of working, and injured to such an extent in her hull as to be kept afloat with difficulty. The narrative of the events of this perilous navigation must, however, be postponed, until I shall have given the proceedings of the other vessels of the squadron, while tracing out the position of the icy barrier, and following along the newly-discovered continent.

ICEBERG.

CHAPTER X.

CONTENTS.

PROCEEDINGS OF THE VINCENNES FROM THE TWENTY-SECOND OF JANUARY—DISAPPOINTMENT BAY—WATERING ON THE ICE—DIAGRAMS OF THE ICE-ISLANDS—THEIR UTILITY—VIOLENT GALE AND SNOW-STORM—NARROW ESCAPE FROM STRIKING THE ICE—THE OPEN SEA REACHED—RETURN OF FINE WEATHER—VINCENNES STANDS AGAIN TO THE SOUTH, AND REACHES THE ICY BARRIER—PINER'S BAY—SOUNDINGS IN THIRTY FATHOMS — ANOTHER VIOLENT GALE — REPORT OF THE MEDICAL OFFICERS—OPINION OF THE WARD-ROOM OFFICERS—DETERMINATION TO PROCEED WITH THE CRUISE—ITS EVENTS UP TO THE FOURTEENTH OF FEBRUARY—LANDING ON AN ICEBERG—SPECIMENS OF ROCKS OBTAINED—INQUIRY IN RELATION TO THE FORMATION OF ICEBERGS—THEIR SEPARATION FROM THE LAND—THEIR PROGRESS—FURTHER EVIDENCE IN RELATION TO THE ANTARCTIC CONTINENT—ESTIMATE OF THE RATE AT WHICH THE FLOATING ICE MOVES—THE VINCENNES BEGINS HER RETURN TO THE NORTH.

CHAPTER X.

ANTARCTIC CRUISE—CONTINUED.

1840.

In taking up the narrative of the disaster sustained by the Peacock, with which the preceding chapter closes, the Vincennes and Porpoise were left on the 22d of January.

On that day the Vincennes passed the place through which the Peacock entered, as has been related, on the 23d, and found no opening. To judge from the manner in which the ice moved during the time the Peacock was enclosed in it, I am inclined to ascribe the alternate opening and closing of the passage into the bay, to a tide setting along this coast. In support of this opinion it is sufficient to state, that the strength of the winds experienced on board that vessel was at no time sufficient to account for the manner in which the ice was found to move.

About thirty miles to the westward of this point, the Vincennes passed a remarkable collection of tabular icebergs, for whose existence I can account in no other manner than by supposing them to be attached to a rocky islet, which formed a nucleus to which they adhered. It was quite obvious that they had not been formed in the place where they were seen, and must, therefore, have grounded, after being adrift.

On the 23d January, after passing around this group of icebergs, the sea was found comparatively clear, and a large open space showed itself to the southward. Into this space the course of the Vincennes was immediately directed. While thus steering to the south, the appearance of land was observed on either hand, both to the eastward and westward.

Pursuing this course, we by midnight reached the solid barrier, and all approach to the land on the east and west was entirely cut off by the close packing of the icebergs. I was, therefore, reluctantly compelled to return, not a little vexed that we were again foiled in our endeavour to reach the Antarctic Continent. This was a deep indentation in the coast, about twenty-five miles wide: we explored it to the depth of about fifteen miles, and did not reach its termination. This bay I have called Disappointment Bay: it is in latitude 67° 04′ 30″ S., longitude 147° 30′ E. The weather was remarkably fine, with a bracing air: the thermometer in the air 22°, in the water 31°.

The next day, 24th, we stood out of the bay, and continued our course to the westward. About noon, to my surprise, I learnt that one of the officers, Lieutenant Underwood, had marked on the log-slate that there was an opening of clear water, subtending three points of the compass, at the bottom of Disappointment Bay. Though confident that this was not the fact, in order to put this matter at rest, I at once determined to return, although forty miles distant, and ordered the ship about, to refute the assertion by the officer's own testimony. This was most effectually done the next morning, 25th, when the ship reached the identical spot, and all were fully convinced that no opening existed. The whole bay was enclosed by a firm barrier of ice, from north-northwest to east-northeast.

The weather proved delightful, with light airs from the southward, and I determined to take this opportunity to fill up the water-tanks with ice. The ship was hove-to, a hawser got in readiness, the boats lowered, and brought alongside of an iceberg well adapted to our purpose.

The same opportunity was also taken to make the magnetic observations on the ice, and to try the local attraction of the ship.

Many birds were seen about the ship, of which we were fortunate in obtaining specimens. The day was remarkably clear, and the same appearance of land was seen that had been witnessed on the 24th. We filled nineteen of our tanks with ice, after having allowed it to remain for some time on deck for the salt water to drain off in part, and it proved very potable.

At about 5 P. M., we had completed our required store of ice, and cast off, making sail to the northward.

In order that no further mistakes should take place as to the openings being passed, I issued an order, directing the officer of the deck on being relieved to go to the masthead, and report to me the exact situation of the ice; and this was continued during the remainder of our cruise among it.

From a Sketch by C. Wilkes U S N

C. A Jewett Sc

VINCENNES IN DISAPPOINTMENT BAY.

In threading our way through the many icebergs, it occurred to me that they might be considered as islands, and a rough survey made of them, by taking their bearings at certain periods, and making diagrams of their positions. This was accordingly done, and every few hours they were inserted on the chart which I was constructing in my progress.

The following is one of the diagrams.

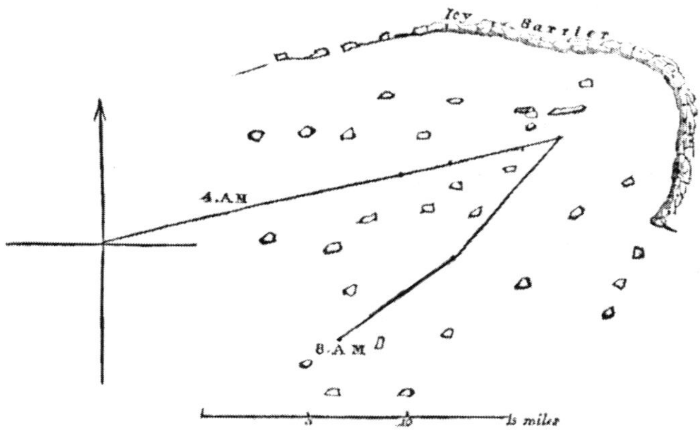

This I found to be very useful, and it gave me confidence in proceeding, for I had a tolerable chart to retreat by in case of need, at least for a few hours, during which time I had reason to believe that there was not much probability of the icebergs changing their relative positions.

The dip observed on the ice was 87° 30′, and the variation 12° 46′ easterly. The compasses were found to be very sluggish, having but little horizontal directive force.

About half an hour after we cast off from the iceberg, a thick snow-storm came up, with the wind from the southeast. Although there were very many ice-islands around us, on our way out, I felt that I understood the ground well, having passed over it twice, and knowing I had a space of a few miles, only thinly sprinkled with icebergs, I hove-to with shortened sail. This was the first southeast wind we had had since being on this coast. I had been disappointed in not finding it from that quarter before; for I had been informed, by those who had navigated in high southern latitudes, that southeast would be the prevailing wind, and would be attended with fine weather. Now, however, with a fair wind, I was unable to run, for the weather was unfavourable.

At 6 A. M. on the 26th, we again made sail, and at 8 A. M. we di-covered the Porpoise, to whom we made signals to come within hail. We found them all well, and compared chronometers.

As it still blew fresh from the southeast, and the weather became a little more clear, we both bore away, running through much drift-ice, at the rate of nine knots an hour. We had the barrier in sight; it was, however, too thick to see much beyond it. Sailing in this way I felt to be extremely hazardous; but our time was so short for the examination of this icy coast, that while the barrier was to be seen, I deemed it my duty to proceed. We fortunately, by good look-outs, and carefully conning the ship, were enabled to avoid any heavy thumps.

On the 27th, we again had the wind from south-southwest. The floe-ice had become so thick, that we found it impossible to get through it in the direction I wished to go, and we were compelled to pass round it. The Porpoise was in sight until noon. The weather proved beautifully clear. A long range of tabular icebergs was in sight to the southward, indicating, as I have before observed, that the coast was near. I passed through these, losing sight of the Porpoise to the northwest about noon, when we were in longitude 142° 40′ E., latitude 65° 54′ 21″ S., variation 5° 08′ easterly.

On the 28th, I found myself completely surrounded by the tabular icebergs, through which we continued to pass. Towards midnight the wind shifted to the southeast, and enabled me to haul more to the southward. At 9½ A. M. we had another sight of the land ahead, and every prospect of nearing it, with a fine breeze. The sight of the icebergs around us, all of large dimensions, was beautiful. The greatest number in sight at one time was noted, and found to be more than a hundred, varying from a quarter of a mile to three miles in length. We took the most open route, and by eleven o'clock had run upwards of forty miles through them. We had the land now in plain view, but the weather soon began to thicken and the breeze to freshen. At noon it was so thick that every thing was hidden, and no observation was obtained. The ship was hove-to, but shortly after again put under way, making several tacks to keep my position, which I felt was becoming a critical one, in case a gale should ensue. I therefore looked carefully over my chart, and was surprised at the vast number of icebergs that appeared on it. At 2 P. M. the barometer began to fall, and the weather to change for the worse. At 5 P. M. a gale was evidently coming on, so we took three reefs in the topsails. It appeared now that certain wreck would ensue, should we remain where we were; and after much consideration, I made up my mind to retrace my way, and seek the open space forty miles distant, taking for a landmark a remarkable berg that had been the last entered on the chart, and which would be a guide to my course out. I therefore stood for its position. The weather was so thick, that it was necessary to run close to it, to be quite sure of recognising it, for on this seemed to depend our safety. About the estimated time we would take to pass over the distance, an iceberg was made (we were within one thousand feet of it) which, at first view, I felt confident was the one sought, but was not altogether satisfied afterwards. I therefore again consulted my chart, and became more doubtful of it. Just at that moment I was called on deck by an officer, who informed me that there were icebergs a short distance ahead! Such proved to be the case; our path was beset with them, and it was evident we could not regain our route. To return was worse, so having but little choice left, I determined to keep on. To encounter these icebergs so soon after seeing the other, was in some respects satisfactory, for it removed all doubts, and showed me that we were not near the track by which we entered. Nothing, therefore, was to be done but to keep a good look-out, and the ship under sufficient way to steer well. My safest plan was to keep as near our former track as possible, believing it to be most free of these masses.

At 8 P. M. it began to blow very hard, with a violent snow-storm, circumscribing our view. and rendering it impossible to see more than two ship's-lengths ahead. The cold was severe, and every spray that touched the ship was immediately converted into ice. At 9 P. M., the barometer still falling and the gale increasing, we reduced sail to close-reefed fore and main-topsails, reefed foresail and trysails, under which we passed numerous icebergs, some to windward, and some to leeward of us. At 10^h 30^m, we found ourselves thickly beset with them, and had many narrow escapes; the excitement became intense; it required a constant change of helm to avoid those close aboard; and we were compelled to press the ship with canvass in order to escape them, by keeping her to windward. We thus passed close along their weather sides, and distinctly heard the roar of the surf dashing against them. We had, from time to time, glimpses of their obscure outline, appearing as though immediately above us. After many escapes, I found the ship so covered with ice, and the watch so powerless in managing her, that a little after midnight, on the 29th, I had all hands called. Scarcely had they been reported on deck, when it was made known to me that the gunner, Mr. Williamson, had fallen, broken his ribs, and otherwise injured himself, on the icy deck.

The gale at this moment was awful. We found we were passing large masses of drift-ice, and ice-islands became more numerous. At a little after one o'clock it was terrific, and the sea was now so heavy, that I was obliged to reduce sail still further: the fore and main-top-sails were clewed up; the former was furled, but the latter being a new sail, much difficulty was found in securing it.

VINCENNES AMONGST ICE-BERGS

A seaman, by the name of Brooks, in endeavouring to execute the order to furl, got on the lee yardarm, and the sail having blown over the yard, prevented his return. Not being aware of his position until it was reported to me from the forecastle, he remained there some time. On my seeing him he appeared stiff, and clinging to the yard and lift. Spilling-lines were at once rove, and an officer with several men sent aloft to rescue him, which they succeeded in doing by passing a bowline around his body and dragging him into the top, He was almost frozen to death. Several of the best men were completely exhausted with cold, fatigue, and excitement, and were sent below. This added to our anxieties, and but little hope remained to me of escaping: I felt that neither prudence nor foresight could avail in protecting the ship and crew. All that could be done, was to be prepared for any emergency, by keeping every one at his station.

We were swiftly dashing on, for I felt it necessary to keep the ship under rapid way through the water, to enable her to steer and work quickly. Suddenly many voices cried out, "Ice ahead!" then, "On the weather bow!" and again, "On the lee bow and abeam!" All hope of escape seemed in a moment to vanish; return we could not, as large ice-islands had just been passed to leeward: so we dashed on, expecting every moment the crash. The ship, in an instant, from having her lee guns under water, rose upright; and so close were we passing to leeward of one of these huge islands, that our trysails were almost thrown aback by the eddy wind. The helm was put up to pay the ship off, but the proximity of those under our lee bade me keep my course. All was now still except the distant roar of the wild storm, that was raging behind, before, and above us; the sea was in great agitation, and both officers and men were in the highest degree excited. The ship continued her way, and as we proceeded, a glimmering of hope arose, for we accidentally had hit upon a clear passage between two large ice-islands, which in fine weather we should not dare to have ventured through. The suspense endured while making our way between them was intense, but of short duration; and my spirits rose as I heard the whistling of the gale grow louder and louder before us, as we emerged from the passage. We had escaped an awful death, and were again tempest-tost.

We encountered many similar dangers that night. At half-past 4, A. M., I found we had reached the small open space laid down on my chart, and at five o'clock I hove-to the ship. I had been under intense excitement, and had not been off the deck for nine hours, and was now thankful to the Providence that had guided, watched over, and preserved us. Until 7 A. M., all hands were on deck, when there was some appearance of the weather moderating, and they were piped down.

The barometer was marked at intervals, for which the reader is referred to Appendix XXVI.

This gale was from the southeast, from which quarter it blew during the whole of its strength; and when it began to moderate, the wind veered to the southward. By noon we felt satisfied that the gale was over, and that we had escaped, although it was difficult to realize a sense of security when the perils we had just passed through were so fresh in our minds, and others still impending. Towards four o'clock, it cleared off, and we saw but few icebergs near us. Our longitude was found to be 140° E., latitude 63° 30′ S., and I again made sail for the ice to the south, to pass over the very route we had just traversed through so many perils.

The wind had now hauled to the southwest. At 6 P. M., we again began to enter among ice-islands. The weather appeared settled; but I had so often been deceived by its fickleness, that I felt no reliance ought to be put in its continuance. A powerful inducement was held out to us, in the prospect of getting close enough to effect a landing: and this rendered us insensible to the dangers.

On the morning of the 30th the sun rose in great brilliancy, and the scene could hardly be realized as the same as that we had passed through only twenty-four hours before. All was now quiet; a brisk breeze blew from the eastward, all sail was set, and there was every prospect that we might accomplish our object; for the land was in sight, and the icebergs seemed floating in quiet. We wound our way through them in a sea so smooth that a yawl might have passed over it in safety. No straight line could have been drawn from us in any direction, that would not have cut a dozen icebergs in the same number of miles, and the wondering exclamations of the officers and crew were oft repeated,—"How could we have passed through them unharmed?" and, "What a lucky ship!" At eight o'clock, we had reached the icy barrier, and hove-to close to it. It was tantalizing, with the land in sight, to be again and again blocked out. Open water was seen near the land to the southwest of us, and a tortuous channel through the broken ice to leeward, apparently leading to it. All sail was immediately crowded; we passed rapidly through, and found ourselves again in clear water, which reached to the shores: the barrier extending in a line with our course, about two miles to windward, and a clear channel to the northwest, about two miles wide, as far as the eye could reach. Seeing this, I remarked to one of the officers that it would have been a good place to drift in during the last gale,—little thinking that in a few short hours it would serve us for that purpose, in still greater need. A brisk gale ensued, and the ship ran at the rate of nine or ten miles an hour; one reef was taken in the topsails, and we stood directly in for the most southerly part of the bay.

This bay was formed partly by rocks and partly by ice-islands. The latter were aground, and on the western side of the bay extended about five miles to the northward of our position.

While we stood on in this direction the gale increased, and our room became so circumscribed that we had not time on any one tack to reduce our canvass, before it became necessary to go about. In this way we approached within half a mile of the dark, volcanic rocks, which appeared on both sides of us, and saw the land gradually rising beyond the ice to the height of three thousand feet, and entirely covered with snow. It could be distinctly seen extending to the east and west of our position fully sixty miles. I make this bay in longitude 140° 02′ 30″ E., latitude 66° 45′ S.; and, now that all were convinced of its existence, I gave the land the name of the Antarctic Continent. Some of the officers pointed out the appearance of smoke, as if from a volcano, but I was of opinion that this was nothing but the snow-drift, caused by the heavy squalls. There was too much wind at this time to tack; I therefore had recourse to luffing the vessel up in the wind, and wore her short round on her heel. At the same time we sounded, and found a hard bottom at the depth of no more than thirty fathoms. I made a rough sketch of this bay, which I have called Piner's Bay, after the signal quarter-master of that name. It was impossible to lower a boat, or to remain longer; indeed, I felt it imperative on me to clear its confined space before the floating ice might close it up.

At $10^h\ 30^m$ we had gone round, and in an hour more we cleared the bay. At noon the wind had increased to a gale, and by one o'clock, P. M., we were reduced to storm-sails, with our top-gallant yards on deck. The barometer had again declined rapidly, proving a true indicator, but giving little or no warning. To run the gauntlet again among the icebergs was out of the question, for a large quantity of field-ice would have to be passed through, which must have done us considerable damage, if it did not entirely disable us. The clear space we occupied was retained until five or six o'clock, when I found the floe-ice was coming down upon us; I then determined to lay the ship for a fair drift through the channel I had observed in the morning, and which I had every reason to believe, from the wind (southeast) blowing directly through it, would not be obstructed until the floe-ice came down. It was a consolation to know that if we were compelled to drift, we should do so faster than the ice; I therefore thought it as well to avoid it as long as possible. Another reason determined me to delay the drifting to the latest moment: I did not believe that the extent of the channel we had seen in the morning was more than ten miles in extent, and at the rate we drifted, the end of it would be reached long before the gale was over. This, like the former gale, was an old-fashioned snow-storm. All the canvass we could show to it at one time was a close-reefed main-topsail and fore-storm-staysail. It blew tremendously, and the sea we experienced was a short disagreeable one, but nothing to be compared to that which accompanied the first gale. From the shortness of the sea, I inferred that we had some current. This state of things continued for several hours, during which we every moment expected to reach the end of our channel. Since the last gale, the whole crew, officers and men, had been put in watch and watch, ready for an instantaneous call, and prepared for rapid movements. The snow was of the same sleety or cutting character as that of the previous day, and seemed as if armed with sharp icicles or needles.

The 31st brought no moderation of the weather. At 1 A. M., a group of ice-islands was reported, and shortly afterwards field-ice close under our lee. We wore ship instantly, and just avoided coming in contact with the latter. Sail was immediately made on the ship, and the scene of the former gale again gone through (which it is needless here to repeat), with this exception, that we were now passing to and fro among icebergs immediately to windward of the barrier, and each tack brought us nearer to it. Between 4 and 5 A. M., our space was becoming confined, and there was no abatement of the gale; I therefore, as it had cleared sufficiently to enable us to see a quarter of a mile, determined to bear up and run off north-northwest for a clear sea. In doing this we passed icebergs of all dimensions and heavy floe-ice. By 8^h 30^m we had run thirty miles, when, finding a more open sea, I judged we had partially cleared the ice. At noon the gale still continued. The lowest reading of the barometer during this gale was 28·59 in.

After lasting thirty hours, the gale, at 6 P. M., began to moderate a little, when we again made sail to the southward. I now felt inclined to seek Piner's Bay again, in order to effect a landing. This would have been a great personal gratification; but the bay was sixty miles distant, so that to revisit it would occupy time that was now precious; and feeling satisfied that a great extent of land wholly unknown lay to the westward, I deemed it my duty to proceed to its discovery, not doubting that if my opinions of its existence were correct, a place equally feasible for landing would be found. Another subject also presented itself, which, for a time, caused me some anxiety, and which I confess was not only unexpected by me, but directly at variance with my own observations on the condition of my crew. As I feel compelled to give a complete detail of our proceedings, I must now revert to this subject.

The following report of the medical officers of the ship was made to me on the day of its date.

U. S. Ship Vincennes,
At Sea, January 31st, 1840.

SIR,—

It becomes our duty, as medical officers of this ship, to report to you in writing the condition of the crew at the present time.

The number upon the list this morning is fifteen: most of these cases are consequent upon the extreme hardships and exposure they have undergone during the last gales of wind, when the ship has been surrounded with ice.

This number is not large, but it is necessary to state, that the general health of the crew, in our opinion, is decidedly affected, and that under ordinary circumstances the list would be very much increased, as the men under the present exigencies, actuated by a laudable desire to do their duty to the last, refrain from presenting themselves as applicants for the list.

Under these circumstances, we feel ourselves obliged to report that, in our opinion, a few days more of such exposure as they have already undergone, would reduce the number of the crew by sickness to such an extent as to hazard the safety of the ship and the lives of all on board.

Very respectfully, your obedient servants,

(Signed) J. L. Fox,

J. S. Whittle,

Assistant-Surgeons.

To Charles Wilkes, Esq.,

Commanding Exploring Expedition.

Although my own opinion, as I have stated, differed from that expressed in the report, I deemed it my duty to ask the opinion of the ward-room officers, and also, in order to procure additional medical advice, restored to duty Acting-Surgeon Gilchrist, who was under suspension. The opinion of the ward-room officers was asked in a written circular, of which the following is a copy.

U. S. Ship Vincennes,

At Sea, January 31st, 1840.

Gentlemen,—

The receipt of the enclosed report of Drs. Fox and Whittle, relative to the health and condition of the crew of this ship, at this time, renders it necessary for me to decide whether it is expedient to push farther south in exploration under the present circumstances.

As you are acquainted with all the circumstances, it is unnecessary to repeat them, except to remark, that your opinion is requested before I decide upon the course to be pursued, in consequence of the strong bias self-interest might give me in the prosecution of our arduous duties. I wish the report returned to me, and for you to communicate your opinion in writing.

I am, respectfully, &c.,

Charles Wilkes,

Commanding Exploring Expedition.

To the Ward-Room Officers,

U. S. Ship Vincennes.

The answers to this letter will be seen in Appendix XXVII.; and it is sufficient here to say, that a majority concurred in opinion with the report of the medical officers. Notwithstanding these opinions, I was not satisfied that there was sufficient cause to change my original determination of passing along to the appointed rendezvous; and after full consideration of the matter, I came to the conclusion, at whatever hazard to ship and crew, that it was my duty to proceed, and not give up the cruise until the ship should be totally disabled, or it should be evident to all that it was impossible to persist any longer. In bringing myself to this decision, I believe that I viewed the case on all sides with fairness, and allowed my duty to my country, my care for those whom it had committed to my charge, and my responsibility to the world, each to have its due weight.

The weather now moderated, and I ordered sail to be made. The 2d of February found us about sixty miles to the westward of Piner's Bay, steering to the southward, and as usual among ice-islands, with the land in sight. The land had the same lofty appearance as before. We stood in until 3 P. M., when we were within two and a half miles of the icy cliffs by which the land was bounded on all sides. These were from one hundred and fifty to two hundred feet in height, quite perpendicular, and there was no appearance whatever of rocks; all was covered with ice and snow. A short distance from us to the westward was a long range of icebergs aground, which, contrary to the usual appearance, looked much weather-beaten. We tried for soundings, but did not get any with one hundred and fifty fathoms, although the water was much discoloured. The badness of the deep-sea line was a great annoyance to us, for deeper soundings would probably have obtained bottom. No break in the icy barrier, where a foot could be set on the rocks, was observable from aloft. The land still trended to the westward as far as the eye could reach, and continued to exhibit the same character as before. Our longitude now was 137° 02′ E., latitude 66° 12′ S.: we found the magnetic declination westerly.

This proved a fine day, so that we had an opportunity of airing the men's bedding, of ventilating the ship, and of getting rid of the ice, with which we were much encumbered. The thermometer varied from 33° to 36°. Our sick-list had increased the last few days to twenty; many of the men were affected with boils, which rendered them comparatively useless; and ulcers, which were caused by the least scratch, were exceedingly prevalent; but their food was good, they had plenty of it, and their spirits were excellent. The high land was seen this afternoon, but the barrier along which we were passing prevented any nearer approach. This evening it was perceptible that the days were becoming shorter, which was a new source of anxiety, for we were often surrounded by numerous ice-islands, which the darkness rendered more dangerous.

Towards evening the weather became unsettled, and the 3d of February was ushered in by another gale, accompanied with snow The barometer fell lower than heretofore, namely, to 28·460 in.; the thermometer stood at 33°. Before the thick snow came on, we had taken the bearings of the ice-islands, and finding we had a few miles comparatively free from them, I determined to await the result of the storm, and made every thing snug to encounter it. The gale continued throughout the day, and although it moderated after 5 P. M., we had some strong squalls, but nothing so violent as those we had already experienced. The ship, in consequence of the snow, became more damp and uncomfortable, and our sick-list was increased to thirty, who were rather overcome by want of rest and fatigue than affected by any disease. To remedy the dampness, a stove was placed on the gun-deck, and fires kept burning in the galleys on the berth-deck, more for the purpose of drying the men's clothes than for warmth. We had no observations this day, but the dead-reckoning gave the longitude 134° E., latitude 63° 49′ S.

The 4th and 5th the weather continued the same. As the winds became lighter thick snow fell, and we were able to see only a short distance from the ship. We contrived by manœuvring to retain our position. On this last day we got a tolerable observation, which gave our longitude as 133° 42′ E., and latitude 64° 06′ S.

The first part of the 6th the same thick weather continued, but towards 4 P. M., it began to clear, when we again made sail, until we saw and took the bearings of the barrier. We found ourselves situated opposite the part of it we had seen three days before. It still had the appearance of being attached to the land, and in one uninterrupted line. Wishing to examine it closely, I hove-to for broad daylight. Many whales, penguins, flocks of birds, and some seals, were reported.

On the 7th we had much better weather, and continued all day running along the perpendicular icy barrier, about one hundred and fifty feet in height. Beyond it the outline of the high land could be well distinguished. At 6 P. M., we suddenly found the barrier trending to the southward, and the sea studded with icebergs. I now hauled off until daylight, in order to ascertain the trending of the land more exactly. I place this point, which I have named Cape Carr, after the first-lieutenant of the Vincennes, in longitude 131° 40′ E., and latitude 64° 49′ S.

On the 8th, at daylight, we again made sail to the southward, and found at 4 A. M. the field of ice had stopped our progress, and the weather was thick. Land was no longer seen to the south, a deep bay apparently making in. We continued our course to the westward along the barrier, until 8 P. M., when we were again brought to. At 7 P. M. we had strong indications of land; the barrier was of the former perpendicular form, and later the outline of the Continent appeared distinct though distant. The night was dark and unpleasant. At noon our longitude was 127° 07′ E., and latitude 65° 03′ S.; variation 14° 30′ westerly.

On the 9th we had the finest day we had yet experienced on this coast; the wind had veered from the east to southwest, and given us a clear, bracing, and wholesome atmosphere. The barrier exhibited the same appearance as yesterday. Our longitude was 125° 19′ E., latitude 65° 08′ S., variation 32° 45′ westerly. The current was tried, but none found; the pot was only visible at five fathoms; the colour of the water a dirty green; the dip sector gave 3′ 15″. I never saw a clearer horizon, or one better defined than we had to the northward. The icy barrier was really beautiful. At midnight we had a splendid display of the aurora australis, extending all around the northern

AURORA AUSTRALIS

horizon, from west-by-north to east-northeast. Before its appearance, a few clouds only were seen in the southeast, on which the setting sun cast a red tint, that barely rendered them visible. The horizon with this exception, appeared clear and well defined. The spurs or brushes of light frequently reached the zenith, converging to a point near it.

Although no clouds could be seen in the direction of the aurora, before or after its appearance, yet when it was first seen, there appeared clouds, of the form of massive cumuli, tinged with pale yellow, and behind them arose brilliant red, purple, orange, and yellow tints, streaming upwards in innumerable radiations, with all the shades that a combination of these colours could effect. In its most brilliant state it lasted about twenty minutes. The gold-leaf electrometer was tried, but without being affected: the instrument, however, was not very sensitive. Being somewhat surprised at the vast mass of cumuli which appeared during the continuance of the aurora, I watched after its disappearance till daylight, but could see only a few clouds: I am therefore inclined to impute the phenomenon to some deception caused by the light of the aurora. The apparent altitude of these clouds was 8°.

On the 10th we were again favoured by the weather; it gave us a fine sunshine, and an opportunity of airing the ship and drying the clothes. All the sick were improving in health.

Running close along the barrier, which continued of the same character, although more broken than yesterday, we saw an appearance of land, although indistinctly, to the southward. The water was of the same colour here as before, and the wind being from the south-southeast, we made some progress, and found ourselves in longitude 122° 35′ E., latitude 65° 27′ S.: the variation had now increased to 44° 30′ westerly. No aurora was seen this night, although it was looked for anxiously.

11th. The barometer had been stationary at 29·080 in. for the last three days: it now began to fall; the temperature of the air was 31°, of the water 32°. The fall of the barometer was soon followed by snow and thick weather. The trending of the barrier had been south-west-by-west, and a good deal of floe-ice had been met with, which we ran through. The sea was quite smooth, and many icebergs were enclosed in the barrier, which was very compact and composed of flat fields. At 10 P. M., I found it too dark to run, and hove-to.

During the 12th we had pleasant weather, and at 2 A. M. filled away. At 8 A. M., land was reported to the southwest. Keeping along the barrier and increasing our latitude, I again had hopes of getting near the land. We passed through great quantities of large floe-ice until 1

P. M., when the solid barrier prevented our farther progress. Land was now distinctly seen, from eighteen to twenty miles distant, bearing from south-southeast to southwest,—a lofty mountain range, covered with snow, though showing many ridges and indentations. I laid the ship to for three hours, in hopes of discovering some opening or movement in the ice, but none was experienced. I tried the current, and found none. The water was of a dirty dark green. We sounded with the wire-line in two hundred and fifty fathoms, and found no bottom. The temperature at that depth was 30½°, of the air 31°. The barrier had in places the appearance of being broken up, and we had decreased our longitude to 112° 16′ 12″ E., while our latitude was 64° 57′ S. This puts the land in about 65° 20′ S., and its trending nearly east and west. The line of the icy barrier was generally uniform, although it was occasionally pierced with deep bays. We saw some icebergs with decided spots of earth on them, which gave me hopes of yet obtaining the object of my wishes. The water was remarkably smooth during this day, and the weather clear, enabling us to see a great distance. Two hours after we bore away, we left the floe-ice, and entered a clear sea to the westward, where we lost sight of the barrier for a time; but in hauling up to the southwest, it was, by 8 P. M., within three miles of us, when we again kept off parallel to its trending. The appearance of land still continued. Shortly after, I hove-to, for the purpose of awaiting the daylight to continue our observations of the land, with little prospect or probability of reaching it, from the immense quantity of ice which continued to form an impenetrable barrier.

13th. At 2 A. M. we made sail to the southwest, in order to close with the barrier, which we found retreated in that direction, and gave us every prospect of getting nearer to it. Our course, for the most part, was through icebergs of tabular form. In the afternoon we had the land ahead, and stood in for it with a light breeze until 6½ P. M., when I judged it to be ten or twelve miles distant. It was very distinct, and extended from west-southwest to south-southeast. We were now in longitude 106° 40′ E., and latitude 65° 57′ S.; the variation was 54° 30′ westerly. The water was very green. We sounded in three hundred fathoms, and found no bottom. The weather having an unsettled appearance, we stood off to seek a clearer space for the night. The land left was high, rounded, and covered with snow, resembling that first discovered, and had the appearance of being bound by perpendicular icy cliffs.

14th. At daylight we again made sail for the land, beating in for it

From a sketch by C. Wilkes, U.S.N. Eng'd by Jordan & Halpin.

VIEW OF THE ANTARCTIC CONTINENT.

until 11 A. M., when we found any further progress quite impossible. I then judged that it was seven or eight miles distant. The day was remarkably clear, and the land very distinct. By measurement, we made the extent of coast of the Antarctic Continent, which was then in sight, seventy-five miles, and by approximate measurement, three thousand feet high. It was entirely covered with snow. Longitude at noon, 106° 18′ 42″ E., latitude 65° 59′ 40″ S., variation 57° 05′ westerly. On running in, we had passed several icebergs greatly discoloured with earth, and finding we could not approach the shore any nearer, I determined to land on the largest ice-island that seemed accessible, to make dip, intensity, and variation observations. On coming up with it, about one and a half miles from where the barrier had stopped us, I hove the ship to, lowered the boats, and fortunately effected a landing. We found embedded in it, in places, boulders, stones, gravel, sand, and mud or clay. The larger specimens were of red sandstone and basalt. No signs of stratification were to be seen in it, but it was in places formed of icy conglomerate (if I may use the expression), composed of large pieces of rocks, as it were frozen together, and the ice was extremely hard and flint-like. The largest boulder embedded in it was about five or six feet in diameter, but being situated under the shelf of the iceberg, we were not able to get at it. Many specimens were obtained, and it was amusing to see the eagerness and desire of all hands to possess themselves of a piece of the Antarctic Continent. These pieces were in great demand during the remainder of the cruise. In the centre of this iceberg was found a pond of most delicious water, over which was a scum of ice about ten inches thick. We obtained from it about five hundred gallons. We remained upon this iceberg several hours, and the men amused themselves to their hearts' content in sliding. The pond was three feet deep, extending over an area of an acre, and contained sufficient water for half-a-dozen ships. The temperature of the water was 31°. This island had been undoubtedly turned partly over, and had precisely the same appearance that the icy barrier would have exhibited if it had been turned bottom up and subsequently much worn by storms. There was no doubt that it had been detached from the land, which was about eight miles distant. The view of the land, ice, &c., taken from this ice-island, is exhibited in the opposite plate, and gives a correct representation of these desolate regions.

Around the iceberg we found many species of zoophytes, viz.: salpee, a beautiful specimen of clio helicina, some large pelagie, and many small crustacea. I made several drawings of them. This day, notwithstanding our disappointment in being still repelled from treading on the new continent, was spent with much gratification, and gave us many new specimens from it.

Finding that we had reached the longitude of 105° E., before the time anticipated, and being desirous to pursue the discoveries further west, I left a signal flying on this berg, with a bottle containing instructions for the other vessels, directing them to proceed to the westward as far as they could, in the time which should remain prior to the 1st of March. At 8 P. M. we joined the ship, and bore away again to the westward, intending to pursue the route pointed out to them.

On the 15th, we passed many icebergs much discoloured with earth, stones, &c., none of which appeared of recent formation. The weather this day became lowering, and the breeze fresh; we double-reefed the topsails, and made every thing snug: the wind was from the southward. At noon this day, we were in longitude 104° E., latitude 64° 06′ S. The sea had been remarkably smooth the last few days, with no swell; and I began to entertain the idea that we might have a large body of ice to the northward of us, for the position where Cook found the barrier in 1773 was two hundred miles further to the north. I determined, however, to pass on in our explorations, hoping they might enable me to join that of Enderby's Land. I deemed it a great object actually to prove the continuity with it if possible; and if disappointed in this, I should at any rate ascertain whether there had been any change in the ice in this quarter, since the time of Cook, which had been done already near his Ne Plus Ultra.

We had a vast number of whales about us this day, as well as penguins, Cape pigeons, white and gray, and small and large petrels. Some seals also were seen.

I was now happy to find the health of my crew had become re-established, and that only a few remained on the sick-list. This, I think, was effected by constant attention to their being warmly clothed.

The icebergs were covered with penguins. Several officers landed on the icebergs to get a few as specimens. On their return, some penguins followed them closely, particularly one, who at last leaped into the boat. It was supposed that its mate had been among those taken, and that it had followed on that account. If this were the fact, it would show a remarkable instinctive affection in this bird.

On the 16th, the barrier of ice trended to the northward, and we were obliged to haul to the northeast, passing through a large number of ice-islands, many of which were stained with earth. In the afternoon a large sea-elephant was discovered on the ice; two boats were sent to effect his capture, and many balls were fired into him, but he showed the utmost indifference to their effect, doing no more than to raise his head at each shot. He contrived to escape by floundering over the ice until he reached the water, in which he was quite a different being. At about 7 P. M., Dr. Fox was despatched in a boat to visit an ice-island that was very much discoloured with clay in patches. He reported that there was upon it a large pond of muddy water, not frozen, although the temperature on board was much below the freezing point. We observed around the icebergs numerous right whales, puffing in all directions. A large quantity of small crustacea, including shrimps, were here seen around the icebergs. These are believed to be the cause that attracts whales to these parts; they also supply the numerous penguins with their food. For several days I observed a great difference in the wind, by day and by night. It had been fresh from the hour of seven in the morning until 8 P. M., when it generally becomes light or dies away altogether. To-day we found ourselves in longitude 99° E., and latitude 64° 21′ S. We to-day made observations throughout the twenty-four hours with Leslie's photometer. These results will be found embraced in the volume of Meteorology.

On the 17th, about 10 A. M., we discovered the barrier extending in a line ahead, and running north and south as far as the eye could reach. Appearances of land were also seen to the southwest, and its trending seemed to be to the northward. We were thus cut off from any further progress to the westward, and obliged to retrace our steps. This position of the ice disappointed me, although it concurred with what was reasonably to be expected. We were now in longitude 97° 37′ E., and latitude 64° 01′ S.; our variation was 56° 21′ westerly, being again on the decrease. To-day we had several snow-squalls, which, instead of being in flakes, was in small grains, as round as shot, and of various sizes, from that of mustard-seed to buckshot. It was remarkably dry, pure white, and not at all like hail. We found the bay we had entered was fifty or sixty miles in depth, and having run in on its southern side, I determined to return along its northern shore, which we set about with much anxiety, as the weather began to change for the worse. Our situation was by no means such as I should have chosen to encounter bad weather in, the bay being sprinkled with a great many large icebergs. Here we met with a large number of whales, whose curiosity seemed awakened by our presence. Their proximity, however, was any thing but pleasant to us, and their blowings resembled that of a number of locomotives.

Their close approach was a convincing proof that they had never been exposed to the pursuit of their skilful hunters. They were of the fin-back species, and of extraordinary size.

Between ten and eleven o'clock at night it was entirely clear over head, and we were gratified with a splendid exhibition of the aurora australis. It exceeded any thing of the kind I had heretofore witnessed; its activity was inconceivable, darting from the zenith to the horizon in all directions in the most brilliant coruscations; rays proceeding as if from a point in the zenith, flashed in brilliant pencillings of light, like sparks of electric fluid in vacuo, and reappeared again to vanish; forming themselves into one body, like an umbrella, or fan, shut up; again emerging to flit across the sky with the rapidity of light, they showed all the prismatic colours at once or in quick succession. So remarkable were the phenomena that even our sailors were constantly exclaiming in admiration of its brilliancy. The best position in which to view it was by lying flat upon the deck, and looking up. The electrometer was tried, but no effect perceived. The star Canopus was in the zenith at the time, and though visible through the aurora, was much diminished in brightness. On this night also the moon was partially eclipsed.

Large icebergs had now become very numerous, and strengthened the belief that the land existing in this vicinity had taken a very decided trend to the northward. I accordingly followed up the northern barrier closely, and passed through the thickest of these bergs, well knowing from our experience that we should have little or no opportunity of seeing the land, unless on the inner side of them. It appeared as though they had collected here from other places, and it is impossible to form an idea of the small space to which we were at times confined. Upwards of one hundred ice-islands could be counted at a time without the aid of a glass, some of which were several miles long. We enjoyed this beautiful sight with the more pleasure, for we had become used to them, and knew from experience that it was possible to navigate through them without accident.

On the 18th, we continued beating to the eastward, and found no end to the apparently interminable barrier. We had a smooth sea, and better weather than I anticipated. At noon, we had retraced our way about forty miles. To-day we again had snow, which fell in the form of regular six-pointed stars. The needles of which these stars were formed were quite distinct, and of regular crystals. The temperature at the time was 28°. The barometer stood at 28·76 in., about three-tenths lower than we had had it for the last twelve days. The wind was easterly.

19th. During this day the barrier trended more to the northeast, and we not unfrequently entered bays so deep as to find ourselves, on reaching the extremity, cut off by the barrier, and compelled to return to within a few miles of the place where we had entered. I thought at first that this might have been caused by the tide or current, but repeated trials showed none. Neither did I detect any motion in the floating ice except what was caused by the wind. Our longitude to-day was 101° E., latitude 63° 02′ S. Some anxiety seemed to exist among the officers and crew lest we should find ourselves embayed or cut off from the clear sea, by a line of barrier. There appeared strong reason for this apprehension, as the smooth sea we had had for several days still continued; we had been sailing as if upon a river, and the water had not assumed its blue colour.

It was, therefore, with great pleasure that, on the 20th, a slight swell was perceived, and the barrier began to trend more to the northward, and afterwards again to the westward. In the morning we found ourselves still surrounded by great numbers of ice-islands. After obtaining a tolerably clear space, the day being rather favourable, we sounded with the deep-sea line eight hundred and fifty fathoms. Six's thermometer gave at the surface 31°, and at the depth of eight hundred and fifty fathoms 35°, an increase of four degrees. The current was again tried, but none was found. A white object was visible at eleven fathoms. The water had now assumed a bluish cast.

We endeavoured to-day to land on an iceberg, but there was too much sea. Shrimps were in great quantities about it, but swam too deep to be taken. The wind again hauled to the westward, which disappointed me, as I was in hopes of getting to the position where Cook saw the ice in 1773, being now nearly in the same latitude. It was less than one hundred miles to the westward of us; and little doubt can exist that its situation has not materially changed in sixty-seven years.

The observations of the squadron during this season's Antarctic cruise, together with those of the preceding year, would seem to confirm the opinion that very little change takes place in the line of ice. It may be inferred that the line of perpetual congelation exists in a lower latitude in some parts of the southern hemisphere than in others. The icy barrier retreats several degrees to the south of the Antarctic Circle to the west of Cape Horn, while to the eastward it in places advances to the northward of that line, which is no doubt owing to the situation of the land. From the great quantities of ice to be found drifting in all parts of the ocean in high southern latitudes, I am induced to believe that the formation of the ice-islands is much more rapid than is generally supposed. The manner of their formation claimed much of my attention while among them, and I think it may be explained satisfactorily and without difficulty. In the first place, I conceive that ice requires a nucleus, whereon the fogs, snow, and rain, may congeal and accumulate; this the land affords. Accident then separates part of this mass of ice from the land, when it drifts off, and is broken into many pieces, and part of this may again join that which is in process of formation. The sketch in Chapter IX. has already given the reader some idea of its appearance in this state.

From the accumulation of snow, such a mass speedily assumes a flat or table-topped shape, and continues to increase. As these layers accumulate, the field-ice begins to sink, each storm (there of frequent occurrence) tending to give it more weight. The part which is now attached to the land remains aground, whilst that which is more remote being in deep water is free to sink. The accumulated weight on its outer edge produces fissures or fractures at the point where it takes the ground, which the frosts increase; thus separated, the surface again becomes horizontal, and continues to receive new layers from snow, rain, and even fogs, being still retained to the parent mass by the force of attraction. The fogs have no small influence in contributing to the accumulation: some idea may be formed of the increase from this cause, from the fact that during a few hours the ice accumulated to the thickness of a quarter of an inch on our rigging and spars, though neither rain nor snow fell. It may, therefore, I think, be safely asserted that these icebergs are at all times on the increase; for there are few days, according to our experience in this climate, in which some mode of precipitation does not prevail in these high latitudes, where, according to our observations, ice seldom melts. The temperature of even the summer months being rarely above the freezing point, masses of a thousand feet in thickness might require

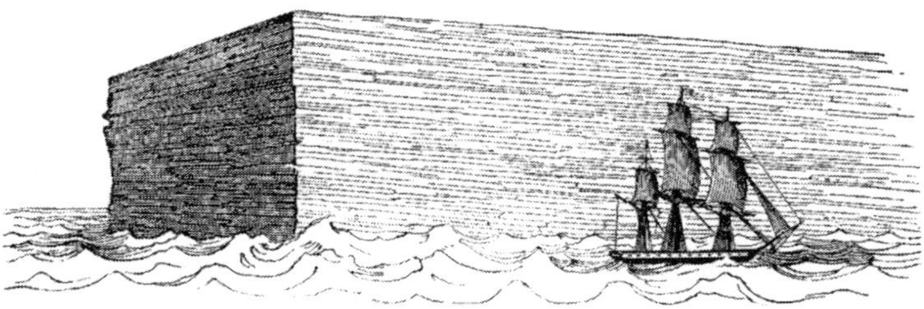

TABULAR ICEBERG.

but few years to form. Icebergs were seen in all stages of formation, from five to two hundred feet above the surface, and each exposed its stratification in horizontal layers from six inches to four feet in thickness. When the icebergs are fully formed, they have a tabular and stratified appearance, and are perfectly wall-sided, varying from one hundred and eighty to two hundred and ten feet in height. These were frequently found by us in their original situation, attached to the land, and having the horizontal stratification distinctly visible.

In some places we sailed for more than fifty miles together, along a straight and perpendicular wall, from one hundred and fifty to two hundred feet in height, with the land behind it. The icebergs found along the coast afloat were from a quarter of a mile to five miles in length; their separation from the land may be effected by severe frost rending them asunder, after which the violent and frequent storms may be considered a sufficient cause to overcome the attraction which holds them to the parent mass. In their next stage they exhibit the process of decay, being found fifty or sixty miles from the land, and for the most part with their surfaces inclined at a considerable angle to the horizon. This is caused by a change in the position of the centre of gravity, arising from the abrading action of the waves.

INCLINED ICEBERG.

By our observations on the temperature of the sea, it is evident that these ice-islands can be little changed by the melting process before they reach the latitude of 60°. The temperature of the sea (as observed by the vessels going to and returning from the south), showed but little change above this latitude, and no doubt it was at its maximum, as it was then the height of the summer season.

During their drift to the northward, on reaching lower latitudes, and as their distance from the land increases, they are found in all stages of decay; some forming obelisks; others towers and Gothic arches; and all more or less perforated: some exhibit lofty columns, with a natural bridge resting on them of a lightness and beauty inconceivable in any other material. The following wood-cut and the tail-pieces of the chapters are sketches of some of them.

While in this state, they rarely exhibit any signs of stratification and some appear to be formed of a soft and porous ice; others are quite blue; others again show a green tint, and are of hard flinty ice.

ICEBERG.

Large ice-islands are seen that retain their tabular tops nearly entire until they reach a low latitude, when their dissolution rapidly ensues; whilst some have lost all resemblance to their original formation, and had evidently been overturned. The process of actually rending asunder was not witnessed by any of the vessels, although in the Flying-Fish, when during fogs they were in close proximity to large ice-islands, they inferred from the loud crashing, and the sudden splashing of the sea on her, that such occurrences had taken place. As the bergs gradually become worn by the abrasion of the sea, they in many cases form large overhanging shelves, about two or three feet above the water, extending out ten or twelve feet; the under part of this projecting mass exhibits the appearance of a collection of icicles hanging from it. The temperature of the water when among the icebergs, was found below or about the freezing point.

I have before spoken of the boulders embedded in the icebergs. All those that I had an opportunity of observing, apparently formed a part of the nucleus, and were surrounded by extremely compact ice, so that they appear to be connected with that portion of the ice that would be the last to dissolve, and these boulders would therefore in all probability, be carried to the farthest extent of their range before they were let loose or deposited.

The ice-islands, on being detached from their original place of formation by some violent storm, are conveyed to the westward by the southeast winds which are prevalent here, and are found, the first season after their separation, about seventy miles north of the barrier. This was inferred from the observations of both the Vincennes and Porpoise, the greatest number having been found about that distance from the barrier. That these were recently detached is proven by their stratified appearance; while those at a greater distance had lost their primitive form, were much worn, and showed many more signs of decay. Near the extreme point of the barrier visited, in longitude 97° E., latitude 62° 30′ S., and where it begins to trend to the westward, vast collections of these islands were encountered. From this point they must pass to the northward during the next season, partly influenced by the current, and partly scattered by the prevailing winds, until they reach the sixtieth degree of latitude, when they encounter the easterly and north-easterly streams that are known to prevail, which carry them rapidly to the north.

Our data for their actual drift, though not altogether positive, are probably the best that can be had, and will go far towards ascertaining the velocity of their progress to lower latitudes; our observations also furnish some estimate of the time in which they are formed. On our way south, we did not fall in with ice-islands until we reached latitude 61° S. The Peacock was the first to return, and nearly upon the track by which we had gone south; the last seen by her was in 55° S. The Vincennes, on her return fifty days later, saw them in 51° S. The Porpoise, about the same time, in 53° S. The observation in the Vincennes gives a distance of ten degrees of latitude, or six hundred miles to be passed over in fifty days, which would give about half a mile an hour; or, taking the Peacock's observations, a more rapid rate would be given, nearly three-fourths of a mile. Many icebergs were met in the latitude of 42° S., by outward-bound ships to Sydney, in the month of November; these, I learned, were much worn, and showed lofty pinnacles, exhibiting no appearance of having ever been of a tabular form. These no doubt are such as were detached during a former season, and being disengaged from the barrier, would be naturally, early the next season, drifted by the easterly current as well as the westerly wind, and would pursue the direction they give them. They would therefore be driven to the northeast as far as the southwest winds prevail, and when these veer to the westward would receive an easterly direction. It is where these winds prevail that they are most frequently found by the outward-bound vessels,—between the latitudes of 40° and 50° S.

Respecting the period of time required for the formation of these ice-islands, much light cannot be expected to be thrown on the subject; but the few facts derived from observations lead to some conclusions. Many of them were measured, and their altitude found to be from fifty to two hundred and fifty feet; eighty distinct stratifications were counted in some of the highest, and in the smallest thirty, which appeared to average a little more than two feet in thickness

Supposing the average fall of snow in these high latitudes to be an inch a day, or thirty feet a year, the largest icebergs would take more than thirty years to form. They were seen by us in all the stages of their growth, and all bore unequivocal marks of the same origin. The distance from the land at which they were forming, fully satisfied me that their fresh water could only be derived from the snows, &c.

The movement of the ice along the coast is entirely to the westward, and all the large ranges of ice-islands and bergs were found in that direction, while the eastern portion was comparatively free from it. A difference was found in the position of the floe-ice by the different vessels, caused rather by the wind than by the tide. When the Vincennes and Porpoise passed the opening by which the Peacock entered, it was found closed, although only twenty-four hours had elapsed. It has been seen that the ice had much movement during the time the Peacock was beset by it, and the bay was all but closed when she effected her escape. Another instance occurred, where the Porpoise, in about the longitude of 130° E., found the impracticable barrier a few miles further south than the Vincennes did six or seven days after; but this fact is not to be received as warranting any general conclusion, on account of the occurrence of southeast gales during the intermediate time. The trials for currents have, for the most part, shown none to exist. The Porpoise, it is true, experienced some, but these were generally after a gale. If currents do exist, their tendency is westward, which I think the drift of the ice would clearly prove. The difference between the astronomic positions and those given by dead-reckoning, was of no avail here as a test,* for the courses of the vessels among the ice were so tortuous, that the latter could not be depended upon.

The winds which prevail from the southwest to the southeast occasionally bring clear weather, interrupted by flurries of snow; the north wind is light, and brings thick fogs, attended by a rise of temperature. Extremes of weather are experienced in rapid succession, and it is truly a fickle climate.

The evidence that an extensive continent lies within the icy barrier, must have appeared in the account of my proceedings, but will be, I think, more forcibly exhibited by a comparison with the aspect of other lands in the same southern parallel. Palmer's Land, for instance, which is in like manner invested with ice, is so at certain seasons of

* The fact of there being no northerly current along this extended line of coast, is a strong proof in my mind of its being a continent, instead of a range of islands.

the year only, while at others it is quite clear, because strong currents prevail there, which sweep the ice off to the northeast. Along the Antarctic Continent for the whole distance explored, which is upwards of fifteen hundred miles, no open strait is found. The coast, where the ice permitted approach, was found enveloped with a perpendicular barrier, in some cases unbroken for fifty miles. If there was only a chain of islands, the outline of the ice would undoubtedly be of another form; and it is scarcely to be conceived that so long a chain could extend so nearly in the same parallel of latitude. The land has none of the abruptness of termination that the islands of high southern latitudes exhibit; and I am satisfied that it exists in one uninterrupted line of coast, from Ringgold's Knoll, in the east, to Enderby's Land, in the west; that the coast (at longitude 95° E.) trends to the north, and this will account for the icy barrier existing, with little alteration, where it was seen by Cook in 1773. The vast number of ice-islands conclusively points out that there is some extensive nucleus which retains them in their position; for I can see no reason why the ice should not be disengaged from islands, if they were such, as happens in all other cases in like latitudes. The formation of the coast is different from what would probably be found near islands, soundings being obtained in comparatively shoal water; and the colour of the water also indicates that it is not like other southern lands, abrupt and precipitous. This cause is sufficient to retain the huge masses of ice, by their being attached by their lower surfaces instead of their sides only.

Much inquiry and a strong desire has been evinced by geologists, to ascertain the extent to which these ice-islands travel, the boulders and masses of earth they transport, and the direction they take.

From my own observations, and the information I have collected, there appears a great difference in the movements of these vast masses; in some years, great numbers of them have floated north from the Antarctic Circle, and even at times obstructed the navigation about the capes. The year 1832 was remarkable in this respect; many vessels bound round Cape Horn from the Pacific, were obliged to put back to Chili, in consequence of the dangers arising from ice; while, during the preceding and following years, little or none was seen: this would lead to the belief, that great changes must take place in the higher latitudes, or the prevalence of some cause to detach the ice-islands from the barrier in such great quantities as to cover almost the entire section of the ocean, south of the latitude 50° S. Taking the early part of the (southern) spring, as the time of separation, we are enabled to make some estimate of the velocity with which they move:

many masters of vessels have met them, some six or seven hundred miles from the barrier, from sixty to eighty days after this period, which will give a near approximation to our results heretofore stated.

The season of 1839 and '40 was considered as an open one, from the large masses of ice that were met with in a low latitude, by vessels that arrived from Europe at Sydney: many of them were seen as far north as latitude 42° S.

The causes that prevail to detach and carry them north, are difficult to assign. I have referred to the most probable ones that would detach them from the parent mass in their formation. Our frequent trials of currents, as has been stated, did not give us the assurance that any existed; but there is little doubt in my mind that they do prevail. I should not, however, look to a surface current as being the motive power that carries these immense masses at the rate they move; comparatively speaking, their great bulk is below the influence of any surface current, and the rapid drift of these masses by winds is still more improbable; therefore I conceive we must look to an under current as their great propeller. In one trial of the deep-sea thermometer, we found the temperature beneath, four degrees warmer than the surface. Off Cape Horn, the under temperature was found as cold as among the ice itself; repeated experiments have shown the same to occur in the Arctic regions. From this I would draw the conclusion that changes are going on, and it appears to me to be very reasonable to suppose, that at periods, currents to and from the poles should at times exist; it is true, we most generally find the latter to prevail, as far as our knowledge of facts extends, but we have not sufficient information yet to decide that there is not a reflow towards the pole; the very circumstance of the current setting from the higher latitudes, would seem a good argument that there must be some counter-current to maintain the level of the waters. These masses, then, are most probably carried away in the seasons when the polar streams are the strongest, and are borne along by them at the velocity with which they move: that these do not occur annually may be inferred from the absence of ice-islands in the lower latitudes; and that it is not from the scarcity of them, those who shared the dangers of the Antarctic cruise, will, I have little doubt, be ready to testify; for, although great numbers of them studded the ocean that year, yet the narrative shows that vast numbers of them were left.

The specific gravity of the ice varies very much, as might naturally be expected; for while some of it is porous and of a snowy texture, other islands are in great part composed of a compact blue flinty ice. This difference is occasioned by the latter becoming saturated with water, which afterwards freezes.

On the ice there was usually a covering of about two feet of snow, which in places had upon it a crust of ice not strong enough to bear the weight of a man. Those ice-islands, which after having been once seen were again passed through immediately after a gale, were observed to be changed in appearance; but though for forty-eight hours a severe storm had been experienced, they had not undergone so great a transformation as not to be recognised. They also appeared to have shifted their position with regard to one another, their former bias and trendings being broken up.

During our stay on the icy coast, I saw nothing of what is termed pack-ice,—that is, pieces forced one upon the other by the action of the sea or currents.

On the 21st, the weather became unsettled, with light westerly winds, and we made but little progress to the westward. The barrier, at 6 P. M., was seen trending to the westward. In consequence of indications that threatened bad weather, I deemed it useless risk to remain in the proximity of so many ice-islands; and a strong breeze, with squally weather, having already set in, I took advantage of it, feeling satisfied that our farther continuance in this icy region would not only be attended with peril to the ship, but would cause a waste of the time which was demanded by my other duties; and having nearly three thousand miles to sail to our next port (Bay of Islands), I made up my mind to turn the head of the vessel northward.

I therefore had the officers and crew called aft, thanked them all for their exertions and good conduct during the trying scenes they had gone through, congratulated them on the success that had attended us, and informed them that I had determined to bear up and return north.

Having only twenty-five days' full allowance of water, I ordered its issue to be reduced to half allowance.

I have seldom seen so many happy faces, or such rejoicings, as the announcement of my intention to return produced. But although the crew were delighted at the termination of this dangerous cruise, not a word of impatience or discontent had been heard during its continuance. Neither had there been occasion for punishment; and I could not but be thankful to have been enabled to conduct the ship through so difficult, and dangerous a navigation without a single accident, with a crew in as good, if not in a better condition than when we first reached the icy barrier. For myself, I indeed felt worse for the fatigues and anxieties I had undergone; but I was able to attend to all my duties, and considered myself amply repaid for my impaired health by the important discoveries we had made, and the success that had attended our exertions.

I shall now leave the Vincennes to pursue her route northward, and return to the Porpoise, the result of whose proceedings will be detailed in the following chapter.

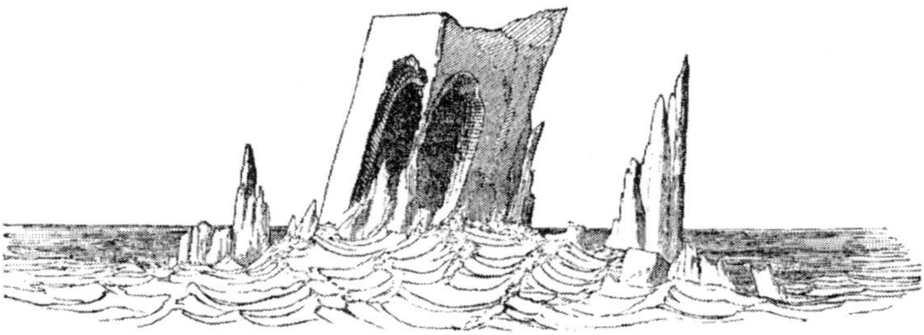

ICE-ISLAND.

CHAPTER XI.

CONTENTS.

PROCEEDINGS OF THE PORPOISE FROM THE TWENTY-SECOND TO THE THIRTIETH OF JANUARY—FRENCH SQUADRON SEEN—ITS COMMANDER REFUSES TO SPEAK THE PORPOISE—PROCEEDINGS UP TO THE THIRD OF FEBRUARY—GALE—FURTHER PROCEEDINGS TO THE TWELFTH OF FEBRUARY—SPECIMENS OF ROCK OBTAINED—WESTERN LIMIT OF HER CRUISE—RETURN TO THE EASTWARD—PORPOISE STANDS TO THE NORTHWARD—AUCKLAND ISLANDS—PORPOISE ARRIVES AT THE BAY OF ISLANDS—CRUISE OF THE FLYING-FISH—LANDING AT MACQUARIE'S ISLAND—PROCEEDINGS OF THE FLYING-FISH UP TO THE FOURTH OF FEBRUARY—STATE OF HER CREW—THEIR LETTER TO LIEUTENANT PINKNEY—HE RESOLVES TO RETURN—ARRIVAL OF THE FLYING-FISH AT THE BAY OF ISLANDS—EVENTS DURING THE RETURN OF THE VINCENNES—SHE FAILS TO REACH VAN DIEMEN'S LAND—ARRIVAL OF THE VINCENNES AT SYDNEY—PEACOCK FOUND THERE—RETURN OF THE PEACOCK FROM THE ICY BARRIER—SHE MAKES MACQUARIE'S ISLAND—SHE ARRIVES AT SYDNEY—STATE OF THE PEACOCK—HOSPITALITIES RECEIVED AT SYDNEY.

CHAPTER XI.

ANTARCTIC CRUISE—CONTINUED.

1840.

On the 22d January, the Porpoise lost sight of the Peacock, and continued beating to the southwest. The weather was extremely cold; sea-water froze on being a few minutes in the bucket on deck. Some shrimps were caught. The water at 3 P. M. was much discoloured; got a cast of the lead with two hundred fathoms: no bottom; found the current south-by-east three-fourths of a mile per hour. At 4^h 30^m, passed large icebergs, one of which had several dark horizontal veins, apparently of earth, through it; large quantities of floe and drift-ice to the southward; the sea very smooth. A report of high land was made this morning; indeed every thing indicated the proximity of land. The number of seals, whales, penguins, shrimps, &c., had very much increased. The pure white pigeons were also seen in numbers.

23d. Countless icebergs in sight; the sea quite smooth; not the slightest motion perceptible. At meridian, they were in latitude 66° 44′ S., longitude 151° 24′ E., and close to the barrier, which appeared quite impenetrable, as far as the eye could reach from aloft, to the north-northwest and north-northeast, with numberless immense ice-islands entangled and enclosed in it in all directions. The position they occupied seemed an inlet of elliptical shape, with an opening to the north. It was needless to count the many scattering islands of ice distinct from the vast chain; intermingled with field-ice, they studded the gulf like so many islands, of various shapes and dimensions. At 2^h 25^m, a sail was discovered on the lee bow; kept off to communicate, supposing it to be the Vincennes or Peacock. At 2^h 30^m, the

Peacock was made out on the southern board, showing no disposition to communicate; showed our colours, and hauled to the westward.

24th. The day was remarkably fine, such as is seldom experienced in this region. The water appeared much discoloured and of a dirty olive-green colour. At meridian, they again made the field-ice, and tacked to the northward, passing through large quantities of ice-islands; weather looking bad, with occasional light snow-storms.

25th. Part of this day was clear and pleasant, though snow fell at intervals; the field-ice was in sight several times, and many ice-islands of great size and beauty. Penguins were swimming round, and also several shoals of black-fish; a black albatross was shot; towards night the weather became very thick; they were in longitude 150° E., latitude 65° 56′ S.

26th. Fresh winds blowing from the eastward; during the first few hours, a thick snow-storm; at 4 A. M. it cleared; at six o'clock made a sail; the strange sail fired a gun and made signal, when we bore down and spoke her; she proved to be the Vincennes; compared chronometers, and received rate; bore off to the westward under all sail; found the drift and floe-ice very thick, and were with great difficulty enabled to navigate through it; wind fresh, with a long swell from the southwest; at 5^h 30^m, the ice increasing in quantity, found it was necessary to haul off. Lost sight of the Vincennes; weather very threatening. The course during the day proved a very tortuous one; many penguins resting on the ice; their gait is an awkward kind of strut.

Received orders to-day by signal to meet the Vincennes along the icy barrier between the 20th and 28th of next month.

27th. This day proved clear and cold; wind from the southwest; ice forming rapidly on the vessel; at meridian, lost sight of the Vincennes; very many ice-islands in sight; latitude 65° 41′ S., longitude 142° 31′ E. On this day, Lieutenant-Commandant Ringgold determined with the fair wind to pass to the extreme limit of his orders, longitude 105° E.; being of opinion he would thereby save time, and be enabled more effectually to examine the barrier with what he thought would be found the prevailing wind, viz.: that from the westward; in this, however, he was mistaken.

The 28th set in with a light breeze from east-northeast; made all sail; at 5 A. M., wind increasing rapidly, snow falling fast, and weather becoming thick; at six o'clock, made the floe and drift-ice; shortened sail, and hauled off to the northwest, it becoming so thick as to render any advance unsafe; until meridian, very strong winds from the eastward, the brig under close-reefed topsails; at 2 P. M. found it difficult and hazardous to proceed, passing within a short distance of ice-islands, and just seeing them dimly through the obscurity; at three, the brig was hove-to, and Lieutenant-Commandant Ringgold says, in reference to their situation—

"I felt great anxiety to proceed, but the course was so perilous, the extent and trend of the barrier so uncertain, I could not reconcile it with prudence to advance. The frequent falling in with fields of drift-ice, the numerous and often closely-grouped chains of icebergs, were sufficient to point out discretion. The long-extended barrier was encountered in latitude 65° 08′ S.; at twelve to-day our position was 65° 16′ S.; it is easy to perceive the possibility of a trend northerly again, which would have placed us in a large and dangerous gulf, with a heavy gale blowing directly on, without a hope of escape.

"At 8 P. M., blowing very heavy; the snow falling rendered vision beyond a few yards impossible; I have seldom experienced a heavier blow, and towards the conclusion the squalls were severe and frequent."

The barometer at 3 A. M., stood at 28·200 in., the lowest point it reached during the gale. The temperature of the air was 26°.

The severe gale continued during the 29th, with a heavy sea, and snow falling thickly; at 8 A. M. the gale abated, and the clouds broke away; through the day the sun occasionally out; the weather appeared unsettled; the sun set red and fiery; the latitude was observed 64° 46′ S., longitude 137° 16′ E.

On the 30th they stood again to the southwest; at 2 A. M. they made the barrier of field-ice, extending from southeast to west, when it became necessary to haul more to the northwest; the weather becoming thick with a heavy fall of snow, at four o'clock, the wind increasing, compelled them to shorten sail; at 7^h 30^m the ice in fields was discovered close aboard, heading west; at this time hauled immediately on a wind to the northeast, and soon passed out of sight of the ice and out of danger; during the day blowing a gale of wind, and very heavy sea running, passing occasional ice-islands; at meridian, being clear of the barrier, the brig was hove-to under storm-sails, to await the clearing of the weather. In the afternoon the weather showed signs of clearing; the sun coming out, again made sail to approach the barrier; no ice in sight; great numbers of black petrels about.

At 4 P. M. discovered a ship ahead, and shortly after another was made, both standing to the northward; the brig hauled up to the northwest, intending to cut them off and speak them, supposing them to be the Vincennes and the Peacock; shortly afterwards they were seen to be strangers, being smaller ships than our own; at 4^h 30^m the Porpoise hoisted her colours. Knowing that an English squadron under Captain Ross was expected in these seas, Lieutenant-Commandant Ringgold took them for his ships, and was, as he says, "preparing to cheer the discoverer of the North Magnetic Pole."

"At 4^h 50^m, being within a mile and a half, the strangers showed French colours: the leeward and sternmost displayed a broad pennant; concluded now that they must be the French discovery ships under Captain D'Urville, on a similar service with ourselves: desirous of speaking and exchanging the usual and customary compliments incidental to naval life, I closed with the strangers, desiring to pass within hail under the flag ship's stern. While gaining fast, and being within musket-shot, my intentions too evident to excite a doubt, so far from any reciprocity being evinced, I saw with surprise sail making by boarding the main tack on board the flag-ship. Without a moment's delay, I hauled down my colours and bore up on my course before the wind."

It is with regret that I mention the above transaction, and it cannot but excite the surprise of all that such a cold repulse should have come from a French commander, when the officers of that nation are usually so distinguished for their politeness and attention. It was with no small excitement I heard the report of it,—that the vessels of two friendly powers, alike engaged upon an arduous and hazardous service, in so remote a region, surrounded with every danger navigators could be liable to, should meet and pass without even the exchange of common civilities, and exhibit none of the kind feelings that the situation would naturally awaken:—how could the French commander know that the brig was not in distress or in want of assistance? By refusing to allow any communication with him, he not only committed a wanton violation of all proper feeling, but a breach of the courtesy due from one nation to another. It is difficult to imagine what could have prompted him to such a course.

At 6 P. M. the weather again was thick, with the wind southeasterly; field-ice again in sight; it commenced snowing and the French ships were lost sight of. At 8 P. M., they passed in sight of large fields of ice and ice-islands; at 10^h 30^m, the snow falling so dense and the weather so thick, that it was impossible to see the brig's length in any direction; she was hove-to, to await a change of weather.

The beginning of the 31st the gale continued; at 7 A. M. moderating, they again made sail to the westward; in half an hour discovered a high barrier of ice to the northward, with ice-islands to the southward; at 10 A. M., they found themselves in a great inlet formed by vast fields of ice, which they had entered twelve hours previously; the only opening appearing to the eastward, they were compelled to retrace their steps, which they effected at 8 P. M., passing some ice-islands which they recognised as having been seen the evening before. They now found themselves out of this dangerous position, and, passing the point, kept away to the westward. Lieutenant-Commandant Ringgold judged it prudent to heave-to during the night, on account of the darkness.

February 1st. The immense perpendicular barrier encountered yesterday was now in sight, trending as far as the eye could reach to the westward; it was of tabular form, from one hundred and fifty to one hundred and eighty feet in height, of solid compact ice, resembling a long line of coast; wind moderate from the southeast,—a brilliant blink extending along and elevated above the barrier. At 4 P. M., they arrived at the end of this barrier, and found it trending off to the southward, seeming as if numbers of icebergs had been broken from the barrier by some mighty force, exceeding in numbers any thing that had yet been seen, and extending as far south as could be distinguished, interspersed with much drift and floe-ice. On the southern horizon sixty-four ice-islands were counted, exclusive of many near them, and those that were not distinguishable from the barrier.

The current was tried here, and found setting southeast nearly a mile an hour. Pigeons around in numbers, also whales and large flocks of penguins.

The nights now evidently lengthened, thus adding to the cares and anxieties attendant on this navigation. It was fortunate that the prevailing winds were from the southeast and southwest, or coming off the ice. If they had blown from the northward, they would have been attended with danger, and might have proved fatal to the vessel.

2d. At meridian, in longitude 130° 36′ E., and latitude 65° 24′ S. They were prevented from proceeding farther to the southward by the impenetrable icy barrier. At this time they had one hundred large ice-islands in sight, without counting any of the smaller bergs, which were innumerable; saw great numbers of penguins and some seals (Phoca proboscida). The current was tried here, and found setting as yesterday, and at the same rate.

At 8 P. M., were obliged to retrace their steps to the northward, the weather becoming thick, with light snow. At eleven, constant and thick snow-storm, and unable to see any distance; the gale continuing, lay-to under a close-reefed main-topsail. The vignette, from a sketch by Mr. Totten will give some idea of her situation.

PORPOISE IN A GALE.

3d. A gale from southeast, heavy sea rising; occasionally passing ice-islands and field-ice. The gale continued throughout the day, but moderated towards midnight; the sea was heavy, the weather thick, and the brig completely covered with ice and snow. The barometer fell to 28·040 in. Temperature of the air 32°.

4th. Although the wind was moderate, yet it was so thick and foggy as to preclude bearing up. Towards meridian it cleared sufficiently for them to bear up and continue their examinations. To day the current was found west-northwest, three quarters of a mile per hour.

On the 5th they had a beautiful day,—no climate or region, Lieutenant-Commandant Ringgold remarks, could have produced a finer: this gave them an opportunity of thoroughly drying every thing and ventilating the vessel, which was much required; standing to the northward, in order to make a long board to the westward; the longitude 127° 08′ E., latitude 63° 22′ S.; few ice-islands in sight, and those appeared much worn, showing marks of rapid decay, with isolated pieces,—some standing erect, while others were inclined, resembling fragments of columns and broken arches. This night there was a brilliant display of the aurora australis: at eleven o'clock there was perceived in the northern horizon a luminous arched cloud, at 15° of altitude, extending from northwest to northeast; the stars were partially obscured in the direction of the clouds; the pale flashes or coruscations vanishing very suddenly, were succeeded by spiral columns or streamers, converging with great velocity towards the zenith; brilliant flashes would again issue forth from the remote parts of the cloud, succeeded in quick succession by perpendicular rays emanating from the cloud, having the shape of a rounded column or basaltic-shaped cylinder, which in contrast with the dark cloud showed in broad relief. As the cloud seemed to rise, the scene became a most interesting one, from the varied and oft-changing coruscations: finally the arc assumed a contracted and elliptical form, vivid streamers bursting forth as if from a corona, converging all towards the zenith, until they were lost in the coming day. The magnetic needle did not show any disturbance. The barometer stood stationary during its continuance. The sympiesometer indicated a slight fall. At the time there was no wind; the stars were brilliant, and all visible.

6th. During this day they had light winds; pursued their course to the westward; wind from the southward. In the afternoon they had light flurries of snow, and at times hail; the sea perfectly smooth, and few icebergs in sight. Longitude 125° 32′ E., latitude 63° 34′ S.

During the 7th, the winds variable; at eight tacked to the southward, in order to close in with the barrier; the wind again hauling, tacked; the number of icebergs increasing; all those seen for the few days past have appeared variously shaped, much worn and fractured, some evidently overturned, and immense arches or caves washed in them; they were totally distinct from those seen to-day.

8th. A brisk breeze from the southward, which carried them on rapidly to the westward. At meridian, discovered compact fields of ice, with many stupendous ice-islands enclosed within it; the ice appeared more broken than any hitherto seen, with many fragments of icebergs resembling spires and broken columns. Altered their course to clear the barrier, and by two o'clock they had extricated themselves. Penguins, whales, brown pigeons, and the black albatross, were seen near the barrier. In the afternoon the snow fell in beautiful shining spiculæ, resembling stars, usually of six, but sometimes of twelve points: they varied from one-eighth to one-sixteenth of an inch in diameter.

The barrier was occasionally seen, and the ice-islands began again to assume a tabular form; towards the close of the day, very many whales, penguins, &c., seen. Longitude 116° E., latitude 64° 01′ S.

On the 9th, fresh breezes from the southeast; at 10 A. M. made the barrier again, the weather being favourable; at 4 P. M. standing along the barrier, through drift-ice, with countless icebergs in sight; good observations were obtained, placing them in longitude 112° 41′ E., and latitude 64° 55′ S. At 10 P. M., some few appearances of the aurora australis in the northern sky, light coruscations streaming upwards, but quite faint, and only for a very short period; many stars and several constellations were traced without difficulty. The sea was smooth; lowered a boat to try the current, but found none. The dip was 83° 30′.

On the morning of the 10th the weather cleared off, and gave them an opportunity of ventilating the vessel; closed in with the field-ice for the purpose of obtaining a supply of water, and the boats were despatched to take in ice; the longitude was found to be 110° 34′ E., latitude 65° 12′ S.; the field-ice here was found to be interspersed with many large ice-islands and bergs. At five o'clock the boats returned with ice. The current was found to be setting north-northeast, five fathoms an hour; the weather continued clear and healthful; made the field-ice ahead and on the lee bow; shortly after, cleared it. The twilight in the southern horizon presented a beautiful appearance, a bright salmon colour radiating from the sun, throwing its tints over the whole sky, tinging the few cirro-stratus clouds that were in the northern quarter, and giving a soft colour to the immense ice-islands that were slumbering along the barrier, and aiding to lend to the scene its peculiar character of silence, solitude, and desolation.

The weather was clear and pleasant on the 11th, with a light wind from the southeast; many penguins and whales were seen. The icebergs were numerous, and some of great beauty, with almost regularly turned arches, and of the most beautiful aqua-marine tints. Longitude was 106° 10′ E., latitude 65° 28′ S.

During the morning of the 12th, running along high broken fields of ice, with a light breeze from the southward; weather overcast; discovered a large piece of ice of a dark brown colour floating by, resembling a piece of dead coral; lay-to, and sent a boat to bring it alongside; obtained from it several pieces of granite and red clay, which were frozen in; the ice was extremely hard and compact, composed of alternate layers of ice and snow; the strata of snow was filled with sand. The icebergs near at the time presented signs of having been detached from land, being discoloured by sand and mud. A number of white procellaria were obtained. The ice-islands again appeared in great numbers. At 3 P. M. hauled up, steering westerly into a very deep inlet or gulf, formed by extensive fields of ice. Believing from the indications of the morning that land could not be far off, in approaching the head of this inlet, several icebergs had the appearance of being in contact with the land, having assumed a dark colour from the clay and sand blown upon them; the whole group around seemed as if in the vicinage of land; sounded with two hundred fathoms: no bottom; also tried the current, but found none. Towards night, it becoming thick with snow, they continued under snug sail, intending to examine more closely the barrier and inlets in the morning.

13th. At 3 A. M. they again made sail to the westward, with wind from the east; at six o'clock they had snow-squalls, rendering it unsafe to proceed, and impossible to make any discovery. A few hours afterwards the weather cleared a little; made sail again to the northwest. At meridian overcast, with a stiff southeast breeze; at 1^h 30^m, approached to within pistol-shot of the barrier, observing much of the dark dirty ice interspersed with the field-ice; kept along it very closely, tracing the barrier northerly; observed a large black object on the ice; shortened sail, and despatched a boat: it proved to be a large mass of black, red, and mixed-coloured earth, resting upon a base of snow and ice, situated some fifty yards back from the margin of the field-ice, and was found to be red earth, mixed with granite and sandstone. Penguins were also procured alive. At 3 P. M. they again followed the trend of the ice in a northwesterly direction; a vast field, of uninterrupted extent, seemed moving along to the westward, the large icebergs containing dark and discoloured masses, with frequent strata of the same description. They were still at a loss to account for these frequent signs of land; discoloured pieces of ice seemed mingled with the general mass; they were often seen along its margin, and appeared as though the icebergs had been turned over, presenting collections as if from the bottom. Great numbers of sperm whales were seen this day. At 8 P. M. they passed out northwardly with a light breeze and smooth sea, through an extensive chain of icebergs, which seemed grouped off the western point of the barrier: upwards of one hundred of them were counted, several of which were very much discoloured. The sunset was brilliant, bright crimson tints illuminating the icebergs, and producing a beautiful effect.

On the 14th, Lieutenant-Commandant Ringgold, having passed a few degrees beyond his instructions, that is, having reached longitude 100° E., and latitude 64° 15′ S., now commenced his return, in order to examine those places in the barrier which he had been prevented from doing on his way west.

15th. Continued their course to the eastward. Lieutenant-Commandant Ringgold frequently refers to the happy and cheerful condition of his crew, and their freedom from all disease.

On the 16th and 17th, they were employed in getting to the eastward, passing many worn and shattered bergs. On the evening of the latter day, they had another exhibition of the aurora australis, extending from north-northwest to east; it was of a light straw-colour, but very indistinct; the luminous bank was at an elevation of 30°. The light in the northwest was most distinct, radiating from a nucleus above the horizon towards the zenith, where it formed a beautiful halo. It was not of long duration. Many ice-islands and bergs in sight: upwards of two hundred; nearly all of a tabular form,—the sides of many of them beautifully excavated by the waves, presenting innumerable Gothic arches, extending often to a considerable distance into the body of the ice.

Their position on the 18th was in longitude 114° 17′ E., latitude 62° 37′ S. Flocks of black-birds were very numerous, but not near enough to be taken.

On the 19th and 20th, proceeding to the eastward. On the 20th, they had but few ice-islands in sight, although they were seventy miles further south than on the 18th, when the largest number ever seen by them at one time was visible; having reached the longitude of 120° E., they again steered south, to make the barrier. The current was tried, but none found.

The 21st proved stormy, with strong breezes from the southeast, and much snow and rain; which covered the brig with ice. Field-ice was seen ahead, when they again stood to the eastward, longitude being 121° 30′ E., latitude 65° 15′ S. On this night they experienced a heavy gale, during which the barometer fell to 27·50 in., where it remained during part of the 22d. The squalls were very severe, accompanied with snow, sleet, hail, and heavy seas; they had now reached longitude 122° E., and latitude 64° 09′ S.

February 22d, being Washington's birthday, the colours were hoisted, and the crew received an extra allowance. Lieutenant-Commandant Ringgold took this occasion to express to them his satisfaction for the manner in which they had performed their duties during the present cruise, and that their conduct would be duly represented to the Commander of the Expedition, and the government.

On the 23d the weather was again thick, with snow and mist.

On the 24th they had reached longitude 126° E., and latitude 64° 29′ S. On this day they again sighted the barrier; when, having completed what he deemed a full execution of his instructions, Lieutenant-Commandant Ringgold determined to put the brig's head north, —which was accordingly done.

Strong winds and gales continued for the next three days. On the 27th they again found themselves in east variation, in longitude 138°

E., latitude 60° 08′ S. The white albatross had now again become common.

On the 29th, they had a beautiful display of the aurora australis; the whole southern hemisphere was covered with arches of a beautiful straw-colour, from which streamers radiated, both upwards and downwards, of almost a lustrous white; numbers of concentric arches would occasionally show themselves, of a width of a few feet, uniting to form a complete canopy for a moment, and then vanish. The arches extended from east-southeast to west-northwest; the display continued for over two hours; the stars were seen above them. Previous to, and during its continuance, the thermometer indicated a change of four degrees, and the wind shifted to the southward.

On the 1st of March, in latitude 55° S., and longitude 140° E., they passed the last ice-island.

On the 2d, great numbers of pyrosoma of large size were passed.

On the 4th, some faint appearances of the aurora australis were seen.

On the 5th, the Lord Auckland Isles were descried. Mr. Totten, who was officer of the deck, was accidentally knocked overboard by the trysail-boom, but was fortunately rescued without injury. Immense numbers of albatrosses were about. The aurora was again seen in the southern hemisphere.

AUCKLAND ISLES.

On the 7th they anchored in the harbour of Sarah's Bosom, in twelve fathoms water. During their brief stay here, all were actively employed wooding and watering, for which this harbour affords a fine opportunity. Assistant-Surgeon Holmes made several excursions on the largest island, of which he gives the following account:

"I found it very thickly covered with trees, in its less elevated parts; as few of them were of any size, I found no small difficulty in penetrating and making my way through them; in many places it was absolutely impossible. It was only after a long and fatiguing walk, that I succeeded in reaching the summit of that part of the island, near which the brig was anchored, where I found the trees less numerous

A thick growth of underwood and dwarf bushes, intermixed with ferns concealed the surface, rendering it difficult to walk. Even on the places apparently most level, the ground was very unequal, and a single step would sometimes send me nearly up to the neck into a hollow filled with large fern fronds. On the highest parts, the small level spots were covered only with moss, and a description of tall grass, and in places also a kind of grain grew abundantly. The ground was dry every where, all the water being found in the streams, which were numerous and pure. Near the summit, the ground was perforated in all directions, probably by birds, who rear their young in these holes. Many of the birds, principally procellaria, were sitting on the ground: they made no effort to escape, but suffered themselves to be taken without any attempt at resistance.

"The forest was full of small birds, of three or four different species, which were perfectly fearless; one little fellow alighted on my cap as I was sitting under a tree, and sang long and melodiously; another and still smaller species, of a black colour spotted with yellow, was numerous, and sang very sweetly; its notes were varied, but approximated more nearly to the song of our blackbird; occasionally a note or two resembled the larks. Hawks too were numerous, and might be seen on almost all the dead trees, in pairs. Along the sea-coast were to be seen the marks of their ravages upon the smaller birds. The sea-birds were very numerous on the opposite side of the island, sitting upon the cliffs or hovering over the islet."

On the western side of the Auckland Island, the under-brush and young trees are exceedingly thick. Dr. Holmes remarks, that it was impossible to penetrate; that he was occupied fully an hour in making his way for a hundred yards, where to all appearance a human step had never before trodden. There was not a vestige of a track; old trees were strewn about irregularly, sometimes kept erect by the pressure on all sides. Some trees were seen upwards of seventy feet in height, although the generality were only from fifteen to twenty; every part of the island was densely covered with vegetation; the soil, from the decomposition of vegetable matter, had acquired considerable richness; specimens of all the plants were collected. The botany of these islands is nearly allied to that of New Zealand, and will be found treated of in the Botanical Report, to which I would refer. Some species resembling the tropical plants were found here, viz., the coffeaceous plants.

These islands have in many places the appearance of having been raised directly from the sea; the cliffs consisted of basalt, and were generally from fifty to ninety feet perpendicular.

The Auckland Islands are the resort of whalers for the purpose of refitting and awaiting the whaling season, which occurs here in the months of April and May. Near the watering place a commodious hut has been erected by a French whaler. Near by was another in ruins, and close to it the grave of a French sailor, whose name was inscribed on a wooden cross erected over it. Some attempts at forming a garden were observed at one of the points of Sarah's Bosom, and turnips, cabbage, and potatoes, were growing finely, which, if left undisturbed, will soon cover this portion of the island; to these a few onions were added. Besides the birds, the only living creature was a small mouse, one of which Dr. Holmes caught: it made no attempt to get out of his way, and seemed to have no fear when taken; being consigned to a pocket, he soon contrived to escape. Many of the smaller islands of this group were visited; they closely resemble the larger one. Penguins were numerous and of a variety of colours.

These isles have a picturesque, wild, steep, and basaltic appearance: the highest peak was estimated to be eight hundred feet; the smaller has a less elevation: the general aspect of the land resembles the region around Cape Horn. The Harbour of Sarah's Bosom is not the most secure; that of Lawrie's is protected from all winds, and has a large and fine streamlet of water at its head. The rocks are covered with limpets, and small fish of many varieties are caught in quantities among the kelp. The crew enjoyed themselves on chowders and fries. No geese were seen, and the only game observed were a few gray ducks, snipes, cormorants, and the common shag. The land birds are excellent eating, especially the hawks; and on the whole, it is a very desirable place at which to refit.

On the 9th of March they had finished, and were prepared for sea, but the weather was threatening and caused them to delay. The magnetic dip was found to be 73° 47′ 30″ S.

A whaler, under Portuguese colours, but commanded by an Englishman, arrived, and anchored in Lawrie's Cove, to await the coming of the whales! The night proved stormy; the wind at 10^h 30^m from the northeast, blowing very heavy in puffs. Towards noon it moderated, and at 2 P. M. they got under way, with a light breeze from the northwest, and stood to sea.

The latitude of Sarah's Bosom is 50° 38′ S.; the longitude 165° 28′ E.

On the 12th no current was found; latitude 49° 27′ S., longitude 168° 13′ E. The weather experienced from this port to New Zealand was very similar to that in passing from Cape Horn to Valparaiso: northerly winds with mist and fog prevailing, with a heavy sea. On the 17th they fell in with the whale-ship Mary and Martha, of Plymouth, Coffin, master, who informed them that there were at least one hundred whale-ships cruising in the neighbouring seas; of these, several were seen. This will give some idea of the number of vessels employed, and how great a capital is engaged in this business.

On the 18th they had a gale from north-northwest, which lasted through the day, moderating at sunset. They were in latitude 43° 02′ S., longitude by chronometer, 175° 24′ E. The barometer sank to 29·30 in. A current was experienced setting northwest, in the direction of Cook's Straits.

On the 20th, in latitude 41° 00′ S., longitude 177° E., the current was found setting northeast-by-north, half a mile per hour. On the 22d and 23d they experienced a heavy gale from the southeast, when they were in longitude 179° 35′ E., and latitude 37° 52′ S.; during the morning of the latter day the wind hauled to the south-southwest; the barometer, at 3 A. M., stood at 29·10 in.; the weather cleared, with the wind at southwest.

On the 26th, they reached and anchored in the river Kawa-Kawa in the Bay of Islands, off the American consul's, about three miles above its mouth. Many vessels were passed lying at anchor off the town of Kororarika. Here they found the tender Flying-Fish; all well.

The cruise of the latter will now be taken up from the 1st of January, on which day she parted company with the Vincennes, in consequence of having carried away a gaff, and being obliged to shorten sail, in doing which their jib-stay got adrift, and carried away the squaresail-yard before it could be secured. The vessel was in the mean time exposed to a heavy sea beating over her, and at midnight they were compelled to heave-to. They then steered for the first rendezvous, Macquarie Island, where they arrived on the 10th, in the afternoon, and saw the Peacock, but it becoming thick, they were not seen by that ship.

On the 11th, Acting-Master Sinclair landed for the purpose of placing a signal on the island, agreeably to instructions. The landing was found difficult and dangerous, and their description of the island agrees with that heretofore given of it from the notes of Mr. Eld, as being dreary and inhospitable. Large numbers of penguins, and small green and yellow paroquets were seen. Near where they landed, they saw about twenty huge sea-elephants basking on the rocks, which did not seem to heed them; when disturbed, they would only throw their carcasses over, open their mouths, utter a loud growl, and go to sleep again; no measurement was taken of them, and one which was killed could not be taken in the boat. The soil was soft and spongy, yielding to the pressure of the feet. The staff and signal being planted, they returned on board, and now passed the surf without difficulty.

On the 12th, they put away for the next rendezvous, Emerald Isle. They reached its position on the 14th, but nothing was seen of it; the weather was thick.

On the 16th, they kept off to the southward, with the wind from the southwest, accompanied with sleet and snow. In latitude 61° S., longitude 164° E., they saw the first ice. The next day, the 19th of January, the water was very much discoloured; got a cast of the lead in ninety fathoms: no bottom; passed a number of icebergs that were all flat on the top, with perpendicular sides.

On the 21st they made the icy barrier, in longitude 159° 36′ E., and latitude 65° 20′ S. From the number of icebergs and the frequency of snow-squalls, they found great danger in running through them, although the water was quite smooth.

On the 22d the weather proved pleasant, and they followed the trend of the ice. The ice-islands still showed flat tops and perpendicular sides, and there were a number of birds, seals, and whales around them; they were at noon in longitude 158° 27′ E. On this day they were close by an iceberg, from the main body of which a large mass fell with a noise like thunder; the snow flying into the air resembled smoke, and the swell produced by the immersion of the fragment caused the schooner to roll water in on her deck. A number of large penguins were in sight, differing from any they had heretofore seen.

On the 23d the weather was pleasant, and they had light winds from the southward and westward. Longitude 157° 49′ E., latitude 65° 58′ S. They continued coasting along the ice in search of an opening. At 8 P. M. they discovered several dark spots, which had the appearance of rocks, and on approaching the margin of the ice, they could make them out to be such with their glasses, but they were situated too far within the field-ice for a boat to get near them. This day being fine, an opportunity was afforded of drying the deck and clothes, and searing the seams with a hot iron. The vessel had been very wet, and her decks leaked badly, notwithstanding the thorough calking and repairs she had received at Sydney: the crew were almost constantly wet, below as well as above deck.

On the 24th they were obliged to steer again to the northward, in consequence of making the barrier ahead. Sea-lions were seen on the ce. They continued to follow the barrier, which trended north-northeast; the compasses were very sluggish. On the 26th and 27th the weather became bad, with the wind to the northward and westward. accompanied by a heavy fall of snow: in the evening of the latter day, the wind hauled to the southward and westward, and brought clear weather. The 28th passed with clear weather, and several seals were about them.

The 29th was thick and snowy, with a northeast wind; passed through quantities of drift-ice, and by 2^h 30^m, it had become so thick as to render a continuance of their course perilous; at 7 P. M. they again made the solid barrier, when it was blowing a stiff gale; at 9^h 30^m discovered the ice ahead, and on both beams; wore round to the northward and eastward, to retrace their steps; it was not long before they discovered a chain of ice-islands ahead, apparently connected by solid ice; about midnight a passage was discovered between two icebergs, through which they passed. It was now blowing a heavy gale, and having gained the open sea, they attempted to reef the foresail, but were unequal to the task (four of the men being on the sick-list), and were compelled to lay-to under the whole sail, which caused the vessel to labour very much, as well as to leak a great deal, and endangered her safety by making her fly into the wind, and get a sternboard in a high sea.

On the 30th, in the morning, the gale abated, and the weather became more pleasant than they had experienced for a number of days. They had reached the longitude of 150° 16′ E., latitude 65° 15′ S. On this day they again passed into blue water.

31st January was thick with snow; a north wind and heavy sea.

1st of February, they were running among ice, until they sighted the barrier, when they again hauled to the northward; a moderate gale blowing, with thick weather and a heavy sea, they were obliged to heave-to.

On the 2d and 3d, they were coasting the ice. In the afternoon of the 3d they again had bad weather, which made it necessary to bring to; surrounded by bergs and drift-ice; the latter, in case of striking, would have seriously injured the tender. The icebergs seen on these days, had the appearance of recent formation; the tops flat, the sides perpendicular, and not worn by the action of the sea.

On the 4th, the gale continued, and the sea had risen to an extraordinary height; the weather was so thick that an iceberg could not be seen further than twice the length of the vessel. The tender was under too much sail, which caused her to labour dreadfully, in consequence of which she leaked in such a manner as to make it necessary to keep the pumps going almost continually. When they were stopped for a short time to rest the men, the water increased so as to reach the cabin-floor: the water came through the seams forward in such quantities as to wet every bed and article of clothing on the berth-deck. This was a great addition to the labour and discomfort of the crew, now reduced by sickness to four men, and the strength of these much impaired by previous sickness, excessive labour, and almost constant exposure. To relieve their situation as much as possible, Lieutenant Pinkney ordered them to make use of the cabin in common with the officers. To ease the pitching of the vessel, a quantity of coal was shifted aft; but although this was a partial relief, yet as she had too much sail on her, which they had been unable to reduce at the commencement of the gale, it was not sufficient to make her easy.

On the 5th, the gale began to abate, when the crew, through one of their number, presented a communication to Lieutenant Pinkney, of which the following is a copy.

(COPY.)

We, the undersigned, the crew of the Schooner Flying-Fish, wish to let you know that we are in a most deplorable condition: the bed-clothes are all wet; we have no place to lie down in; we have not had a dry stitch of clothes for seven days; four of our number are very sick; and we, the few remaining number, can hold out no longer; we hope you will take it into consideration, and relieve us from what must terminate in our death.

(Signed) A. MURRAY. THOMAS DARLING.
JOHN ANDERSON. JAMES DANIELS.
F. BEALE. JOSEPH.
JAMES DARLING. JOHN H. WEAVER.

TO LIEUTENANT PINKNEY,
U. S. Schooner Flying-Fish.

On the receipt of this appeal, Lieutenant Pinkney addressed an order to the officers, a copy of which follows.

U. S. Schooner Flying-Fish,
Lat. 66° S., long. 143° E., Feb. 5th, 1810.

GENTLEMEN,—

You will furnish me with your opinion, and the reasons which induced that opinion, of the propriety of any longer endeavouring to accomplish that part of the accompanying order, which refers to penetrating to the south.

I am, respectfully, &c.,
R. F. PINKNEY,
Lieutenant-Commandant.

TO ACTING MASTER GEORGE T. SINCLAIR.
PASSED MIDSHIPMAN WILLIAM MAY.
PASSED MIDSHIPMAN GEORGE W. HARRISON.

COPY OF REPLY.

U. S. Schooner Flying-Fish,
Lat. 66° S., long. 143° E., Feb. 5th, 1840.

SIR,—

Agreeably to your order of this date, we, the undersigned officers, have to express our most thorough conviction, that the condition of this vessel's *crew*, and the vessel, loudly demand an immediate return to milder latitudes.

The causes of this opinion are these: that the crew of this vessel, consisting of fifteen persons (four officers and eleven men), even if well, are entirely inadequate to her safe management; but five are now confined to sick beds (one a servant), one of them is in a very critical state of health, and three others dragging out upon duty, complaining, and under medical treatment. Out of four, nominally performing duty, one of them, the cook, is totally unfit to a turn at the helm, and another cannot be trusted without the closest watching; indeed, so deficient in force are we, that in the gale of yesterday and the day before, and on a previous occasion, when it became extremely necessary to reef the foresail, the men were so deficient in physical strength as to make it impossible to accomplish it.

The crew's apartment is in the most deplorable state, leaking like a sieve, all their beds being wet, their clothes on them being so, even to their under flannels, for *one week*, and without a dry change on hand, and no prospect of having one; so miserable is their situation, that at length you have been compelled to allot them the cabin, in common with us, for the purpose of cooking, eating, and sleeping.

Furthermore, sir, in the gale now abating we find that nearly constant application to the pump is barely sufficient to keep the water from flooding the cabin-floor, evidently having started a leak; notwithstanding this, the condition of the crew is more imperative, much more so in this, our recommendation, for a return to the northward; in fact, we would cheerfully continue to the southward, if we had a proper crew.

Lastly, understanding that the crew, through one of their body, have waited upon you, and, by written application, also stated their inability to live through these hardships much longer, and begging your return.

We are respectfully, your obedient servants,

(Signed) GEORGE T. SINCLAIR,
Acting Master
WILLIAM MAY,
GEORGE W. HARRISON,
Passed Midshipmen

LIEUT. COM. R. F. PINKNEY,
Commanding U. S. Schooner Flying-Fish.

Lieutenant Pinkney, in accordance with this opinion, and his own conviction of the necessity of an immediate return to milder latitudes, as the only means of restoring the sick, and preserving those on duty, who were then incapable of managing the vessel without the assistance of the officers, deemed it his duty to steer for the north, which he accordingly did.

The 6th and 7th continued thick, with occasional squalls. On the 8th, the weather again broke up, when they had several hours of sunshine, which proved of great benefit to the sick, who were enabled to come again on deck, and enjoy the milder weather, affording so great a contrast to that of the Antarctic circle. They had reached the longitude of 139° 45′ E., latitude 61° S. At 11 P. M. the aurora was seen; it was first visible in the southeast quarter, in spots resembling pale moonlight, extending to the zenith, from whence it diverged in rays, some of which reached the horizon, but the greatest number terminated at an altitude of twenty-five or thirty degrees. On the 9th, the aurora was also seen in the west, in vertical rays of pale yellow light, commencing about five degrees above the horizon, and extending to an altitude of thirty degrees. After a short time it disappeared, and was again seen in the zenith, radiating in lines to the northeast and west, reaching to within ten degrees of the horizon. The wind was from the southward. Temperature 34°. The following five days they had thick weather, and nothing occurred until the evening of the 14th, when they again had a display of the aurora; the coruscations were frequent and brilliant, but did not exhibit any different form, until after midnight, when it appeared in arches, reaching nearly to the horizon, at from 45° to 73° of altitude, and composed of short perpendicular lines, blending at one moment into a sheet of misty light, and then breaking out into brighter lines, some of which were broad. It then again shifted to the zenith, with radiations extending in every direction, in straight and wavy lines. The changes were incessant, but not shooting.

On the morning of the 15th, they again had a display of the aurora. It first appeared in the southern heavens, at an altitude of 45°, flashing to the zenith, where it disappeared. After midnight it was again visible in the southern quarter, at about 30° of altitude. It finally centered in a bright spot, which changed into a crescent, with the rounded side to the northward. From this, feathery-edged rays of pale orange-colour branched off in every direction, over which the prismatic colours seemed to flit in rapid succession. The rays would sometimes fold into one another like a fan, and reach the horizon in one direction, while in another they were drawn up to the zenith, again to burst forth in repetitions, until lost in daylight. On the 19th, the aurora again appeared in an arch of 15° altitude.

They passed the last icebergs in latitude 55° 30′ S., longitude 145° 30′ E.

On the 22d they spoke a French whaler from Hobart Town, who expressed much surprise at finding so small a vessel in such high latitudes. The captain sent a boat on board, and invited them to "soup" with him.

On the 23d they made the southern island of New Zealand. On the 1st of March they experienced a most violent gale. The wind, about noon on the 29th of February, hauled to the southward and eastward, and by midnight it blew a gale, hauling to the eastward, until about 8 P. M., when its violence moderated. Their latitude was 40° S., longitude 178° 30′ E. For several days previous to this, a noise was heard about the heel of the main-mast; an examination was had, and the conclusion arrived at that it worked in the step, the wedges in the partners having been driven without obviating it. On the 9th of March they arrived at the Bay of Islands, where they found the gentlemen who had gone there to pursue their researches in natural history waiting our arrival.

AURORA AUSTRALIS

The Vincennes was left on the 21st of February on her way north On the night of the 22d, we had a beautiful and novel appearance of the aurora australis. The sketch of it which I made will in some measure convey an idea of it.

Black clouds were passing rapidly over the sky; an orange glow of light seemed to cover the heavens, emanating from a point, over which flitted rays of the prismatic colours, directed towards the horizon, lighting up both edges of the clouds, and throwing them into bold relief. The rays seemed to dart simultaneously towards the horizon, on reaching which they would seem to be gathered, as if by magic, towards the centre, and slowly vanish, to reappear again and fold up.

Strong gales from the west-northwest with snow-squalls continued until the 27th, with thick misty weather. Numerous ice-islands were passed during this interval. The last iceberg seen, was in the latitude of 53° S., and longitude 120° 25′ E., the temperature of the water was 46°.

On the 28th, we found our variation 1° easterly, in the longitude of 131° 50′ E., latitude 50° 30′ S.; and in attempting to get a deep-sea sounding of eight hundred and fifty fathoms, we lost our Six's thermometer by the wire parting. The sea was a deep blue; the temperature 45°. We found a current setting west-northwest three-fourths of a knot per hour. The white object was seen at the depth of fifteen fathoms.

On the 1st of March we had reached the latitude of the Royal Company's Isles, and I continued to run in nearly the same parallel for eight degrees of longitude, without seeing any signs of the supposed land. Having sailed far to the eastward of their supposed position, I again hauled to the northward to proceed to Hobart Town, Van Diemen's Land, to fill up our water. We now saw a sail, the first during sixty days, which made us feel as if we were returning to a habitable part of the globe. This night we had a brilliant display of the aurora australis, resembling that seen on the 9th of February, with this difference, that it was seen to the southward, extending from east-southeast to west-southwest.

On the 5th of March the wind headed us off our course to Hobart Town; I then determined to proceed direct to Sydney, and thus be enabled to communicate as speedily as possible with the United States. The consideration of getting intelligence respecting the other vessels, also led to this determination. I felt, in truth, forebodings that all was not well, from not having met any of the vessels at the appointed rendezvous, along the icy barrier; and I was anxious for their safety, after the severe gale of the 28th of January.

Having reached a lower latitude, the weather had now become pleasant, and we could dispense with our winter clothing,—a relief which the whole of the crew seemed to enjoy. It was the reverse with me; I had a feeling of exhaustion and lassitude that I could not account for, and the least exertion caused me much fatigue.

On the 9th, we reached the latitude of Cape Howe, and were seventy miles to the eastward of it. We there experienced a rise in the temperature of the water: six degrees in less than an hour.

On the 10th, when off Cape Jervis, and about forty miles to the eastward of it, we again changed the temperature from 68° to 73°, as we steered in for the land to the northward, but on hauling to the eastward it again fell to 68°. A strong southerly current has been long known to exist along this coast; and I feel well satisfied that the thermometer is a good guide in making the passage from the southward. The coasting vessels, as I was informed at Sydney, had frequently made long passages from Van Diemen's Land, and South Australia, which I have but little doubt is owing to the prevalence of this minor Gulf Stream, the position of which the use of the thermometer will clearly indicate. This current will be noticed particularly in the chapter on currents; its width no doubt varies with the season.

On the 11th of March, at noon, we passed the Heads of Port Jackson, and took a pilot. We were, as a body, in better condition than when we left Sydney three months before.

In an hour afterwards we dropped our anchor in Farm Cove, off Fort Macquarie. Our reception was flattering; scarcely was our anchor well down before many of our friends came on board to bid us welcome; and we felt tenfold that kind hospitality which on our former visit we had first become acquainted with. They appeared to rejoice in our success as if we had been their countrymen.

During our absence from Sydney, many improvements had taken place. The storehouses for the deposit of grain on an island in the harbour were in rapid progress; the new Government-House nearly completed, and the foundation of an Exchange laid; besides this, many improvements in town that were then in progress, had been completed; and the rapidity with which these works had been accomplished, strongly reminded me of similar operations at home.

The country was looking quite green and pretty; indeed, the sail up the noble harbour was truly beautiful; it wore quite a different face from its former parched appearance, the rains having been abundant during our absence.

Observations were obtained for the rates of our chronometers and the magnetic needles again experimented with.

On overhauling my ship, the fore-topmast was found to be slightly sprung.

It was with great pleasure I learned the safety of the Peacock; for that vessel had occupied my thoughts more than the others, on account of the condition in which she left Sydney. All on board of her were well, and the vessel was undergoing repairs in Mossman's Cove, one of the many which this harbour forms. These coves may be termed wet-docks, affording as they do every facility for the repair of vessels of any size. They are more like artificial than natural basins, and are secure against any wind. There is no port in the world that offers so many natural advantages as Port Jackson, for a great naval power. We had many things to relate to each other; among others, the particulars of the accident that befell the Peacock, that has already been noticed. The return of that vessel to this port now claims our attention.

On the 28th of January, their sick-list had increased to thirteen, more in consequence of the fatigue the men had undergone, than from any disease.

On the 29th, they experienced strong gales from the northwest, which continued to increase until midnight, after which the weather moderated. The ship during this gale was in latitude 61° 20′ S., and longitude 154° 09′ E. This gale is remarkable, in consequence of its blowing in a contrary direction to that which the Vincennes experienced on the same day; while the former had it from the northwest, the latter had it from southeast. Their distance apart was four hundred and fifty miles, in a northeast direction.

On the 1st of February the weather was stormy until towards evening, when it moderated and cleared off, with the wind to the northwest, and gave them a view of the aurora australis lighting up the southern portion of the horizon. Rays were thrown out in different directions, some reaching an altitude of 30°, others of 40°, whilst others again almost spanned the heavens.

On the 2d, they had another display of the aurora, but contrary to that of the previous day, it was first seen at an altitude of 70°, diverging towards the horizon, from east-southeast to the southwest-by-west, before it disappeared. The point from which the rays diverged reached the zenith.

On the 4th they made Macquarie Island, and shortly after passing it, experienced another gale from northwest to southwest, which caused them much anxiety for their rudder, which thus far had answered well, although great attention was necessary to prevent strain upon it.

Strong gales yet continued. On the 5th, they had a faint display of the aurora.

On the 7th of February, the weather had become less boisterous, and having reached latitude 49° S., longitude 155° 23′ E., the aurora australis again appeared. It was first seen in the north, and gradually spread its coruscations over the whole heavens; the rays and beams of light radiating from nearly all points of the horizon to the zenith, where their distinctive outlines were lost in a bright glow of light, which was encircled by successive flashes, resembling those of heat lightning on a sultry summer night; these formed a luminous arc in the southern sky, about 20° in altitude, from the upper part of which, rays were continually flashing towards the zenith; light showers of rain finally shut it out from view. On the same night, between one and three, the aurora burst out from the southwestern horizon, streaming up and concentrating in the zenith, and attended with quick flashes of every variety of tint. The wind was moderate from the southwest, and a squall of hail passed at the time. In latitude 47° S. they first encountered phosphorescence in the water. On the 17th they made the land of New South Wales, and continued to experience a variety of weather until the 21st, when they arrived off, and anchored within, the Heads of Port Jackson.

The next day they proceeded up the harbour, and anchored off Sydney Cove. The ship was much shattered, but her officers and crew all in good health. Here they were kindly received, and no time was lost in proceeding to make the necessary repairs. The collector was kind enough to give them permission to land every thing that might be necessary, when and where they pleased. The powder and fire-works were received into the public magazine, and when called for were politely sent in a government boat, free of expense. The railway for merchant-vessels was found too light to trust the Peacock upon it; Mossman's Cove, on the north shore, was then resorted to, not only as a convenient place for making the necessary repairs, but as affording more security for the crew against the crimps and rum-shops.

The day after my arrival, I visited the Peacock, in order to examine into her condition, and could not withhold my astonishment that she had been able, after undergoing such damage, to reach a distant port. The visible injuries have already been stated, in speaking of her accident. On their arrival at Sydney, it was found that her stem had been chafed to within one and a half inches of her wood ends, and much strained throughout. After a full examination of the circumstances, I feel it a duty I owe to Captain Hudson, as well as to his officers and crew, to state that I am well satisfied, that his coolness, decision, and seamanship, with the good conduct of his officers and men in the perilous situation in which they were placed, are worthy of the highest encomiums. The preservation of the ship and crew, and her subsequent navigation to a distant port, reflect the highest credit upon her commander and upon the service to which he belongs.

Sydney was now much crowded with people, and several balls were given, to which we had the honour of an invitation. That of the St. Patrick Society was attended by the chief people in the neighbourhood of Sydney, including the governor and most of the officers of the crown. It was given in the new court-house, and was a handsome and well-conducted entertainment. Two military bands were in attendance; quadrilles and country dances followed each other in rapid succession; rooms were provided for cards, refreshments, teas, lemonade, &c.; and towards the close of the evening, the company was ushered in to an elegant supper, which was partaken of standing.

I was struck with the beauty and general appearance of the ladies, though I was informed that many of the belles were absent. The style of the party was neither English nor American, but something between the two. I scarcely need remark that we were all much gratified and pleased. The hospitality and kindness shown us were of that kind that made us feel truly welcome.

Our last week at Sydney was spent in a round of pleasure, and the attention we met with being entirely unexpected, was doubly gratifying to us.

ICEBERG.

CHAPTER XII.

CONTENTS.

DEPARTURE FROM SYDNEY—PASSAGE TO NEW ZEALAND—ARRIVAL AT THE BAY OF ISLANDS—MEETING WITH THE SCIENTIFIC CORPS—THEIR PASSAGE FROM SYDNEY—BAY OF ISLANDS—RIVERS WHICH FALL INTO IT—FACE OF THE COUNTRY—ACTIVE VOLCANO—HOT SPRING OF TAIAIMI—CRATER OF POERUA—DR. PICKERING'S VISIT TO HOKIANGA — MISSIONARY ESTABLISHMENT AT PAHIA — KORORARIKA — ENGLISH POLICE MAGISTRATE AND ACTING GOVERNOR—TREATY OF CESSION TO ENGLAND—CONDUCT OF THE AMERICAN CONSUL—INSTALLATION OF THE LIEUTENANT-GOVERNOR—OPINION OF THE CHIEFS IN RELATION TO THE TREATY—ARRIVAL OF ENGLISH REVENUE OFFICERS—LAND CLAIMS—BURTHENSOME TAXES AND TARIFF—THEIR EFFECT ON AMERICAN COMMERCE—EXPENSE OF THE NEW GOVERNMENT—CASE OF JOHN SAC—HIS LETTER TO MR. WALDRON—FURTHER REMARKS ON THE TREATY OF CESSION—VIOLENT GALE—ITS EXTENT AND ROTARY CHARACTER—FOREIGN RESIDENTS—HIGH PRICE OF LAND—MISSIONS—TABOO—PAS, OR FORTIFIED TOWNS—DWELLINGS—TOMB—DRESS OF THE NATIVES—THEIR STOREHOUSES—THEIR FOOD—THEIR ARMS AND ORNAMENTS—KING POMARE—MAUPARAWA—CHARLEY POMARE—POMARE'S WARS—CEREMONY OF HIS RETURN—HIS MEANNESS—POPULATION OF NEW ZEALAND—VISIT TO WANGARARA—POLITENESS OF KO-TOWATOWA—WANGARARA BAY—CHARACTER OF THE NEW ZEALANDERS—THEIR PERSONAL APPEARANCE—TRADITION IN RELATION TO THEIR ORIGIN—TATTOOED HEADS—CANNIBALISM—CONDITION AND PROSPECTS OF THE NATIVES—NATIVE DANCES—MUSIC—CHATHAM ISLAND—CHART OF THE BAY OF ISLANDS—MR. COUTHOUY'S PASSAGE FROM SYDNEY—HIS ACCOUNT OF MOUNT EGMONT—OF PORT COOPER—WARS OF ROBOLUA—PORT LEVY—KORAKIBARURU—PIGEON BAY—CAPE CAMPBELL AND SNOWY PEAKS—CLOUDY BAY—ROBOLUA—HABITS OF THE NATIVES, AND PREVAILING WINDS AT CLOUDY BAY—CLIMATE OF NEW ZEALAND—DISEASES—SOIL—CULTIVATION—VEGETABLE PRODUCTIONS—TIMBER—CANOES—QUADRUPEDS—BIRDS—COMMERCE.

Drawn by A. T. Agate

G. H. Cushman Sc.

VIEW IN NEW ZEALAND

CHAPTER XII.

NEW ZEALAND.

1840.

Having replenished our stores of provisions, we took, with much regret, a final leave of our friends at Sydney. The Vincennes weighed anchor, and at 3 P. M. on the 19th March, we discharged our pilot, and bade adieu to these hospitable shores. The Peacock, not having completed her repairs, was left at Sydney for a few days, with orders to follow us to Tongataboo.

On reaching a distance of thirty miles from the coast, we again found a difference of three degrees in the temperature of the water, and experienced the effects of a strong current towards the south. The wind was from the northward and eastward.

On the 23d we spoke the French whale-ship Ville de Bordeaux, in want of provisions, which we supplied her. She had been out three years, and had on board four thousand barrels of oil. The crew was reduced to bread and water, and the vessel was apparently in a bad condition in other respects.

On the 25th, in latitude 34° 24′ S., longitude 160° 26′ E., we experienced a current setting to the south at the rate of twenty miles in twenty-four hours.

On the 26th the current set east-southeast at the rate of twelve miles per day.

The wind on the 27th hauled to south-southeast by the east, and became a fine breeze.

On the 29th, we made the North Cape of New Zealand. The current for the two previous days had been setting north-northwest and the temperature of the air varied during our passage from Sydney from 63° 3′, to 76° 4′; that of the water from 70° to 72°.

At daylight on the 30th, we made Cape Brett, and after groping our way through the dark, into the Bay of Islands, anchored at 10 P. M. in the Kawa-Kawa river, opposite the residence of Mr. Clendon the American consul. Here I had the satisfaction to find the Porpoise and Flying-Fish, and receive the reports of their cruises, which will be found in Appendix XXX.: they were all well on board. The former vessel had arrived a few days, and the latter about three weeks, before us. We were also gratified with the receipt of letters from the United States. Every exertion was made to shorten the duration of our stay in New Zealand, and the necessary instruments were landed without delay.

Here also we met all the scientific gentlemen,—who, as has been stated, had been left at Sydney when the squadron sailed upon the Antarctic cruise,—anxiously awaiting our arrival.

They had been forced to remain inactive at Sydney, in consequence of a change in the destination of the vessel in which they had first taken their passages, and, by this vexatious delay, had not only been prevented from pursuing further researches in New South Wales, but had lost time that might have been advantageously employed in New Zealand. They finally succeeded in finding an opportunity of reaching the Bay of Islands, in the British brig Victoria.

After leaving Sydney in this vessel, a sea was shipped, which, besides doing other mischief, entered at the cabin-windows, and filled the chronometer-box with salt water; in consequence of which the master considered it necessary to put back, in order to exchange the injured time-piece for another. She accordingly anchored again in Port Jackson.

On the 7th February, they had a beautiful exhibition of the aurora australis: the coruscations were of a straw-coloured light, reaching nearly to the zenith in the southern sky, and lasting from seven until ten o'clock. A noddy lighted on the brig, and remained on board many days; so tame was it that it even suffered itself to be handled.

On the 16th, when they had performed about half the passage, they had another exhibition of the aurora, much like the former; after which they experienced a gale of wind of five days' duration. On the 21st, they were enabled again to make sail, and, on the 23d, they made the North Cape. A gale then came on from the eastward, and they had a narrow escape from shipwreck while running down the land. On the 24th, they dropped anchor at Kororarika, about three miles above which place they found the United States Consul, Mr. Clendon, at Ornotu Point.

From the splendid panorama of Mr. Burford, I had pictured the Bay of Islands to myself as a place of surpassing beauty, and I could not but feel gratified at the idea of paying it a visit: it did not, however, realize my expectations. It might, with more propriety, be called the Bay of Inlets. The best idea that can be given of its geographical features is, to liken it to an open hand with the fingers spread apart. The land is much indented with bays, or arms of the sea, running up among hills, which are nearly insulated. The distance between the two capes (Brett and Point Pocock) is ten miles, and there are several secondary bays facing this opening. Four rivers flow into them, the Kawa-Kawa, Kiri-Kiri, Loytangi, and Waicaddie, into which the tide flows a few miles, after which they become small streamlets, varied by some waterfalls. There are many minor indentations, which render it impossible to move any distance without a boat; and it is often necessary to make a turn of five or six miles around an inlet or marsh in going to a place, which might be reached in one-tenth of the distance by water.

The land has the appearance of barren hills without accompanying valleys, and there is so little level ground that terraces are cut in the hills to build the cottages on. The whole view is any thing but picturesque, and there is little to meet the eye except bare hills and extensive sheets of water. Some fine views are, however, to be met with from the elevated ridges, which afford occasional glimpses of the bay, with its islets.

Many of our gentlemen were struck with the resemblance of this land to that of Terra del Fuego. Black islets and rocks, worn into various shapes, are found, as in that country, at all the points in the bay through which a boat can pass. These rocks are of a basaltic character. About the Bay of Islands the rock is compact and argillaceous, showing little or no stratification, and is for the most part covered with a layer of stiff clay, two or three feet thick, the result of its decomposition. The hills about the Bay of Islands are generally from three to five hundred feet high, but some of those at the head of the bay reach one thousand feet. The district about the Bay of Islands, and the northern portion of the island, may be styled volcanic; for, in addition to rocks of undoubted volcanic origin, all the others had in a greater or less degree undergone the action of fire. Our naturalists were informed that the valley of the Thames was of a different character, although many persons represented the whole island as volcanic. The ridges in the northern part of the island were not thought to rise more than two thousand feet. The Rev Mr. Williams, missionary at Pahia, has crossed the island from Port Nicholson to Taaranga, during which journey he passed a district from which the snow was absent only four months in the year. This region is in the neighbourhood of the high peak of Mount Egmont, said, in the Sydney Almanac, but upon what authority is not stated, to be fourteen thousand feet high. Mr. Williams described the route as exhibiting volcanic phenomena on a large scale, among which were quantities of pumice, extending entirely across the island, and an extensive plain, which had sunk in one place, and disclosed a bed of that substance, three or four hundred feet in thickness; he likewise spoke of geysers or jets of boiling water.

The only volcano that was known to be in action, was one on a small island in the Bay of Plenty, on the east coast.

The embedded minerals in the rock about the bay are quartz, iron, and iron pyrites.

The hot spring of Taiaimi was visited, but it is described as rather an emission of gas than of water. It is situated in a small basin, and forms a lake of three or four acres in extent; near the edge of this lake, gas is constantly bubbling up, usually through the water, to which it gives the appearance of boiling; and gas also issues from the surrounding land for an extent of several acres. The water was found to be warm, but did not scald. The neighbouring ground was destitute of vegetation, and appeared as if the surface of the earth had been artificially removed. Sulphur was abundant, and there was also a slight incrustation of alum. The water was strongly impregnated with iron, was much discoloured, and in smell and taste not unlike pyroligneous acid. A quantity of gas was brought away, but the bottle met with an accident before it could be analyzed. It is not inflammable, and had it been of a deleterious nature, the fact, (from the quantities emitted,) could not fail to have been perceived. It had no smell, and appeared not to differ from atmospheric air. The natives attribute medical virtues to these waters.

Twelve or fifteen miles to the westward of the Bay of Islands, near Taiaimi, there are several small extinct craters, rising about five hundred feet above the surrounding country. One of them is called Poerua, and is remarkable for the regular figure of its cone when seen from the eastward. Its western side is cut through by a deep gorge. The interior is covered with large forest trees and huge blocks of lava, while the exterior is clad in ferns of low growth. The diameter of the crater is about half a mile. The plain which surrounds the cone is composed of an uncommonly rich soil, strewed with lava, which the natives collect in heaps, in order to obtain space for cultivation. The lava does not extend far from the cone, and even in the interior, rock seldom appeared, but where it was seen it proved to be vesicular lava. The soil in the neighbourhood of the craters is richer, looser, and more fit for cultivation than in other places.

Dr. Pickering made a visit to Hokianga, on the western side of the island, and found that it had more of the forest character than the eastern. He took the direct road to Waimati, which is fifteen miles from the Bay of Islands. The river Waitanga was very high, and one of the chiefs, a large and muscular man, seemed to take particular interest in getting them across safe and dry; but notwithstanding his stature and all his care, he could not prevent a slight immersion.* The Doctor arrived at Waimati at 4 P. M., and was kindly welcomed by Mr. Davis, the Methodist missionary, to whom he had a letter of introduction. It was not without surprise that he found here a water-mill in operation, which the guides took care to point out with no little exultation. This, together with the fences, and well cultivated fields, were the works of the missionaries. He remained with Mr. Davis for the night, who advised his proceeding direct to Hokianga; but the guides who had hitherto accompanied him were ignorant of the route, and another became necessary.

The next day they passed over the flank of Te-ahooahoo, a volcanic cone, and the most prominent elevation in this region. A little farther on, a fine lake was passed, about three miles in length. At nine miles from Waimati, the wooded region was entered, which extended to Hokianga. Just before crossing the Hokianga river for the first time, the Baron de Thierry was met with, who was exceedingly polite. The road after this became difficult, it being necessary to cross the river repeatedly, and to follow the stream for some distance. The usual manner of crossing here is to be carried. The guides, under various pretexts, prevented them from reaching Hokianga, and they were compelled to stop four miles short of it, at a chief's called Tooron, of rather doubtful character.

Tooron, with his family, had worship both morning and evening, as is customary with converted natives, he himself officiating. The accommodations were none of the best. An open shed, with fire and blanket, were, however, sufficient to insure a good night's rest. Tooron was liberally paid, and so well pleased, that he said he was determined to carry his guests over the river himself. The road was any thing but good, being miry, and filled with roots of trees, so that

* On the banks of the Waitanga, the adult inhabitants, to the number of twenty, were collected in a circle, each armed with a musket, and several had been met on the way, all armed. The cause of this unusual occurrence was not known. They are very fond of fire-arms, and on welcoming any one, particularly a chief, all the people of the village assemble and salute him with a number of rounds, in proportion to his rank.

their attention was wholly engrossed in seeking a good foothold. The river was again repeatedly crossed. On the way they met natives loaded with baskets of peaches, the season for which had arrived. They freely offered their fruit, for which tobacco was returned. Before noon, they arrived at Baron de Thierry's house, where they were hospitably received by his lady. This house is situated at the head of tide-water on the Hokianga river, about thirty miles from its mouth, and boats can ascend as far as this place. There is no village at the mouth of the river, but many whites reside at different points on its banks. There is a bar between the headlands at its mouth, which will admit only of small vessels entering.

Our travellers had intended to return the next day, but one of their guides, by the name of Pooe, was missing. He had been allowed to take up his quarters at a short distance, on condition of his being ready for an early start; on inquiry, however, they were informed that Pooe had said he did not intend to go back until Monday, which was two or three days off. They departed without him, but before reaching Tooron's, Pooe again joined them, having a piece of pork, which one of his friends had furnished for the Doctor's supper.

Mr. Davis's was reached at dark, and the same warm greeting experienced as before. The next day they reached the Bay of Islands, at Pahia.

Pahia is the principal missionary establishment of the Episcopal Church. It is pleasantly situated on the bay, opposite Kororarika, and is the residence of all those attached to the mission, and their printing-presses are there. It is too much exposed to afford a good harbour for shipping, but as it is the most favourable side for communication with the interior, the advantages and disadvantages of its position are nearly balanced.

Kororarika is still the principal settlement, and contains about twenty houses, scarcely deserving the name, and many shanties, besides tents. It is chiefly inhabited by the lowest order of vagabonds, mostly runaway sailors and convicts, and is appropriately named "Blackguard Beach."

The appointment of the police magistrates was one of the first acts under the new order of things. Mr. Robert Shortland, the first police magistrate, after the illness of Governor Hobson, styled himself acting governor, and a more ridiculously pompous functionary could scarcely be imagined. He paid a visit to the vessel in which some of our gentlemen had made the passage from Sydney, and demanded the reason why the mail-bag had not been sent to the new government postmaster. The master of the vessel replied, that he thought it his duty, not having been informed of any change, to deliver them to the old postmaster, until he should be directed otherwise by Governor Hobson. This pompous functionary, in an improper tone as well as manner, exclaimed, "I wish you to know that I am governor now!" In the words of one of the gentlemen, "had he been the viceroy of the Indies, he could not have made his inquisitions in tones of loftier supremacy."

Some of our gentlemen arrived at the Bay of Islands in time to witness the ceremonies of making the treaty with the New Zealand chiefs. I mentioned, whilst at Sydney, the arrival of H. B. M. frigate the Druid, with Captain Hobson on board, as consul to New Zealand. It was well understood that he had the appointment of Lieutenant-Governor in his pocket, in the event of certain arrangements being made. His arrival at the Bay of Islands, in H. B. M. ship Herald, seemed to take the inhabitants, foreigners as well as natives, by surprise. A few days afterwards, on the 5th February, a meeting was called at the dwelling of Mr. Busby. The meeting was large and numerously attended by the chiefs. Many arguments and endeavours were used to induce them to sign a treaty with Great Britain, all of which were but little understood, even by those who were present, and had some clue to the object in view. Great excitement prevailed, and after five hours' ineffectual persuasion, the meeting broke up, every chief refusing to sign or favour Captain Hobson's proposition, which was in reality nothing more or less than a cession of their lands, authority, and persons, to Queen Victoria. Among the arguments made use of, he stated that unless they signed the treaty, he could do nothing more than act as consul! Nothing having been effected, the meeting was broken up, and the following Friday appointed for a second. Tobacco and pipes were given them before they departed, which restored their good humour, and they went away shouting.

In the mean time, Mr. J. R. Clendon, an Englishman acting as American consul, the missionaries, and many interested persons residing there, or about becoming settlers, were made to understand that their interest would be much promoted if they should forward the views of the British government. Every exertion was now made by these parties to remove the scruples of the chiefs, and thus to form a party strong enough to overreach the rest of the natives, and overcome their objections. About forty chiefs, principally minor ones,—a very small representation of the proprietors of the soil,—were induced to sign the treaty. The influence of Mr. Clendon, arising from his position as the representative of the United States, was among the most efficient means by which the assent, even of this small party, was obtained. The natives placed much confidence in him, believing him to be disinterested. He became a witness to the document, and informed me, when speaking of the transaction, that it was entirely through his influence that the treaty was signed.

The Lieutenant-Governor installed himself, confirmed the appointments of a host of government officers, and the whole machinery, that had been long prepared, was put in motion. Proclamations were issued by him, extending his authority over all the English residents on both islands! and it was considered by the Englishmen as good as law, though far otherwise by the other foreigners. After this, the Lieutenant-Governor proceeded to the district of the Thames River, or Hauaki, in the Herald, for the purpose of procuring a similar cession of the country; but before this could be consummated, he was attacked with paralysis, and the Herald was obliged to depart for Sydney.

So far as the chiefs understand the agreement, they think they have not alienated any of their rights to the soil, but consider it only as a personal grant, not transferable. In the interview I had with Pomare, I was desirous of knowing the impression it had made upon him. I found he was not under the impression that he had given up his authority, or any portion of his land permanently; the latter he said he could not do, as it belonged to all his tribe. Whenever this subject was brought up, after answering questions, he invariably spoke of the figure he would make in the scarlet uniform and epaulettes, that Queen Victoria was to send him, and "then what a handsome man he would be!"

Those who are not directly benefited by the change, cannot but view it as a disastrous circumstance for the natives, which will seal their doom, and make them the prey of the hosts of adventurers who are flocking in from all parts, some to be engaged as public officers, and to fatten on the coming revenues, and others as speculators. During our stay, a cutter arrived from Sydney, with a number of revenue officers, magistrates, and other minor dignitaries.

New Zealand continued under the authority of New South Wales until September, 1840, when it became a separate colony. One of the first acts of the new government has been, by proclamation, to require all those who have acquired lands by purchase from the natives, to exhibit their vouchers, and to show how much land they had purchased, and the price paid. At the same time, a committee was appointed to examine these claims. A few statements made by this committee, will show how the spirit of speculation has been at work in New Zealand Up to October, 1841, they reported that five hundred and ninety-one claims had been entered by two hundred and eighty individuals; of these, there are four hundred and thirty-five claims, amounting to thirteen millions nine hundred and twenty thousand four hundred and eighty-two acres. The remaining one hundred and fifty-six claims are not defined by ordinary landmarks, but are limited by degrees of latitude and longitude, and computed in square miles instead of acres. The last description of claims are considered, at a moderate calculation, to be double the amount of the four hundred and thirty-five claims, so that in round numbers, the claims already sent in to the commissioners may be estimated at forty millions of acres. For four hundred claims, affidavits have been made, and the total value of goods and money paid by these claimants is thirty-four thousand and ninety-six pounds.

For one hundred and ninety-five claims, no value is stated; but if paid for in the same ratio, the amount will be nearly forty thousand pounds, or about one penny for three acres. The whole surface of the two islands does not contain more than eighty thousand six hundred square miles, or fifty millions of acres, and the largest part of them has not yet been sold by the natives, viz., the Waikati district, Rotorua and Taupo, in the interior, as well as the whole of the eastern coast of the northern island; so that it will be difficult to find a space wherein to locate these enormous claims.

Laws have likewise been promulgated and imposts levied, harassing to foreigners, (Americans and others,) and most destructive to their commercial pursuits, while they offer the most marked protection to those of British subjects! This would seem not a little unjust to those who have been resident, and extensively engaged in commerce, before England took possession, and whilst New Zealand was acknowledged as an independent state. It has, among other things, been enacted, that all goods imported and remaining on hand on the 1st of January, 1840, the time of British assumption, shall pay duties; that all lands are to be considered as belonging to the Queen, even those purchased of the chiefs prior to the treaty, while the purchasers shall be only entitled to as many acres as the amount paid to the chiefs will cover at the rate of five shillings per acre. The government in addition reserves to itself the right to such portions as it may require. Many of these purchases were made from the native chiefs, prior to the treaty, in good faith, and for an equivalent with which they were well satisfied, and so expressed themselves.

The destructive effect of these laws on American commerce will be great, particularly as those engaged in mercantile pursuits find themselves called upon to pay heavy duties on their stocks. Americans are not permitted to hold property, and, in consequence, their whaling establishments on shore must either be broken up altogether, or transferred to other places, at a great loss of outlay and capital. Our whalers are now prevented from resorting to the New Zealand ports, or fishing on the coast, by the tonnage duty, port charges, &c.; are denied the privilege of disposing of any thing in barter, and obliged to pay a duty on American articles of from ten to five hundred per cent. The expenses of repairs have so much increased, that other places must be sought for the purpose of making them. The timber and timber-lands are exclusively claimed as belonging to Her Majesty. Thus have our citizens been deprived of a fishery yielding about three hundred thousand dollars annually in oil.

Governor Hobson's proclamation will be found in Appendix XXXI.

The expenses of this new government were estimated for the year 1841 at £50,922 3*s.* 4*d.*, sterling, which is about equal to £10 for each man, woman, and child; for the whole foreign population on all the islands, is not supposed to be more than five thousand. The great precipitancy with which the islands were taken possession of, is said to have been owing to the fears entertained that the French intended forming a colony on the southern island in like manner.

After my arrival I gave the men liberty. Among the first who obtained it was John Sac, a native of New Zealand, and of the neighbourhood of this bay. His native name was Tuatti, and he was a petty chief. He had been some time absent from his country, and had sailed in the Expedition from the United States, was an excellent sailor, a very good fellow, and had been enthusiastic in the praise of his country and countrymen. According to him, there was nothing like New Zealand; and under this feeling he hired a canoe to take him on shore, for which his countryman charged him three dollars, although half a dollar would have been an exorbitant price. He landed at Tibbey's, and being desirous of going to his friends, wished to engage a canoe to take him about ten miles up one of the rivers, the Kawa-Kawa, where they resided. For this conveyance he was asked £2, nearly a month's pay. Poor John could not submit to this extortion, and was found sitting on a log, greatly mortified, depressed, and incensed at such treatment.

After John returned on board, he made a proposition to Mr. Waldron, in a letter, to purchase the island which he called Motugee, with the territory of Muckatoo, belonging to his father and family, and expressing his belief that they were all opposed to the encroachments of the English, and were determined not to part with their land to them.

Although the land about the Bay of Islands is much cut up by indentations, yet from this circumstance it affords many pretty views, which have in some respects an appearance of an advance towards civilization, that one hardly expects to find within the scope of the residences of these savages.

One of the many sketches Mr. Agate made, will serve to convey an idea of their beauty, as well as a distant view of their pas.

VIEW IN NEW-ZEALAND

At the time of my visit, which was, as has been seen, immediately after Captain Hobson's arrival, and the signing of the *treaty*, or cession, it was evident that full seven-eighths of the native population had the same feelings as are found expressed in this note. The circumstances that have occurred at New Zealand fully prove the necessity of having American citizens as our consuls abroad. Mr. J. R. Clendon, our consul at New Zealand, an independent state, and the only representative of a foreign power, whose interest was at stake, was consulted by some of the most powerful and influential chiefs, who had refused to sign the treaty or cession to Great Britain. They came to Mr. Clendon for advice, how they should act, and he admitted that he had advised them to sign, telling them it would be for their good. He himself signed the treaty as a witness, and did all he could to carry it into effect; but, in doing this, he said, he had acted as a private citizen, by request of the Governor, thus separating his public duties from his private acts. At the same time he buys large tracts of land, for a few trifles, and expects to have his titles confirmed as Consul of the United States. This is not surprising, and any foreigner would undoubtedly have pursued the same course; for his personal interest was very great in having the British authority established, while the influence he had over the chiefs was too great not to attract the attention of the Governor, and make it an object to secure his good-will and services.

The prospects of these islanders are, in my opinion, any thing but pleasing, and the change by no means calculated to insure their happiness, or promote their welfare. It seems to have been brought about by a rage for speculation, and a desire to take possession of this country, in order to secure it from the French. The idea that it was necessary to extend the laws of New South Wales over the island, in order to protect the natives, and break up the nest of rogues that had taken refuge there, is far from being true. No such necessity existed, for there was no difficulty in having any one apprehended by sending officers for the purpose, or offering a reward.

The New Zealand Land Company have been the secret spring of this transaction, and under the shelter of certain influential names, the managers have contrived to blind the English public. It will scarcely be believed that the New Zealand Land Company had disposed of several thousand shares of land before they purchased an acre. Some three or four thousand emigrants, who had purchased allotments, left England on their way to take possession of them, just after the agent. Upon their arrival they could obtain no satisfactory information respecting their allotments, and were left in a destitute condition, to spend the few earnings they had left, and to endure all the privations to which people landed in a new country are subject.

Even of those allotments that have been given out, many are not susceptible of cultivation. It is scarcely to be believed that the high names which stand at the head of this Company could have been informed of the true state of things; yet it is generally supposed in this part of the world, that it is by their exertions and influence that the British government has been induced to take forcible possession of the territory of an independent state, which New Zealand undoubtedly was. However this may be, the speculators have succeeded in their object, and the country will now be retained by England, even if a military power should be necessary. Should the New Zealanders resist, and they are a warlike race, yet acting against European discipline, they will readily be overcome. They are not unlike grown children, and may be more easily ruled by kindness, and by satisfying the wants of the chiefs, than by force. The population will soon disappear before the whites, for the causes that have operated elsewhere are to be seen in action here, where the savage is already sinking imperceptibly before the advances of civilization. While philanthropy, real or pretended, is ransacking the globe to find subjects for its benevolence, it seems a little surprising that scarcely a voice has been raised in Parliament against this act of usurpation.

On the 29th of February, 1840, there was a violent gale at the Bay of Islands, said by the missionaries to have been the severest they had experienced, with perhaps the exception of one which took place shortly after their arrival. Many vessels suffered great damage. The Thorn, of Sag Harbour, which sailed a few days before, bound home, was obliged to put back, and in consequence of the damage received, was condemned as unseaworthy, as was also the Tuscan, an English whaler. The barque Nimrod arrived, having lost her topmast, and several coasters were missing, supposed to have been lost. Most of the vessels lying off Kororarika dragged their anchors, but they suffered less from not being much exposed; the Harriet was driven ashore at Tipoona, a few miles to the eastward, near Point Pocock. This vessel parted her cables during the night, and the next morning was found a complete wreck. The crew barely escaped with their lives. Besides these disasters on the water, those on the land were also great: fences were carried away, houses deluged, grounds overflowed, wharves injured, and the extensive embankment of the missionary establishment at Pahia nearly demolished. The tide rose six feet, during the night of Saturday, beyond its usual mark, which caused most of the damage.

This gale was experienced at the Thames on board H. B. M. ship Herald, one hundred and forty miles to the south; also by the Flying-Fish, off Cook's Straits, and by the barque Achilles, to the north. Mr. Hale was a passenger in the last named vessel, and took barometrical observations and notes during the continuance of the gale.

From the observations, it appears that the change took place at the two northern and two southern positions, in opposite directions, proving that the gale was a rotary one, and that its centre must have passed between the Bay of Islands and the river Thames. The greatest force of the gale was between the hours of 1 and 3 A. M., on the 1st of March. At the Bay of Islands, a calm was observed by Mr. Dana and others, which lasted fifteen minutes, after which the wind rapidly hauled round to the westward, and blew with increased violence. On board the Herald, the barometer fell to 28·75 in., and from the fact of the gale having been experienced first to the northward and eastward, it is certain that it came from that quarter, and passed over New Zealand in a southwest direction: the width of the track was about five hundred miles. The particulars of the preceding observations will be found in the Meteorological Report.

Foreign residents have established themselves in many places, and on all the inlets or arms of the Bay of Islands their cottages are to be seen, occupying the points and coves.

On the north, the British resident, Mr. Busby, has built a large and commodious cottage, and commenced laying out his grounds in town lots for the future city of Victoria, of which there was a public sale previous to our arrival. All the lots were, I believe, purchased on speculation, for after seeing the locality, one must be convinced that it offers no advantages for more than a village, if indeed for that. More to the westward, is situated Pahia, the mission establishment. For commercial purposes, the south or Kororarika shore offers the greatest advantages, having the deepest water, and being the most sheltered from the stormy winds.

The extent to which speculation has raised the prices of land in this neighbourhood is almost incredible. Mayew's Point, the first above Kororarika Bay, has on it a few storehouses, which are rented for six hundred pounds ($3,000) a-year.

Mr. Clendon, the American consul, for about three hundred and twenty-five acres, of which only fifty are level, has received thirty thousand pounds from the British government, reserving to himself the remainder, one hundred acres. He bought the whole for a trifle a few years ago. The location is a pretty one, on a hill about three hundred feet high, and is, perhaps, the most commanding spot on these waters. The neatness of his cottage and of the grounds about it adds much to its pleasing appearance.

The introduction of a Sydney police at Kororarika has been of service to that place, for they have dealt in a summary manner with the vagabonds who formerly frequented it.

A Roman Catholic bishop is established here, who has a chapel, and it was said, was making many converts; but it was supposed that the principal inducement to conversion was the liberality with which he and his associates bestowed gifts and presents upon those who joined in their prayers and received the cross.

Besides the Episcopal mission, under the Reverend Mr. Williams, formerly a lieutenant in the British navy, there is a Wesleyan mission at Hokianga, which is highly spoken of. Many reports have been put in circulation by the evil-disposed, in relation to these missions; but as far as my observations went, they seemed exemplary in their duties; they were also occupied in farming, in which native labourers were employed. Mr. Williams having a large family growing up, many of them obtained farms, and are now in the successful occupation of them. There is no doubt the hue and cry against the father, that the mission had obtained all the best land from the natives, arose from this cause. Some circumstances were remarked, from which it was evident that the interests of the natives were looked after by the missionaries, who protected their lands and induced them not to sell to the emigrants, who would otherwise have found them only too ready to part with them.

It is true that the situation of these missionaries of the Church of England is different from that of any we had heretofore seen, and equally so that they do not appear to have succeeded as well in making proselytes as those in the other Polynesian islands; but I am persuaded that they have done and are still endeavouring to do much good. They are, however, separated, as it were, from their flocks, and consequently, cannot have that control over their behaviour that would be desirable. Many scenes, therefore, take place at the pas or strongholds, that might be prevented if the missionaries mingled more with their converts.

Mr. Williams was kind enough to have divine service at the house where our naturalists stayed,—Mr. Tibbey's. I was not a little surprised when I heard that Mr. Williams had refused any opportunity to our philologist to inspect a grammar of the New Zealand language, that was then going through the press. I mention the circumstance as remarkable, from being the only instance of the kind that occurred to us during the cruise; and it cannot be easily imagined what could have been the cause of his refusal, for a very short period after our departure it would be published, and there could have been no fear of his being forestalled by us.

Among the natives the taboo is yet law, though endeavours are making to introduce other laws among them. It was told me, on good authority, that there had been a trial for murder by a jury of chiefs at or near Hokianga, under the direction of a white man, but there was great reason to believe that the person did not receive that impartial justice which a duly organized court would have assured him. The evidence was said to have been deficient, but the current belief being against him, he was notwithstanding shot.

The natives, we were told, were not a little surprised at the summary way in which justice, or rather punishment, is dealt out by the magistrate of Kororarika.

Their taboo laws are very strict, and carefully observed, even among those who are considered Christians. The chief, Tomati refused to enter the house of a person whom he took Mr. Hale to visit for if he had entered, it would have become tabooed; and the native law, which does not permit any man to enter a house in which a chief has resided, even temporarily, would have compelled him to abandon his dwelling. Women alone are allowed to enter the houses of chiefs. An instance of this was witnessed at the pa of Pomare, and another where we attempted to purchase the prow of a canoe. This prow, which was elaborately carved to represent some non-descript animal, with a human head, having the tongue protruded, was accidentally seen in an out-of-the-way storehouse, and was somewhat mutilated; it had belonged to the late chief Kiwikiwi, and was tabooed in the first degree. Overtures were made to the widow of Kiwikiwi for its purchase. It was evidently considered very sacred, for none of the natives would touch it, or even enter the storehouse in which it was kept. Notwithstanding all its sacredness, it was sold, after a little chaffering, for six dollars. The first price asked was two pounds, but the widow could not resist the chance of its sale. After the bargain was concluded, no native could be found willing to incur the penalty of the taboo, by carrying it. When the transportation was accomplished, a new and unexpected difficulty arose: it could not be carried across the water in a canoe, as it was against taboo to do it. The threat of making them refund the money, and take back the *ihu* or nose, so worked upon the covetousness of old Kawiti, the chief, that he consented to remove it, and also promised to come the next day and paint it red, after the native fashion. This he punctually performed, using a kind of red earth mixed with water. This is represented in the tail-piece at the end of this chapter.

The taboo is always resorted to, to protect their kumara-patches, and the fear of breaking it was strongly shown by the intrusion of Mr. Tibbey's goats into the kumara-patch of Pomare, near his pa. No one could be induced to go in to drive them out, for fear of punishment; and a message was sent to the chief to allow them to be expelled. After the permission was given, the natives could not be induced to enter by any other place but that where the goats had broken through.

The natives, for the most part, have their permanent residence in towns, or what are here termed "pas," which are generally built on high promontories, or insulated hills, and fortified in a rude fashion, with a palisade of upright stakes, about ten feet high: the houses or huts are all built closely together.

Pomare's pa being near our anchorage, was frequently visited. It contained about three hundred huts. There was a main entrance through the palisade, near which are two posts, the tops of which are carved into distorted representations of the human figure.

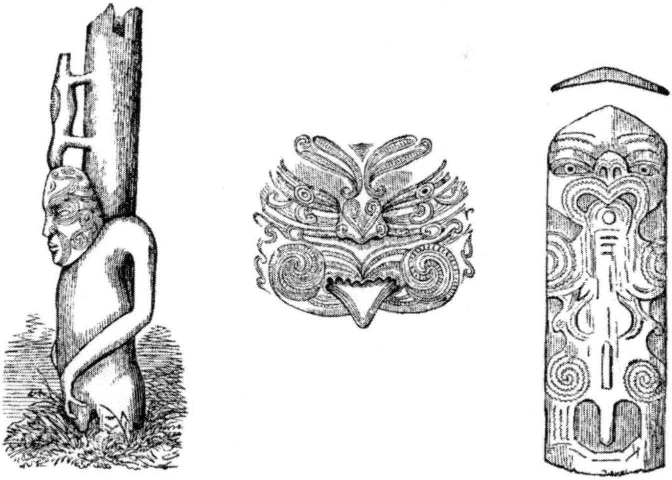

DRAWINGS OF NEW ZEALAND CARVING.

Within the main enclosure are other enclosures, each containing five or six houses, with alleys of two feet wide, that traverse the town. Their houses are very simply constructed: four corner-posts are driven into the ground, and left from two to five feet above the surface; in the centre line two or three strong posts are firmly set in the ground, to support the ridge-pole of the roof; on the posts is placed and lashed a horizontal beam for the rafters to rest upon, and smaller poles are lashed to the posts, at one foot apart, from the ground up; on these the roofing is worked: the material used in thatching is the rush (Typha latifolia), or our common cattail. The manner of making the roof is to tie the materials on the horizontal strips or poles, setting the larger ends on the ground, and driving them close against each other, generally with the fist, and so on until all is closed in, leaving doorways under the eaves, at the gable-ends; the *rappooing* is then cut square off at the upper horizontal beam or plate-piece, and the roof is put on, made of the same material, and generally thatched with it or fern. The roofs have usually but little pitch, which gives a squat look to the houses. Mats are generally hung up at the doorways, but some have doors made of pine; they are low, obliging one to stoop or creep, in entering. Around their houses they have usually peach trees growing, but nothing else is cultivated about them.

The furniture consists of mats, a few baskets and trinkets, an old chest to lock them up in, an iron pot, and a double-barrelled gun, generally of the best maker.

POMARE'S HOUSE

Pomare's house was about twenty feet long by twelve broad; from five to eight feet high. The mode of construction was the same as above described, with the exception that the rafters were flat and ornamented with arabesque work, drawn with soot or black pigment. The posts were likewise carved; but from the dirt and filth with which they were covered, it was difficult, if not impossible, to decipher them. It is said that the New Zealanders have improved in the art of building since they were first visited, but they are still in this respect far behind any of the islanders we have visited.

Four of our gentlemen, before my arrival, had paid Pomare a visit, and made him some presents, which, so far from satisfying his cupidity, only made him more covetous. On receiving a watch-chain, he asked for the watch; and could not be induced to exhibit a dance, unless each person presented him with a shilling. This exaction was submitted to, though they were disgusted and disappointed with the greediness he manifested. The dance proved very similar to those seen among the Samoans and Tahitians, with the same tossing of the arms and legs,

TOMB OF A NEW ZEALAND CHIEF

and various contortions of the body, performed by a number of men and women. The only music was that of the voice, two or three singing in a high monotonous key. The dance was, however, seen to disadvantage by candlelight.

On the top of the hill is a sacred enclosure, or Kianga-taboo, in which is erected the tombs of the chiefs. A few days before our visit one was interred here. The vignette represents the tomb.

This tomb is formed of a small canoe, cut across through the middle, and the two parts joined face to face, forming a hollow cone, about seven or eight feet long. The corpse is placed inside, in a sitting posture, and would remain there a year, after which the bones would be carried up the river, and as Charley Pomare expressed it, would be "thrown away any where."

The tomb is painted red, and ornamented with feathers on each side, from the ground to the top; it is covered with a small shed, to protect it from the weather, and enclosed all around with a fence. The funeral ceremonies were not witnessed, but, from the description of the natives, were very noisy, and accompanied with firing of many guns,—a general practice on all public occasions. Their faces and arms bore evident marks of their having been engaged in the ceremony, being covered with scratches which they had inflicted on themselves.

The pas of the natives are not in reality strong places, but are little more than insulated and commanding situations. Pomare makes some show of warlike instruments, in the formidable array of three ten-pounders, all of them in bad condition, though looked at and spoken of by the natives with no small pride and conceit. The natives, in time of peace, do not live constantly in these pas, but are mostly occupied at their plantation-grounds; for which reason only a few men were seen lounging about in front of their houses. The women were generally engaged in making and plaiting mats, or cooking, and the men seemed the greater idlers.

Their native dress consists of mats of various kinds, made of the native flax (Phomax), which are braided by hand, and are, some of them, finer than carpeting, while others are as coarse as our corn-leaf mats. The latter were worn by the women while at work, tied around the hips, and sometimes over the shoulders. They carry their children on the back, like our Indians.

The men were more luxurious in their dress, having fine mats, nearly as large in size as our blankets, ingeniously and beautifully wrought, and sometimes embroidered. Both of these kinds are still worn, though they are gradually disappearing, and the dress is becoming more European, or rather Tahitian. The women now often wear loose slips of calico, drawn about the neck, which are any thing but becoming, while the men have coarse clothing, sometimes a dirty white blanket, at others, different parts of European dress. The blanket is

NEW ZEALAND WOMAN AND CHILD.

worn in the same manner as the native kakahu. They never think it necessary to use clothing for a covering; it is worn more from pride and ostentation than any thing else; and not unfrequently a native may be seen decked out in a coat and vest without any covering on his nether limbs, and occasionally with a pea-jacket and no shirt. That which gives a foreigner a peculiar disgust to the persons of the New Zealanders, is their filth, which also pervades their houses. They seldom, if ever, bathe themselves, or wash their clothes, which are usually worn until they drop off from age. They occasionally anoint their skins with fish-oil, and of course cannot be expected to keep themselves clean.

To their houses, the description of Cook still applies: they are small, low, begrimed with soot, besmeared with grease, and are filled with filth. As yet, their furniture has received no addition from their intercourse with the whites, except the huge sea-chest and iron pot: the former to deposit their valuables in, and the latter for cooking. It was remarked by us all, how few of the grotesque figures, so much spoken of by voyagers, were to be seen. There appeared to be little carving recently done, in comparison with former times. They are said to have improved in the construction of their houses; but there is still great room for improvement, before they can vie with any of the other islanders we have visited. Their food consists principally of the potato, fish, kumara, or sweet potato, Indian corn, and fern-root, which is found throughout the country. The kumara is much smaller and inferior in quality to those grown in the other Polynesian isles. Here it is a small watery root, and is generally disliked by foreigners. It is preserved in houses constructed for the purpose, to prevent the depredations of the rats. These are built on four posts, which are scraped exceedingly smooth, and are only entered by a single slanting post. The roots are also suspended beneath these houses in large baskets.

Fish are taken with hooks and nets, and are dried and laid by for use. They also eat a clam, which they call *pipi*. Hogs and poultry are raised in abundance, for their own use and the supply of ships. They have, as I before stated, peaches, as well as many small berries, and in a few years they will have all the fruits of the temperate zone introduced by settlers. They formerly ate their fish raw, or cooked with the kumara, after the Polynesian fashion, in the ground, with hot stones; but now they use an iron pot, in which all their food is boiled together. They have a great fondness for rice, with sugar or molasses. They do not want for food, for their country is well supplied with wild roots, which in case of necessity or scarcity can be resorted to. They also make a pleasant beverage, resembling spruce-beer, which they call *wai-maori*.

The greatest changes which have taken place in their customs are the introduction of the use of fire-arms, and the adoption of whale-boats instead of their canoes. The latter are without an out-rigger, and differ in this respect from the boats of all the other Polynesians south of the equator. They have also adopted the square sail (which generally consists of a blanket), in place of the triangular one common to all Polynesia.

The ornaments of the New Zealanders are few; those of the men, who are chiefs, generally consist in an elaborate tattooing, that gives a striking appearance to the face; the regularity with which it is done is wonderful. They all have their ears bored, and have small rings in them, made of jade or shark's-teeth, tipped with sealing-wax, or small bright-coloured feathers. Around the necks of the chiefs and their wives is hung their "heitiki," made of a stone of a green colour, which is held very sacred, and which, with their "meara,"—a short cleaver or club,—is handed down from father to son. The heitiki has some resemblance to a human figure, sitting with crossed legs. This stone is procured from the southern island, near the borders of a small lake, which receives its name from the stone, being called Tewai Pounamu or the Green-stone Water. From the name of this stone, Cook, by mistake, gave the name of Tavy Poenammoo to the southern island. It is also supposed that Captain D'Urville's name of Ika-na-maw (meaning, the fish out of Mawi), given by him to the northern islands, may also be the name of some place on the northern side of Cook's Straits. Those who are acquainted with the natives and their language say, that they have no native name for either of the islands, or any part of the country, and have adopted into their language the names given by the whites, with modifications to suit their tongue.

It was a long time before Pomare would consent to his wife parting with the heitiki which she wore, and that belonging to himself (his atua) he would not allow us to take off his neck, even to look at. Our consul interpreted for me a singular story that the southern natives had invented, relative to these stones: "That they were found in a large fish, somewhat resembling a shark, which they were obliged to capture and kill for the purpose of obtaining them. When first taken from the stomach of the fish, the stone is soft, but from exposure becomes hard, and must be wrought in its soft state." This story was related by Pomare. The smaller stones were about three inches in length, and the larger ones about five inches.

Pomare is a fine-looking man, and is handsomely tattooed. He is six feet in height, and well formed, with the exception of his feet and legs. His dress was any thing but becoming: a blanket was tied about his neck, and hung ungracefully about his person, leaving his right arm free; beneath this he wore a shirt and loose pair of drawers, descending to his knees; the rest of his person and his feet were bare. In his hand he usually carries a short cloak of dogskin, called *topuni*, *shupuni*, or *patutu*. These short cloaks are, in shape, not unlike those of the knights in ancient times; they are about three feet long, being formed of common cloth, mat, or sewed dogskin, dressed with the hair on. Pomare's dress was surmounted by a blue naval cap, with a gold-lace band. The tattooing may give his features somewhat of a fierce aspect, and serve to disguise the expression, yet I cannot but believe that his true feelings are developed in it. His face indicates any thing but a kingly character. Perhaps his reputation for business may have something to do with the impression his physiognomy produced. He told me he had two wives, but it is generally believed that thirty would be nearer the truth. The favourite one usually accompanies him; she is highly spoken of for her good sense, and

Drawn by A T Agate

V Balch Sc

KO-TOWA-TOWA

A NEW ZEALAND CHIEF

Pomare is said to place much confidence in her judgment. She was the best-looking native I saw in New Zealand, but would not be called handsome elsewhere. The missionaries have not yet been able to produce any effect upon Pomare or the family connected with him. Pomare's chief warrior is Mauparawa, who has been persuaded to remain with him, although a native of Hauaki, on the river Thames.

Mauparawa is a much finer-looking man than Pomare,—in appearance a very Hercules; but the effects of dissipation are beginning to be perceived in his powerful frame. He has long been a favourite with the whites, who admire him for his prowess. Many of his followers came with him to join Pomare, of whom few are now left; for in an expedition last year he lost almost all of them: having landed on Aoteu or Barrier Island, he was overpowered and badly wounded, barely escaping with life. One of his acts of daring took place in the last feuds with the Kororarikans, by whom he was much detested. Wishing to put a disgrace upon them and show his contempt, he one night took his canoe, and with six of his followers left Pomare's pa or stronghold for Kororarika, the heart of his enemies' strength. He landed there in the midst of his foes, whom he found fast asleep. Drawing up his canoe on the beach, he went to the house of a white man, whom he awoke, and ordered him to give himself and followers some spirits. threatening him, in case of refusal, with instant death. They took their spirits quietly, desiring the man to say to the Kororarikans in the morning, that Mauparawa had been there in the night, with some insulting message; but before leaving, it occurred to him that the man would not have the courage to tell of his visit: he therefore determined to leave his own canoe, (which was very well known,) and take a whale-boat in its stead. All of which was done merely to throw a slur upon his enemies, at the risk of his own life.

Another person of some note, is a cousin of Pomare, called Charley Pomare, the son of the former ruling chief of that name. Hoia, the brother of the king, appears to be a stupid fellow. Charley Pomare was very talkative, and although young, appears well-informed in the history of the island, and is quite intelligent. In his accounts, he dwells particularly on the extensive ravages committed by Shougi, who I believe was taken or went to Europe. After his return, finding he had lost influence in his tribe, in order to regain it, he committed some of the most barbarous cruelties that have ever disgraced these islands, and made his name terrible among the tribes. Most of these, before his wars, had from three hundred to one thousand warriors, but only a few now remain in some of those who were formerly powerful and independent, and who being from their weakness unable to contend by themselves, have become incorporated with other tribes. The reason that the natives give for this diminution is, that Shougi had killed them all. His conquests embraced nearly all the northern part of the north island, whose warriors he then united, and led against the people of the south, about Hauaki, on the river Thames. With these he waged a long and bloody war, and extended the name of Ngapuhi, which properly belongs to the people about the Bay of Islands, as far south as Kiapara. His death, which happened a few years since, was a great relief both to his followers and foes.

The last war took place in 1837, about two years before our arrival. It was, in all probability, the last native contest that will be waged. It was caused by the disappearance of a woman of Otuiha, whom the tribe of Kororarika were suspected and accused of having killed and eaten. Formidable preparations were made, and the allies on both sides called in; the people of Kororarika being aided by the forces from Hokianga. The principal battle was fought in a piece of marshy ground between Waikereparu and Otuiha. Here Pomare, better known by the name of Charley, then quite a boy, led the forces of Otuiha, while those of Kororarika were marshalled by Pi, a great chief of Hokianga; and the fight was terminated by Charley first shooting Pi, and then the second chief, who was endeavouring to save the body, with his double-barrelled gun. The heads of the warriors were cut off, and preserved as trophies, while their bodies were left on the ground. They were not eaten, though the Hokianga people are said to be cannibals. This latter imputation, however, should be received with caution, as the information was derived from their enemies.

From all I could learn, Pomare is not deemed very courageous, and was not himself engaged in the fight. He is looked upon as quite avaricious, and as a great coward: he is much addicted to liquor. It will, perhaps, excite surprise to learn how he came to exercise the influence he does over his countrymen; it is entirely owing to his eloquence, by which he is enabled to lead them any where. When Charley was asked the cause of his uncle's influence, he said that Pomare could lead the people wherever he chose; and to the question as to why he himself was not king, he answered, "Oh, that is maori" (country fashion).

Some of the gentlemen visited the pa of Pomare, for the purpose of witnessing his return from a visit to one of his allies. The canoe was seen coming up the bay, paddled by forty-five natives, and on the side of the hill all the people of the pa were collected, shouting, waving their garments, and firing muskets, to welcome their friends. When the chief touched the shore, a curious scene ensued. All the boatmen seized their paddles, and ran some distance along the beach, where they halted, and formed themselves into a compact body, in martial array. Those of the pa did the same, and were stationed in front of the canoe; the former party then returned, and when near, the latter made simultaneously, ten or twelve leaps directly upward, waving their paddles over their heads, and giving at each jump, a hard guttural sound, like *hooh.* The two parties then changed positions, when the boatmen went through the same motions, after which the whole mingled together. This ceremony was supposed to represent that used on the return of a war-party. Pomare was found shortly afterwards seated in front of his house, surrounded by his people, who were busily engaged in preparing a great feast, for which he was giving directions, and which shortly took place, accompanied by much merry-making.

The chief, Pomare, on one occasion paid a visit to the gentlemen of the squadron at Mr. Tibbey's, with some fish for sale, and for which he had been fishing several hours. He first asked a shilling for them. which was handed to him, when he immediately raised his price to two shillings, and when this was refused, he went away in high dudgeon, and complained to me on my arrival, that he had not been treated well. Many instances of the same kind occurred.

Mr. Hale induced Hoia, Pomare's brother, to give him a list of the various clans of the great Yopaki tribe, which under Shougi had formerly been the terror of all New Zealand. From this and other authorities, the number of the tribes were given at one hundred and five, in which were comprised upwards of sixty thousand fighting men. Those who are more acquainted, and have the best opportunities of knowing, state the population at less than three hundred thousand; there are others who rate the population from thirty to forty thousand. A mean between the two estimates would be nearer the truth. From the information I received, I am satisfied that it cannot be great. The population of both islands is said to amount to from one hundred and forty to one hundred and eighty thousand, and the whole of this number are on the north island, with the exception of three or four thousand who are on the southern island. It is remarkable that every tribe has a name peculiar to itself, and distinct from the district which it inhabits: thus the natives of Kororarika are called Yaitawake; those of Hauaki (the river Thames), Ngaitawake; and with few exceptions these names begin with the syllable of Nga or Ngati—most commonly the latter. These names are thought to have reference to clanship. The members of each tribe appear to be all connected by the ties of consanguinity.

Some of our naturalists made a visit to a town called Wangarara, situated near the coast, about thirty miles to the southward of Cape Brett. They passed up the Waicaddie river eleven miles to Waicaddie Pa. Here they found a missionary station occupied by a Mr. Baker; but none of the family were at home. The old chief of Waicaddie was very indignant, and treated them quite uncivilly, because they were going to Wangarara. After procuring a guide, they set out on foot for that place. The distance is twelve miles, which they accomplished by sunset. The road lay over mountains. The village of Wangarara consists of four or five miserable huts, or what would more properly be designated kennels, made in the rudest manner, and thatched with fern-leaves. In order to enter these, they were obliged to crawl on their hands and knees. The furniture of the chief's house consisted of a few mats, two or three fishing nets, and an old chest. A fire was smoking in the centre to keep out the musquitoes, and the resemblance to a smoke-house was striking; or, perhaps, the latter would have suffered by the comparison. The accommodations in this hut were rather confined and crowded; for besides themselves, there were three runaway sailors as guests. They, therefore, gladly accepted the invitation of the chief Ko-towatowa, who was on a visit here, to accompany him to his hut, at the mouth of the bay. They went with him in his fine large canoe, and reached his residence late in the evening, where they found themselves much more comfortably accommodated, having clean mats and a good supper of pigeons and potatoes. This was Ko-towatowa's principal farm. His pa is situated a few miles up the bay, on a rocky point, and contains one hundred and fifty houses. It was, at the time of their visit, nearly deserted, in consequence of the attention demanded by their crops; and this is the case with nearly all the other pas at this season.

This part of the country is flat, and has a good soil; and here Ko-towatowa raises most of his potatoes and kumaras, which are larger and better than those raised at the Bay of Islands. They also raise a good supply of Indian corn, and are at no loss for food, which was evident from the quantities of dried as well as fresh fish which was seen.

A great difference was perceived between the natives of this place and those of the Bay of Islands. The former have had little or no communication with foreigners, their manners are more simple, and they have little or no idea of the conventional value of money. The people of this place appeared more virtuous and happy, and a number of young women were seen, good-looking, sprightly, and full of animation.

NEW ZEALAND GIRL.

They here saw the old chief of Wangarara, grand-uncle to Ko-to-watowa. He was very feeble, with white hair, and clad in an old dogskin robe. He was observed to sit all day on a small mound of dirt and pipi-shells; having lately lost a relation, he, according to custom, is tabooed for the season. He does not help himself, and is not allowed to touch any thing with his hands; his grand-daughter, a sprightly girl, waits upon him; and it was pleasing to witness the watchfulness she evinced in attending to his wants, often filling and lighting his pipe, and holding it in his mouth while he smoked. Notwithstanding the promising appearance of Ko-towatowa's house and premises, it was found swarming with fleas and other vermin. Ko-towatowa is a member of the Episcopal Church, and daily performed worship in his native tongue. After their morning meal, they began their rambles, but had not proceeded far before they were met by a large party of natives, who kept saying to them, "*walk about one hilling*," by which they soon understood that they were required to pay one shilling for the privilege of walking on the beach and picking up shells; on Ko-towatowa's being appealed to, he soon dispersed them. On a hill, near this place, Mr. Drayton found a beautiful specimen of Bulimus Shougii.

Wangarara Bay is a deep indentation in the coast, to which it runs parallel, and is separated from the ocean by a narrow belt of high and rocky land. It is said to have good anchorage for a distance of six miles from its mouth. The entrance is very deep, free from danger, and about one mile wide: it is a much safer port than the Bay of Islands. A vessel might pass by its entrance without suspecting that a harbour existed. Provisions of all kinds are much cheaper and better than at the Bay; and although the natives are aware of this difference, yet not being able to transport their provisions there, they are content to dispose of them for a less price.

Their kind friend Ko-towatowa took them back to Wangarara, stopping on the way at his pa, where he presented them with quantities of peaches, which had been tabooed to his people. At Wangarara they again found their guide, and the two old chiefs,—the elder of whom was called Kawau, and the other, a little younger, Ruahenna: both of them have the character of being great rascals. The contrast between them and Ko-towatowa was very much to their disparagement. With some reluctance they ordered a pot of potatoes to be boiled; but when night came, they positively refused entrance into their huts unless each gave a shilling, to which Ko-towatowa sternly objected, saying that they were his guests, and should not pay. A quarrel between the chiefs ensued, and the only way it was prevented from going to extremity, was to slip the money quietly into old Kawau's hand; after which, peace was restored, and they retired for the night, where they were effectually tormented by the fleas and vermin. Ko-towatowa, on taking leave of them, refused any compensation for his services; but a pressing invitation to pay them a visit at the bay was accepted.

They returned by the same route, and by noon reached Waicaddie Pa. It contains about two hundred houses, and is situated between two small fresh-water streams. This is the most cleanly and extensive town in the neighbourhood of the Bay of Islands. Mr. Baker, of the Episcopal Mission, has settled here; he has many acres of land, and comfortable dwellings, farms extensively, and has about twenty head of cattle, with good pasture for them. The natives also possess some cattle. By night they reached their lodgings.

One who has long known the New Zealanders, and on whose judgment reliance may be placed, gives them credit for intelligence and generosity, and says that they are hospitable and confiding to strangers, persevering where the object concerns themselves, strongly attached to their children, and extremely jealous of their connubial rights. A violation of the latter is punished with death, not only to the parties themselves, but sometimes extended to the near relatives of the offenders. They are crafty, but not overreaching in their dealings, covetous for the possession of novelties, although trustworthy when any thing is placed under their immediate charge, but not otherwise over-honest.

A transient visiter would hardly give them so high a character, and would, I think, have an unfavourable opinion of the race. He might, however, award to them intelligence; but they appear vindictive, and, from a number of facts, must be treacherous. One cannot be long among them, without discovering that they are adepts in trickery, and suspicious in their dealings. These bad qualities they may have acquired from the number of low whites that are among them. They seem destitute of any of the higher feelings, such as gratitude, tenderness, honour, delicacy, &c. They are extremely indolent and dirty, disgusting in their habits, and carry on the infamous practice of traffic in women, which even the highest chiefs are said to be engaged in, openly and without shame. The vice of drunkenness does not exist among them to any degree, and it is not a little astonishing that the bad example set them should not have been more followed. They are extremely proud and resentful of any insult, to avenge which the whole tribe usually unites. As an instance of this, we may cite the conduct of Ko-towatowa, whose hospitality to one of our parties has been recorded. At the invitation of the gentlemen who had been indebted to him for attentions, he visited them at Tibbey's, when an untoward circumstance occurred, which had well-nigh ended in an open affront. As they were seated in the porch of Tibbey's house, one of their thoughtless visiters, by way of affording amusement to the company, played off upon Ko-towatowa a boyish trick, by burning him on the nose with a cigar. This produced great anger in the chief, who would have at once punished the rudeness, but through the timely interference of the bystanders, he became appeased, but required some atonement for the insult offered him; a half-dollar was given him, but he said he would accept only half, as he did not want to be paid for it, but merely desired a token that it had been atoned for. In the opinion of all, he rose much above the silly trifler who had been the perpetrator of the joke.

The natives are peculiarly sensible to any insult of this kind. A short time before our arrival, a mischievous white boy, staying with our consul, had placed a small brass kettle on the head of an old chief, which caused some amusement to the bystanders. The chief at the time did not show any signs of being offended. He had always been well disposed and peaceable towards the whites, and was known to have a strong partiality towards the family. On going to the pa, however, he mentioned the circumstance to his tribe, which produced a great excitement among them. They assembled and advanced in a body to the dwelling, to require satisfaction for the affront offered, and although they were told and convinced it was done in playfulness, they required atonement; and this being refused, they took all the clothes that were hanging to dry on the lines, and every thing they could find about the premises. They even took the shoes and clothes off a sick boy, who was lying in the veranda. Their rapacity was only stopped by the courage of the mistress of the house, who, being unable to check their proceedings by remonstrances, threw a billet of wood at the principal chief. This bold act astonished him, and from admiration of her courage, caused them at once to desist, saying she had a big heart, which is their figurative term for a courageous person. Insults given in this accidental way, have been known to occasion the most deadly feuds. They have, however, great command of temper when insulted. As an instance of this, an anecdote was related to me of some chiefs having become offended at the Episcopal missionaries in consequence of some transaction respecting lands, in which they conceived themselves wronged. The offended parties proceeded to Pahia in order to demand redress; but on their arrival there, the missionaries were absent, and although the whole property was at their mercy, there being no one on the premises but females, they did not harm any thing, and declined to enter into any explanation until they had seen the missionaries. Taking their seats quietly at the gate, they awaited their return, which did not take place for some hours after, when they demanded an explanation of the supposed wrong, and atonement for it; and being satisfied, they departed without any molestation or injury whatever. It will, in all probability, be said, that such patience was in consequence of the parties complained of being missionaries; but that could not well have been the case, for they are by no means popular with the natives, and the reason is, that the missionaries show very little regard for their own countrymen, which, in the eyes of a New Zealander, is a great crime.

From all I could gather, I am inclined to believe them an observant people, and that they would become an industrious one, were it less easy to provide themselves with the necessaries of life. They show much energy of character in their warlike pursuits, on which their whole minds seem yet to dwell. The spontaneous productions of their soil furnish them so easily with all that is required for their food and clothing, that there is no sufficient incitement to industry.

The New Zealanders are above the middle size, well formed, and athletic; they vary in colour from a chestnut to a light copper; they have black hair, very thick and curly, which many suffer to grow long, while others crop it close. I saw few with whiskers, and their beards were light. The forehead is high, sloping backwards; the nose frequently aquiline and prominent; the eyes are black and piercing, but rather small; the tattooing gives a hardness of outline to the chiefs that is not so observable in the common people; they want, however, the softness of the rest of the Polynesian family, of which they are a part, not having the full muscles, or soft contour of face, which we had hitherto observed among the groups we visited. They are as indolent as the other cognate races, but more capable of undergoing fatigue.

The following is one of their traditions respecting their origin. The first natives came from Hawaiki, situated towards the east, in several canoes, and the names of some of the principal men, were Tanepepeke, Tanewitika, Taneweka, Rongokako, Kopaia, Kornanpoko: the canoes in which they came were called Kotahinui, Kotearawa, Kohorouta, Takitima. They settled first at Kawia, on the western coast; then near Maketu, Turanga, and Ahuriri, at the east cape. The natives, it may be as well to remark, say that this story is all nonsense, yet the similarity of the foregoing names with those of the people of Savaii, in the Samoan Group, is striking. This, connected with the story, which we shall hereafter quote, of the introduction of the kumara in canoes, taken together, would appear to afford very strong reason for the conjecture that they were derived from the same source. In their native traditions there appears to be some idea of a creation, having a general resemblance to that of the other nations of Polynesia.

The trade in native curiosities is not quite so great as it used to be, particularly in tattooed heads. So great at one time was the traffic in the latter article, between New Zealand and Sydney, that, in 1831, it was prohibited by law. In Governor Darling's administration of the colony, the chief Shougi is supposed to have made large sums by it, and there are some persons who, in part, impute his wars to his desire of gain; for, having been in England, he became acquainted with the value set upon them, and the demand for them. It is generally thought that many of the heads thus sold have been prepared by the white runaway convicts, who have learnt the mode of doing this from the natives. They are still to be obtained, though great precaution is used in disposing of them. A missionary brig, lying at the Bay of Islands, had many curiosities on board, in the possession of the steward; and after the buying of mats, &c., had been finished, he invited our officers to step down to his little store-room, under the forecastle, where he had a curiosity which could not be brought out. After this mysterious enunciation, they followed him to the bottom of the ladder; he then told them he was about to put his fate into their hands, believing that they were too much men of honour to betray him. He then proceeded to inform them that he had two preserved heads of New Zealand chiefs, which he would sell for ten pounds. He could not venture, he said, to produce them on board the brig, but if they would appoint a place, he would bring them. The penalty for selling them was fifty guineas, and he conjured them to the most perfect secrecy. These proved to be beautiful specimens, and now form a part of our collections. So effectually has the fine prevented this traffic, that it is an extremely difficult matter to obtain a head; they are as rare now as they have been common heretofore; and the last place in which it could have been expected to find them, would have been on board a missionary vessel.

The New Zealanders are still cannibals, although in the districts where the missionaries reside, they have done much to put a stop to this practice. After the arrival of our gentlemen, an instance occurred of a chief having killed a boy about fourteen years of age, as a medicine for his son, who was sick; and as this prescription did not effect a cure, a girl about the same age was to be served up, but the timely interference of the missionaries prevented it.

The present condition of the New Zealanders is inferior to that of some of the other Polynesian nations. There is, as in other places, little or no occasion for labour; the industry of a few weeks is all that is needed to supply them with food for the year; their traffic in pigs and other supplies to whalers and traders is quite sufficient to procure their necessary supply of clothing. It is said their moral condition has much improved of late, and that they are becoming sensible of the advantages of civilized life. In the former direction there is still great room for improvement, and the latter, I should think, as yet far above their ideas of honesty and of the obligations they owe to those about them. Perhaps those who have become somewhat attached to the Christian religion may be a little improved, but the only instance that we can recall to our recollections is that of the chief Ko-towatowa. The chiefs, however, in general show a growing disposition to acquire comforts about their dwellings, and in comparison with the other natives, are almost cleanly in their persons. Industry is also making progress in the cultivation of their plantations. If I could believe it possible that the dwellings of the lower classes of the people had ever been more filthy, or their persons less cleanly, I would more readily credit that some improvement had taken place. Numbers are said to be able to read and write their own language, having been taught by the missionaries, and then have afterwards been known to take a pride in instructing others, and to display a great eagerness in the acquisition of farther knowledge; but they are far, very far behind, in the rudiments of education, the natives of other groups where the missionaries have been established, although, as respects natural capacity, they may probably rank higher.

There is much that is worthy of notice in the missionary operations here. They seem to have pursued a different course from that followed at the other groups, and appear to begin by teaching the useful arts, and setting an example of industry. This has given rise to much remark. The missionaries of the Episcopal Church appear to keep aloof from the natives, and an air of stiffness and pride, unbecoming a missionary in most minds, seems to prevail. They have a chapel at Pahia and one at Tipoona, but very few persons attend; their native and Sunday schools have also very few scholars; and they appear to be doing but little in making converts. Most of the natives, however, have morning and evening prayers, but their practices and characters show any thing but a reform in their lives. The missionaries hold large tracts of land, and about the Bay of Islands the Church Mission (Episcopal) may be said to have the entire control of the property. At the missionary establishment at Pahia they have a printing-press, and have printed some parts of the Scriptures. They are now printing a New Zealand grammar. In the native traditions, there appears to be some idea of a creation, having a general resemblance to that of the other nations of the Polynesian groups. The first god was Maui, who fished up the earth out of the sun; afterwards a great flood came which covered the land, and then the waters were dried up by another god, who set fire to the forest. From the accounts and observations of all, it may be safely asserted that the natives have no religion. Some few apparently follow the form of it, and call themselves professing Christians; but the majority or greater number of the natives have none, either Christian or pagan. When undergoing tuition by the missionaries, they are said frequently to stop and ask for a present for having said their hymn, and it is said, I know not with what truth, that the Catholic missionaries have been in the habit of giving them some small token in the shape of crosses, which the natives look upon as a sort of compensation.

At Kororarika, as has been stated, there is a Roman Catholic chapel, and it is the residence now of the Bishop of the South Sea Catholic Mission. Some singular anecdotes are related of the natives, of their first joining one denomination and then another, receiving little articles as presents from each; indeed, it is said that there are few of them but conceive they ought to be paid for saying their prayers, or attending mass. At Hokianga there is also a Methodist or Wesleyan

Mission, which is generally considered the most active, and is doing a great deal of good.

The native pas are generally scenes of revelry and debauchery. My crew soon got tired of their visits to that of Pomare, and complained much of the dishonesty of the natives. Pomare and his suite paid the ship a visit a few days after our arrival, for the purpose of obtaining his quota of presents. I received him and all his retinue with kindness, and made him several presents, among which was a fowling-piece; but he had, in going round the ship, seen one of Hall's patent rifles, that loaded at the breech; and nothing would satisfy him but to exchange the gun I had given him for one of these. He surprised me by at once comprehending its facility of use, and its excellent manufacture. After a great deal of importunity, I consented to the exchange, but found that he was inclined, after having once succeeded, to beg every thing that struck his fancy. In this he was followed by the other chiefs, among the rest by Hoia, his brother. To the latter, I gave an old cocked-hat, which pleased him exceedingly, and I was not a little amused to see him wearing it, and dressed in a tight coat and vest, with bare legs, exhibiting one of the most ridiculous figures imaginable, although in his own opinion the beau ideal of elegance. Pomare went about the ship begging for military caps with gold bands, and was extremely importunate until he found that nothing more could be obtained. I by no means admired his appearance on this visit; for, although of good proportions, tall, and well made, he is awkward and parrot-toed. His height and manner of walking make this defect more apparent, and he wants that dignity which is sometimes seen in a savage of our country. The New Zealanders, however, struck us as having a closer resemblance to our North American Indians than any others we had yet met with among the Polynesian nations. I was surprised to see how little respect was paid to the orders of Pomare by his followers, and was told that there is little authority acknowledged by those who are free. His slaves and wives are those who must sustain the burden of his wrath; their lives are at his disposal, and with them his will is law; they seem, however, to be treated kindly. Pomare is said to be entirely under the control of his favourite wife, of whom I have heretofore spoken. She is a far more respectable person than her husband, and was the most intelligent native I met with.

Wishing to see their war-dances, I requested Pomare to gratify us with an exhibition, which he consented to do. The ground chosen was the hill-side of Mr. Clendon, our consul's place, where between three and four hundred natives, with their wives and children.

assembled. Pomare divided the men into three parties or squads, and stationed these at some distance from each other. Shortly after this was done, I received a message from him, to say that they were all hungry, and wanted me to treat them to something to eat. This was refused until they had finished their dance, and much delay took place in consequence. Pomare and his warriors were at first immovable: but they in a short time determined they would unite on the hill-top, which was accordingly ordered, although I was told they were too hungry to dance well. Here they arranged themselves in a solid column, and began stamping, shouting, jumping, and shaking their guns, clubs, and paddles in the air, with violent gesticulations, to a sort of savage time. A more grotesque group cannot well be imagined; dressed, half-dressed, or entirely naked. After much preliminary action, they all set off, with a frantic shout, at full speed in a war-charge, which not only put to flight all the animals that were feeding in the neighbourhood, but startled the spectators. After running about two hundred and fifty yards, they fired their guns and halted, with another shout. They then returned in the same manner, and stopped before us, a truly savage multitude, wrought up to apparent frenzy, and exhibiting all the modes practised of maiming and killing their enemies, until they became exhausted, and lay down on the ground like tired dogs, panting for breath. One of the chiefs then took an old broken dragoon-sword, and began running to and fro before us, flourishing it, and at the same time delivering a speech at the top of his voice. The speech, as interpreted to me, ran thus: "You are welcome, you are our friends, and we are glad to see you;" frequently repeated. After three or four had shown off in this way, they determined they must have something to eat, saying that I had promised them rice and sugar, and they ought to have it. Mr. Clendon, however, persuaded them to give one of their feast-dances. The performers consisted of about fifteen old, and as many young persons, whom they arranged in close order. The young girls laid aside a part of their dress to exhibit their forms to more advantage, and they commenced a kind of recitative, accompanied by all manner of gesticulations, with a sort of guttural husk for a chorus. It was not necessary to understand their language to comprehend their meaning, and it is unnecessary to add, that their tastes did not appear very refined, but were similar to what we have constantly observed among the heathen nations of Polynesia. Their impatience now became ungovernable, and hearing that the rice and sugar were being served out, they retreated precipitately down the hill, where they all set to most heartily, with their wives and children, to devour the food

This to me was the most entertaining part of the exhibition. They did not appear selfish towards each other; the children were taken care of, and all seemed to enjoy themselves. I received many thanks in passing among them, and their countenances betokened contentment. Although they were clothed for the occasion in their best, they exhibited but a squalid and dirty appearance, both in their dress and persons.

No native music was heard by any of our officers, and they seem to have little or none in their composition. In their attempts to sing the hymns, chaunts, or old psalm-tunes, they entirely failed to produce any thing like a resemblance. The pitch of their voices when speaking, is higher than that of Europeans, (the French excepted,) and that of the women was not a tone above, which gives additional coarseness to their character. Both sexes have but little intonation in conversation, and there are no tones heard which would indicate sympathy of feeling.

Chatham Island, which will probably soon be connected with the English colony of New Zealand, is now considered as a nest of rogues, and several vessels have been robbed there. Its inhabitants have a tradition that they are derived from New Zealand, whence their progenitors came about a century since, having been driven off in their canoes by a storm, and that on landing they had changed their language. The change consisted in reversing the ordinary construction of their phrases, and the syllables of words, as, for Hare-mai, Mai-hare; and for Paika, Ka-pai. The natives of Chatham Island are not tattooed, do not wear clothing, and are said to be more intelligent than their progenitors. They were conquered a few years ago by a party of New Zealanders from Port Nicholson, who had been driven out by the Kapiti tribes, under the celebrated Rauparaka.

An examination of the charts of the Bay of Islands was made, and some additional soundings added; the meridian distance, measured by our chronometers from Sydney, gave the longitude of the point opposite Mr. Clendon's wharf, 174° 07′ E.; its latitude was found to be 35° 17′ S. The dip and intensity observations were also made here, and will be found registered with those results in the volume on Physics.

Mr. Couthouy, who was left sick at Sydney, took passage in a vessel to Tahiti, and passed through Cook's Straits, touching at several of its anchorages. To his observations I am indebted for the following information relative to the southern part of these islands.

The first point they made was the Sugar Loaf Islands and Mount Egmont. The charts published by Clintz at Sydney, give also the height of this mountain as fourteen thousand feet, but this was believed to be erroneous,* for only a small portion of the top was covered with snow. The day previous to their making land, they had been set to the northward by current about twenty miles in fourteen hours.

They next passed through Cook's Straits to Port Cooper, on the north side of Banks' Peninsula, where they anchored. This harbour is sheltered, except from the northerly winds, and is much frequented by whalers, who resort thither to try out the whale-blubber. The beach is in consequence strewn with the bones of these monsters. On going on shore, a party of three natives and their wives were found in a state of wretchedness and degradation,—their only clothing being an old blanket, disgustingly dirty, besmeared with oil and with a reddish earth which had been rubbed from their bodies, and a coarse mat of New Zealand flax; they depended for subsistence on a small potato-patch, and smoked fish; they lived in low huts formed of stakes, covered with mats, and thatched with grass in the rudest manner: their condition was but little better than that of the Fuegians. A fellow-passenger, who had seen the oldest man left of the tribe, stated that these were the remnants of a tribe that, but a dozen years before, could muster six hundred fighting men; they were all cut off, about ten years since, by the noted chief Robolua, residing near Cook's Straits. The old man appeared deeply affected whilst dwelling on the history of his people. The cupidity of the whites in this case, as in many others, had brought about, or was the cause of, this deadly attack; the particulars were as follow.

The master of an English vessel, by the name of Stewart, (the same person from whom the small southern island takes its name,) was trading along the northern island, and fell in with the chief, Robolua, who was then meditating an excursion to the south. Feeling confident that if he could come upon his enemies unawares their defeat was certain, he offered Stewart to load his vessel with flax, if he would transport him and his warriors to the place he wished to attack. The contract was readily entered into by Stewart, and the warriors were taken on board, and landed on various parts of the coast, where the inhabitants, taken by surprise, were butchered without mercy. Not less than fifteen hundred persons were cut off at this and the adjoining harbour of Port Levy, or Kickurarapa. This Stewart is said to be still living on the northern island of New Zealand.

Many specimens of shells were obtained here, and a few presents, consisting of pipes and tobacco, were made to the remnant of this once powerful tribe. Two of their fellow-passengers intended to land

* I have seen other authorities, which give its height at eight thousand feet.

here for the purpose of establishing themselves, but the place offered so little inducement that they determined to proceed to Port Levy, a larger harbour to the eastward, where the natives informed them that refreshments could be had in plenty. The next day they anchored in it, and found it somewhat similar to Port Cooper, but more open. In the afternoon a party went on shore, and returned with sixty-four brace of pigeons, and three black parrots. The former were in great abundance and very large, some of them weighing twenty ounces: the colour of their backs was a dull slate, passing into bronze on the neck and wings; the head was very black, the breast white, deepening into a reddish brown on the belly; the bill and feet of a bright red. The parrots were quite black, about the size of a crow, and remarkable for two rose-coloured wattles at the lower mandible, like the common fowl. They also killed a species of pica, called *cuga* by the natives, about the size of a blackbird; it was of a dull black, with greenish reflections on the back, and on each side of the neck was a single white feather, which curled forward and upward.

Here they became acquainted with Charley, or Karakiharuru, the chief proprietor of Port Cooper, Port Levy, and Pigeon Bay. Notwithstanding these extensive possessions, neither himself nor his followers were better clad, housed, or superior in any respect to those already described. As for Charley himself, he appeared in a striped shirt, pea-jacket, and trousers, the cast-off clothing of some sailor. From having made the voyage to Sydney, Charley fancied he had seen the world, and took great pains to show his knowledge and excite the admiration of those about him. The captain of the vessel obtained from him about twenty bushels of potatoes, at the rate of a pound of tobacco for a basket containing about a peck; he besides offered to sell one-third of his dominions or estate for a new whale-boat. Charley had on the usual heitiki or neck ornament. The only account he could give of the locality of this green stone was, that it was found to the southward, in a large bed between two mountains. Among other things in Charley's possession, was an enormous wax doll, dressed in the height of the Parisian fashion, which had been presented to him by the officers of a French expedition that had touched there, some time previously,—rather a droll occupant of a dirty New Zealand hut.

About Port Levy the land rises nearly twelve hundred feet high: the soil is every where exceedingly rich, but its value for agricultural purposes is diminished by its steepness; it would be impracticable to use cattle in ploughing. The land in all parts of the peninsula exhibited the same character: a succession of steep hills, intersected by deep and narrow ravines, clothed with a thick forest, except where they terminate on the coast, and form a tolerably level spot of a few acres in extent, available for cultivation. The forest consisted of an abundance of fine timber, principally the Kaurie pine, from one hundred and twenty to one hundred and thirty feet in height, and seven to eight feet in diameter. The fern was thick in patches, but in no great variety; some scandent and parasitic plants were met with, and a great number of flourishing ones observed; but Mr. Couthouy having no means for the purpose, was not enabled to secure any specimens. He remarked that the vegetation appeared much more luxuriant and diversified than that of any country he had seen since leaving Brazil. The soil is a rich black loam, composed of vegetable mould and decomposed basalt; the structure of the rocks decidedly columnar, exposing at the summit of the hills large masses of compact dark gray basalt, containing numerous crystals of olivine, pyroxine, and other volcanic minerals. At the base of the hill, the rock was frequently a coarse cellular lava, and the beach was covered with boulders of all these varieties.

They next stopped at Pigeon Bay, but remained there only a few hours; the passengers who were in search of a position to establish themselves, found this quite as unfavourable as either of the two previous places.

In passing to the northward, towards Cape Campbell, the coast is high and broken, with no level land in the vicinity of the sea; but notwithstanding its abruptness, they found only fourteen fathoms of water at a distance of four miles from the shore, with sandy bottom. They had a fine view of the snowy peaks, called the "Lookers On," about twenty miles to the southward. These are supposed to be nearly as high as Mount Egmont, and tower up in sharp peaks, covered with snow for fifteen hundred feet from the summit. The land along this part of the coast is very rugged, is apparently unsuited for any kind of cultivation, and has no harbours. Off Cape Campbell, a line of rocks was seen extending to the eastward about a league, which do not appear on the charts; they are partly above and partly below water.

They then anchored in Cloudy Bay, which, contrary to the representation of the charts, proved a good anchorage. The wind here sweeps down the gullies in strong squalls, but the water is at all times smooth. There are five whaling establishments in Cloudy Bay, each employing from twenty to thirty hands, chiefly New Zealanders. The kind of whale taken here is principally the right whale, and the quantity of oil collected the previous year was four thousand five hundred barrels, which was sold on the spot to Sydney dealers, at forty pounds the tun. In addition to this quantity, five thousand five hundred barrels were taken in the bay, by whale-ships, principally Americans, from which some idea of its value to our countrymen may be formed. The establishments on shore have connected with them stores for supplying ships, where articles may be had at one hundred per cent. advance on the Sydney prices; potatoes are sold at thirty dollars the ton, and pork at twelve and a half cents per pound; boards and plank may also be obtained at fifty dollars per thousand; wood and water are purchased of the natives for muskets, powder and ball, blankets, pipes, and tobacco. It is also customary to make a present of two muskets, or an equivalent, to Robolua, the chief, for harbour dues. A Mr. Williams, who was one of the establishment, furnished the above information.

Two American whalers were found here. A number of chiefs came off to the vessel, in the course of the day; they were fierce-looking savages, with coarse matted hair, tattooed visages, and bodies besmeared with red earth and oil; some of them were clad in coarse mats, others in blankets, and all exceedingly filthy; most of them had the heitiki ornament about their necks, and some in their ears, which were also decorated with red and white feathers, and the holes pierced in them were also made the receptacle of their pipes; others had necklaces of human bones, polished,—trophies of the enemies they had slain.

Their manners were uncouth, exhibiting none of that amenity so remarkable in the natives of the other Polynesian groups; yet there was a rude dignity about them, that evinced a consciousness of their rank and consequence. Three or four women came on board, but not one of them could be called good-looking, and they appeared to care less about their appearance than the men.

The noted Robolua made his appearance at the breakfast-table, unannounced and uninvited; he most unceremoniously took his seat next the captain, remarking, "Me, Robolua!" In person, he is above the middle stature, powerfully built, and rather ill-featured. The usual expression of his countenance is not bad, but when enraged, it is truly fiendish, and his small deep-sunk eyes, which betoken cunning, gleam with the ferocity of a tiger. His head is of enormous size, covered with long matted hair, sprinkled with gray; his eyebrows were long and shaggy; he had a bad expression of the mouth, resulting from the loss of his teeth, a circumstance of rare occurrence among these natives. He seemed in feeble health, and his figure was slightly bent by age; he wore a filthy blanket, and over it an old-fashioned plaid cloak, the colours of which, like those of his under garments, were no longer distinguishable. All the chiefs wore their dress so as to cover their left arm, and leave the right bare, which Mr. Williams said was for the purpose of concealing their meara, or stone cleaver, which is constantly suspended to the left wrist, ready, at a moment's warning, for use, and which they take particular care never to expose to view. With Robolua was his principal warrior, Oranga-dieti, a fine specimen of a savage chieftain, about fifty years of age, with a noble though fierce cast of countenance, nearly six and a half feet in height, and as straight as an arrow; his long hair was tied up behind, *à la Grecque,* the knot being secured by two long black feathers stuck through it; altogether he had more the appearance of a chief than Robolua; the latter, from the account Mr. Williams gave of him, owes his ascendency more to his powers of persuasion in council, and his talents for strategy in their system of warfare, than to his warlike achievements; and he seldom risks his person in battle. The chiefs, in their figurative language, say, "The breath of Robolua can turn them round and round, and his tongue is more powerful than any of their weapons." He was originally a petty rangatira (landholder). Of late years his power had very much declined: five or six years ago he could number more than six thousand warriors, but now he has not over four hundred. His rapid rise is imputed to the introduction of fire-arms, for he was long the only chief who possessed any number of them; and the decay of his power is attributed to the acquisition of this weapon by others, and the inactivity arising from his advancing age. Several of the natives who were met here could read, and a portion of the Testament was seen in their possession; two women in particular were desirous of showing their accomplishments, and remarked that the missionary religion was not made for New Zealanders; it was too good for them. Drunkenness and dishonesty prevail, by their own confession, among the white men, who are at times entirely beyond the control of their masters; they all have native wives, who are taken and discarded at pleasure.

The whalers stated that the prevailing winds at Cloudy Bay in summer and the beginning of autumn, from November to March, are from the southeast and northwest, which usually succeed each other at short intervals; during the rest of the year, winds from south round to west are more frequent, and bring with them wet weather.

The general information which we obtained, and which has not been included in the preceding portions of the chapter, is as follows:

The climate of New Zealand is extremely changeable; but although it may be considered as the cause of many diseases among the natives it is, perhaps, the best suited to a European constitution of any in the South Seas. A large quantity of rain falls during the year, but I was unable to obtain any record of its exact amount. The temperature at Kororarika, during the months of February and March, varied from 53° to 78°, and the mean was 64·2°. In the sun the thermometer rose as high as 110°. The principal prevailing winds are from the south-east and west; the former are frequently in squalls, and attended with rain: May and June are the rainy months.

Warm days are often succeeded by cold nights, which give rise to pectoral diseases among the natives, many of whom are affected by phthisis, or swept off by rapid consumptions. They are also liable to rheumatism and pleurisy. European and American residents, who enjoy better food and clothing, and inhabit more comfortable dwellings, are exempt from these complaints. Measles, hooping-cough, and other epidemics, have been introduced from foreign vessels. While we lay at the Bay of Islands, the influenza prevailed on shore and was communicated to our crew. The venereal disease, propagated by their licentious habits of life, and unchecked by medicine, is rapidly reducing the numbers of the natives.

The greater part of the soil of the portion of New Zealand which fell under our observation is too sterile to be profitably employed in agriculture. It consists, in general, of an obdurate yellow loam, capable of bearing little else, after it is cleared of trees and brushwood, than the fern (Pteris esculenta). Where the soil is volcanic, however, it is comparatively fertile; but this description of ground is rare.

Wheat and other grains are raised, and the fruits and vegetables of temperate climates succeed well. The hills are almost bare of vegetation; for after the ground is cleared, the heavy falls of rain sweep the soil from them into the valleys, and wear the hill-sides into gullies. In this manner patches of good land are formed in them, which, however, rarely exceed fifteen or twenty acres in extent. The only continuous level tract of as much as a hundred acres, is on the farm of Mr. Clendon on Manawa Bay. The sterility of the soil is not the only obstacle the agriculturist has to contend with. The fern, of which we have spoken, springs up the moment the forest is removed, and covers the land with a dense vegetation. Ploughing is not sufficient to extirpate it, for it will spring again from the severed roots, and choke the grain. It can only be completely eradicated by removing it by hand and burning it. The ashes are then spread upon the ground, and are found to be a good manure. In this manner the sons of Mr. Williams the missionary at Pahia, are endeavouring to bring a farm they possess into cultivation. Natives are employed in the labour, and they have in this way cleared several acres.

The fern, from its size and strength, is supposed to indicate a fertile soil; but this is not the fact, for I have seen nearly a thousand acres in a body covered with a growth of it six feet in height, where the ground was deemed fit for no purpose but to furnish brick-clay. So densely do the ferns grow, that it is impossible to force a way through them, and the only mode of traversing the country where they abound, is by following the native paths; these pursue the high ground and ridges, and have branches which lead to the neighbouring cultivated spots. The moment the culture of the land is neglected, the fern again makes its appearance.

The clayey soils afford only a scanty growth of grass, which is scarcely fit for pasture, and indeed there appear to be no native grasses. In the more fertile soils, red clover, according to Mr. Brackenridge, does well; and he believes that white clover would succeed on the hills, which are now bare. The climate is favourable to the growth of the foreign grasses.

After the fern has been burnt and the ashes spread, a crop of wheat is raised, and the land is laid down in grass. To give an idea of the produce of land near the Bay of Islands, we may cite the instance of Captain Wright's farm, which is eligibly situated, and is considered as possessing a fertile soil. He had twenty acres in wheat, whose average product was only fourteen bushels per acre.

Among the foreign fruits which have been introduced, are apples, peaches, and grapes. The latter grow best in the volcanic soils, but the climate is considered to be too moist to permit them to attain perfection. The peaches are fine, but the propensity of the natives to pluck them before they are ripe, prevents them from attaining their full flavour. Cape gooseberries are plentiful, but the common description of that fruit, and the currant, have not been introduced. Late writers have given marvellous accounts of the growth of the fruit trees of temperate climates, in New Zealand; but these may be set down as exaggerations calculated to mislead, and intended to subserve speculation. The success of Captain Wright, however, in raising fruit and vegetables, has been great.

Among the native vegetables is the sweet-potato, which they call kumara: it is plentiful.

The missionaries stated that the natives have a remarkable tradition in relation to this root; namely, that it was first brought to the island in canoes of a different construction from their own, and composed of pieces of wood sewed together.

Cook left the common potato, which has been cultivated ever since his visit, and is now plentiful.

The native hemp (Phormium tenax) is a most useful plant; it grows in large quantities, and is applied by them to many purposes, besides being a principal article of foreign trade. It is an important material in the construction of their houses, for which purpose it is made into cords, that are also employed for other more common uses. It is manufactured into fine fishing-lines, which are much prized at Sydney for their strength and beauty.

The manufacture of the hemp is altogether performed by the women, who cut it, and after it has been dried a little, divide it into strips of about an inch in width. The outer green fibres are then scraped off with a piece of glass, or a sharp shell. The inner fibres being thus exposed are easily separated, and the greatest care is taken to keep all the fibres as straight as possible, both in this and the following operations. To this precaution the great strength of the cordage the natives make of it, is owing. After the fibres are separated, they are washed, rubbed, and laid in the sun to bleach.

The vegetation of New Zealand is of a fresher and deeper green than that of New Holland, and has some resemblance to that of Terra del Fuego. According to the missionaries, the ridges, and indeed the greater part of the northern island, are destitute of trees; and the woods, which are confined to the valleys, are for the most part in detached spots. The western part of this island contains more actual forests than the eastern.

It was remarked by our botanists that trees of genera which in other countries grow in the more barren soils, are found in New Zealand in those which are fertile. This is in particular the case with the pine tribe. It also appeared to them, from the position of isolated trees, and the quantity of Kaurie-gum found embedded in the soil, that forests had formerly been more generally spread over the face of the country, than they are at present.

The gum which has just been spoken of, is still produced by the Kaurie pine, which is the finest of the timber-trees of New Zealand. The greatest portion of that which is shipped from the island, is dug from the ground. Small quantities of the latter description have been purchased by our countrymen, and shipped to the United States, where it was manufactured into a varnish. This was of a good quality, and was afterwards sent to New South Wales, and New Zealand, where it is sold for copal varnish.

The Kaurie and Kaikotia pines yield spars which for large ships are not surpassed by any in the world. The trees are generally large, and are easily brought to the coast by means of the numerous streams.

The natives use these trees in building their canoes, which are dug out of a single log. They have no out-rigger, and are in consequence liable to accident from want of stability. Great ingenuity is shown in repairing them. We saw a war-canoe which was sufficiently large to be manned by fifty men; it had a prow extended ten feet upwards, which was elaborately carved and decked with tufts of feathers. The paddles have spoon-shaped blades, by which the canoes are propelled with great swiftness.

No native quadrupeds were found wild in New Zealand. Cattle have been introduced, and thrive. Those which are imported require to be fed, but those raised in the country can provide for themselves, and grow fat by browsing.

Among the birds, are the native nightingale and the tui, also known under the sobriquet of the parson-bird. The latter is a great favourite with the natives.

I saw it only in a cage, and its note did not strike me as pleasing, but several of our gentlemen saw and heard it in the woods; they describe its note as rather louder than that of the bird called by the Samoans "poe," and it is at times said to utter a cry resembling the sound of a trumpet.

The domestic fowl does not appear to have been known before this island was visited by white men.

I made inquiries in relation to the mode in which birds were taken in this country before the introduction of fire-arms, but could not obtain any satisfactory information. I was inclined to think that the natives had no method of doing this in former times.

The great staple articles of trade are flax, spars, and wheat; potatoes and gum are also exported; but the whale-fishery is of more value at present to foreigners than all the productions of the soil. This is carried on from the shores by parties of New Zealanders and foreigners; but they are rapidly destroying this source of wealth, for, as has been stated, their eagerness for present gain leads them to destroy the animals whether old or young, without discrimination.

The whaling establishments of British subjects on the coast are numerous, and the most disgraceful acts are perpetrated by their occupants and by the crews of the whale-ships, who not only use violence against the natives, but against each other. As New Zealand is in the immediate vicinity of the whaling-ground, it is a desirable rendezvous for our whalers; and the American whaling fleet, actively employed on the coast in the spring of 1840, amounted to one hundred sail.

Many spars are now exported to England, where, however, the smaller sticks are not as much esteemed in proportion as the larger ones. Several government vessels have recently obtained spars for the Royal Navy at the trifling cost of a few blankets and muskets. The latter, in particular, are a great inducement to the chiefs, who are willing to devote much labour for the purpose of acquiring the means of rendering themselves powerful. Besides guns and blankets, gunpowder, lead, coarse blue and white cottons, whiskey, rice, sugar, and molasses are the articles most in request. These now bring enormous prices, in consequence of the demand caused by the number of immigrants; but the effect of these prices is to render labour proportionably dear.

NEW ZEALAND IHU AND WEAPONS

APPENDIX.

CONTENTS.

APPENDIX.

I.

REGULATIONS RESPECTING THE TRADE, ETC., WITH THE NATIVES OF THE ISLANDS THAT MAY BE VISITED BY THE VESSELS OF THE EXPLORING SQUADRON.

1st. The purser on board each vessel appointed to take charge of the articles on board, intended to be exchanged for fresh provisions and vegetables for the crew, and for shells, specimens, &c., will regulate the rates of exchange, under the direction of the commander of the vessel, in such a manner that they may be the same on board all the vessels.

2d. No one on board, except the trade-master and his assistants, will be permitted to hold any trade or barter with any natives on board or alongside the different vessels of the squadron.

3d. Any one wishing to obtain articles not wanted for the Expedition, will procure them through the trade-master, under such directions as may be given by the respective commanders.

4th. No natives whatever, either male or female, will be allowed on board any of the vessels (except hostages or interpreters) before seven o'clock in the morning, or after half an hour before sunset, and none will be allowed on board during the day, except when the white flag remains at the mizzen, or at any time when they may interfere with the duties of the ship.

(Signed) Charles Wilkes,
Commanding Exploring Expedition.

U. S. Ship Vincennes,
Tahiti, September 10th, 1839.

ORDERS RELATIVE TO THE WATCHES, ETC., ON BOARD THE VESSELS OF THE EXPLORING SQUADRON.

The following alterations are made in the General Rules and Regulations of the Squadron, and will be obeyed and respected accordingly:

1st. The thirty-second article of the General Rules and Regulations will be stricken out, and the following article be inserted in lieu thereof, as—

Article 32d. The first-lieutenant may grant leave of absence to those officers whose services may not be required during the day, until sunset. All officers will return on board their respective vessels by sunset. Boats will be in waiting agreeably to the arrangements made by the first-lieutenant.

2d. The forty-third article of the Rules and Regulations is amended by striking out the words "and third" in the first line.

3d. The "General Instructions to all Officers," in the same Rules and Regulations, are altered by striking out the words "or Passed Midshipman," and inserting, "except by special direction of the commander of the vessel."

No division of watches will take place, except by permission of the commanders.

The order of the 20th September, 1838, relative to medical officers, is revoked; they will hereafter, when in port, (unless ordered to attend to special duty,) attend to the medical duties of the vessel to which they are attached, agreeably to the usages of the service.

No boat will leave the ship for the shore after sunset, unless by order of the commander.

(Signed) Charles Wilkes,
Commanding Exploring Expedition.

U. S. Ship Vincennes,
Tahiti, September 10th, 1839.

II.

ORDERS FOR LIEUTENANT EMMONS AND DR. GUILLOU.

1st. They will compare the sympiesometer of the Porpoise and of this ship together, and carefully note the difference.

2d. Lieutenant Emmons will provide himself with a pocket sextant, artificial horizon, and pocket azimuth compass, note-book, &c.

3d. Dr. Guillou will provide himself with two thermometers and an hygrometer.

4th. The observations required to be made with the sympiesometer half-hourly, on ascending the heights; the temperature of all springs; the temperature of all the different heights, in the sun and shade; morning and evening, at noon, 9 A. M., 9 P. M., 3 P. M, and 9 A. M.; their watches will be set to ship's time, in order that the observations may be compared.

5th. As many observations relative to the topography of the island as may be in their power; the dimensions of the lake on the mountains, its altitude, and, if any opportunity offers of getting at its depth, to ascertain that also.

6th. The meridian altitude, or that of double altitudes for latitude.

7th. Any peculiar or remarkable atmospheric phenomenon, such as refraction, &c., will be sketched, and the time of appearance noted; the other observations will be made at the same time.

(Signed) CHARLES WILKES,
Commanding Exploring Expedition.

U. S. Ship Vincennes,
Matavai Bay, September 13th.

III.

U. S. Ship Vincennes,
Matavai Bay, September 19th, 1839.

SIR,—

You will proceed with the Porpoise under your command, to the west end of Nairsa, Dean's, or Prince of Wales's Island, in longitude 148° 02′ W., and latitude 15° 05′ 30″ S. You will measure the distance from this point to Krusenstern's Island, both by your log and chronometer observations.

You will circumnavigate Krusenstern's Island, and ascertain the latitude and longitude of its eastern and western points; thence to the small island which you saw before, carefully ascertaining its distance from Krusenstern's; after which you will proceed westerly, to search for Lazareff Island, said to be in latitude 14° 56′ S., longitude 149° 10′ 00″ W. I am disposed to believe, from our examinations, that Krusenstern's Island may have been taken for a part of Dean's Island; that a passage of a few miles may exist, dividing what we now believe to be Krusenstern's Island; if such exist, it is at its western extremity.

From thence you will proceed along the parallel of latitude 15° 30′ S., and keep a good look-out for Recreation and Fugitive Islands, laid down on Arrowsmith's Chart, and until you have passed the longitude of 165° W. There is a reef and island supposed to exist in about 161° W.; this ground you will well examine.

Thence you will proceed to the easternmost of the Navigator's Group, the south part of Manua being in latitude 14° 30′ S., longitude 169° 45′ W., which you will examine, and remain off until my arrival, or farther orders.

You may expect me off them between the 1st and 5th of October.

I am, &c.,

(Signed) CHARLES WILKES,

Commanding Exploring Expedition.

LIEUT. COM. C. RINGGOLD,

U. S. Brig Porpoise.

IV.

TO POMARE IV., QUEEN OF THE SOCIETY ISLANDS.

THE undersigned has been instructed by the President of the United States to visit Tahiti with the squadron under his command, to inquire into and make representations respecting certain occurrences that have taken place; and to assure you that the President and country place full reliance upon the letter addressed by you to the President of the United States, dated September 20th, 1839, a copy of which is enclosed, in which you invite all Americans to visit your ports, and give assurances of protection and friendship; in consequence of which the President has established a consulate at Papieti, for the purpose of watching over an extensive commerce now carried on by American citizens, which results in advantages to both countries; and farther, that the President continues to feel much good-will and interest toward the government and inhabitants of these islands.

1st. That the President, feeling this strong reliance in the promises contained in the said letter, could not but be astonished that the authorities had suffered an occurrence to take place that might be deemed an infringement of the former consular premises. At the same time that he was disposed to overlook that circumstance, he desired that some assurance should be given that such an occurrence, so contrary to the laws and usages of nations, should not be repeated, and to warn them, through me, that it must not occur again.

2d. Relative to the ill treatment of the crew of a whale-boat belonging to an American vessel, in 1836, and the detention of the boat at the Queen's boat-house, and the delay of bringing the offenders to justice. *Such acts* must not occur again.

3d. The consul of the United States will maintain and be ever ready to aid the authorities in the execution and ascendency of their laws, particularly as respects the deportment of American seamen; and every facility offered to other consuls should be extended to him for the discharge of his duties, by assigning for his use a portion of land in a suitable situation to enable him to perform his duties advantageously, particularly as you invited the President to send him, and requested that our vessels might visit your ports.

4th. Relative to the inefficiency of the laws in regard to the apprehension of deserters from the whale-ships, and the absolute necessity of providing a jail or place for their safe-keeping, until they can be sent on board; also, for the appointment of a proper police to watch over and secure any offenders, otherwise it is impossible that the consul can carry into effect the good intentions of the President, that all who violate the laws may be punished and made examples of.

5th. It has been represented that a law formerly existed or still exists, exacting from seamen landing on this island the sum of thirty dollars.

If there is such a law it is extremely oppressive to them, as they are under the protection of, to be taken care of, and sent away by the consul as soon as convenient, agreeably to the laws of the United States. Any such law should be revoked or amended. Seamen elsewhere are exempted from such dues, unless they do not leave in a reasonable time, or intend to become residents.

The President and country entertain the fullest confidence that justice will be exercised at all times to American citizens, and that they will be safely guarded and protected in their rights secured to them by treaty and assurances given them heretofore.

The President will order vessels of war to visit these islands frequently, for the protection of American citizens and commerce, and to see that justice be done to all.

The undersigned requests that this communication may receive the careful attention and consideration of the Queen and chiefs, and that a written answer may be returned to him on his anchoring in the harbour of Papieti, where he intends to proceed with the vessels under his command in a few days, for the purpose of affording the Queen and chiefs an opportunity of visiting the vessels, and of receiving a few tokens of friendship from the President as evidences of his continued regard.

All of which is respectfully submitted.

(Signed) Charles Wilkes,
Commanding Exploring Expedition.

U. S. Ship Vincennes,
Matavai Bay, Sept. 18th, 1839

V.

U. S. Ship Vincennes,
At Sea, October 10th, 1839.

Sir,—

You will proceed to the most western island of the Navigator Group, called Savaii, passing on your route to the northward of Upolu, and leave an officer at Apia, to take charge of and attend to tide-staves hourly.

From thence on your route you will proceed to Manono and Apolima, ascertaining the distance between each of them and Savaii, by the methods pointed out in my surveying instructions; you will particularly examine Savaii for harbours, and will not allow any thing to escape you that will tend to give facilities to commerce and navigation.

Dr. Pickering will join the Porpoise for the purpose of procuring every thing relative to its natural productions. You will land him at the missionary settlement at Savaii, with an officer to attend to the tide-staves hourly, until you leave that island.

Lieutenant Alden, with a boat and crew from this ship, will be put under your orders, for duty.

You will return to the harbour of Pago-pago as soon as you shall have finished, which will not detain your arrival beyond the 20th inst.

If there are any of the crew of the Porpoise who are on the sick-list, and unfit for duty, you will send them, with their bags and hammocks, to this ship.

As many specimens, sketches, &c., as possible, will be obtained as usual by the officers; also observations, &c., noted.

The presents for natives will be exchanged for specimens, also fresh provisions, &c., for the crew.

Wishing you a successful cruise,

I am, &c.,

(Signed) Charles Wilkes,
Commanding Exploring Expedition.

Lieut. Com. C. Ringgold,
U. S. Brig Porpoise.

VI.

U. S. Ship Vincennes,
Harbour of Pago-pago, October 13th, 1839.

Sir,—

You will, with the Peacock and Flying-Fish under your command, proceed to the next island of this group, called Upolu, and enter the harbour of Apia, situated on the northern side, and will remain while the Flying-Fish and your boats are making a survey of it, by circumnavigating it, and making all possible observations, sketches, &c.

The officers sent in the schooner and boats will be careful to examine and survey any harbour or shelter that may be found, making sketches of their approach, and surveys; also observations to determine their situation, &c.

You will also make a careful survey of the harbour of Apia.

Every facility will be afforded to the scientific corps, for making collections and observations on shore; all information possible will be obtained from the missionaries and others, relative to the island; as to the population, productions, soil, number of white missionaries, &c.

You will remain at Apia until you hear from me, or until you have completed the duties required, which I deem five days amply sufficient for. One of the officers of the Porpoise was left to superintend tide-staves, which observations will be continued by you after that officer leaves.

I am yours, very respectfully,

(Signed) Charles Wilkes,

Commanding Exploring Expedition.

Captain Wm. L. Hudson,

Peacock.

U. S. Ship Vincennes,

Harbour of Pago-pago.

Sir,—

You will receive on board Lieutenant Perry, and Mr. Colvocoressis, of the Peacock, with their boats and crews, delivering your boats to the Peacock, and then you will carry into effect the following instructions.

Proceed to the eastern point of the island of Upolu, then commence the survey of its south side, particularly examining it for harbours, and all places that may afford any shelter for vessels; the boats are intended to operate with you, and will trace the shore or reefs by compass, and astronomical bearings. You will be very particular in keeping your deck-board as ordered for surveying duty. The boats will make sketches of all the shores, and you will be particular in getting your observations for latitude and longitude, at the same time taking astronomical bearings, that the point may be determined therefrom. If possible, you will pass between Upolu Island and Manono, making your survey embrace the eastern shore of Manono Island; but if the passage proves dangerous, you will then go round Manono, between it and Apolima, and survey the whole of Manono Island only. You will then proceed to the eastward, along the northern shore of Upolu, to the harbour of Apia, continuing the survey in like manner up to that harbour; here you will again rejoin the Peacock,—when Lieutenant Perry and Mr. Colvocoressis will rejoin the Peacock, and you receive your boats, and remain under the orders of Captain Hudson until further instructions. During the survey, you will ascertain how far soundings exist from the shore. You will proceed to sea at daylight.

I am, very respectfully,
Your obedient servant,
(Signed) CHARLES WILKES,
Commanding Exploring Expedition.

LIEUT. COM. R. F. PINKNEY,
Schooner Flying-Fish.

VII.

ABSTRACT FROM A DIARY KEPT BY WILLIAM FOXALL, PILOT OF THE HARBOUR OF PAGO-PAGO, ISLAND OF TUTUILA, SAMOAN GROUP, 1839.

MONTHS.	WINDS.				CALM DAYS.	CLOUDY DAYS.	RAINY DAYS.	CLEAR DAYS.
	Northward and Eastward.	Southward and Eastward.	Southward and Westward.	Northward and Westward.				
January .	5	13		12	1	5	6	20
February .	3	12		13		4	12	13
March . .	7	16	1	6	1	2	18	11
April . .		22	1		7	4	9	17
May . .	1	19	1	6	4	8	18	5
June . .	1	27	2			4	15	11
July . .	3	22	5	1		1	12	18
August .		25	1	5		9	7	15
September	8	19	3			4	4	22
October .		12						11
	28	187	14	43	13	41	101	143

VIII.

EXTRACT OF A LETTER FROM GEO. BURADER TO THE REV. W. MILLS, RELATING TO A REMARKABLE PHENOMENON OF THE TIDES, AT THE HARBOUR OF PAGO-PAGO, ISLAND OF TUTUILA.

WEDNESDAY, November 7th, 1837, 2^h 20^m P. M. the tide rose suddenly to two feet above high-water mark, spring tide.

At 2^h 30^m, sunk to low-water mark, neap tide.

At 2^h 35^m, rose as high as at first.

At 2^h 40^m, sunk to low-water mark, spring tide; immediately rushed in again with great violence, and with such rapidity, that in two minutes it was as much as three feet higher than it had been before. Receded with equal force, and at 2^h 52^m had sunk a great deal below low-water mark.

At 2^h 55^m rose again as high as before; receded about eighteen inches, and suddenly rushed in again to the same height as at the third time.

At 3^h 3^m it was high, and at 3^h $6\frac{1}{2}^m$ it was much lower than it had been before.

At 3^h 12^m it rose again to the same height as at 3^h 3^m; receded suddenly one foot, and at 3^h $17\frac{1}{2}^m$ rose again nearly as high as at the third time.

At 3^h 20^m, sunk to low-water mark, spring tide.

At 3^h 35^m, rose to high-water mark, spring tide; sunk to half tide, and returned rapidly.

At 3^h 36^m, was as high as at first and second times; fell two feet.

At 3^h 42^m, nearly the same height as at third rising; it then sunk much lower than at any other time.

At 3^h 50^m, several feet below low-water mark; spring tide rushed in again.

At 3^h 54^m, rose full two feet higher than at the third time; fell nearly as fast as it rose, and with equal force.

At 4^h, was down at high-water mark, neap tide.

At 4^h 15^m, rose about half as high as at last time.

At 4^h 20^m, was down to low-water mark.

At 4^h 32^m, rose to same height as at 4^h 15^m.

At 4^h 40^m, a heavy shower of rain fell; the sea at this time at low-water mark.

At 4^h 55^m, rose to the same height as at first time.

At 5^h 3^m, again at low-water mark.

At 5^h 8^m, rose again about one foot above high-water mark, spring tide; after which it gradually sunk to low-water mark, and during the remainder of the evening continued to ebb and flow less frequently than during the afternoon, and not passing the bounds of high and low water.

During the evening preceding the above remarkable event, we had frequent and heavy squalls from the east, which continued till 7 A. M. from which time the day was calm and cloudy, and frequent light showers, but no heavy rains till the one referred to above, at 4^h 40^m, after which it continued to rain more or less till ten o'clock.

On Thursday, the 8th, the tide continued to ebb and flow in an irregular manner, both as to time and height, being sometimes a little below low-water mark, and a little above high-water mark, and at other times ebbing and flowing the same as at neap tide, till 4 P. M., when it rose to the same height as at the first time the day before; fell about half tide, and for a short time appeared nearly stationary.

At 4^h 24^m, rose to the same height as at 4^h.

Sunk a little below high-water mark, and continued to ebb and flow during the evening in the same way as during the morning. During the whole of Thursday the weather was very hot, and the sun shone bright, without clouds, and with little wind.

IX.

COMMERCIAL REGULATIONS, MADE BY THE PRINCIPAL CHIEFS OF THE SAMOAN GROUP OF ISLANDS, AFTER FULL CONSIDERATION IN COUNCIL, ON THE 5TH DAY OF NOVEMBER, 1839.

1st. ALL foreign consuls duly appointed and received in Samoa shall be protected and respected both in their persons and property, and all foreigners obtaining the consent of the government and conforming to the laws, shall receive the protection of the government.

2d. All foreign vessels shall be received into the ports and harbours of Samoa for the purpose of obtaining supplies and for commerce, and with their officers and crews, so long as they shall comply with these regulations and behave themselves peaceably, shall receive the protection of the government.

3d. The fullest protection shall be given to all foreign ships and vessels which may be wrecked, and any property saved shall be taken possession of by the consul of the country to which the vessel belongs, who will allow a salvage or portion of the property so saved, to those who may aid in saving and protecting the same, and no embezzlement will be permitted under any circumstances whatever. The effects of all persons deceased shall be given up to the consul of the nation to which they may have belonged.

4th. Any person guilty of the crime of murder upon any foreigner, shall be given up without delay to the commander of any public vessel of the nation to which the deceased may belong, upon his demanding the same.

5th. Every vessel shall pay a port-charge of five dollars for anchorage and water, before she will be allowed to receive refreshments on board, and shall pay for pilotage in and out, the sum of seven dollars, before she leaves the harbour; and pilots shall be appointed, subject to the approval of the consuls.

6th. No work shall be done on shore, nor shall any natives be employed on board vessels on the Sabbath day, under a penalty of ten dollars, unless under circumstances of absolute necessity.

7th. All trading in spirituous liquors or landing the same is strictly forbidden; any person offending shall pay a fine of twenty-five dollars, and the vessel to which he belongs shall receive no more refreshments. Any spirituous liquors found on shore shall be seized and destroyed.

8th. All deserters from vessels will be apprehended, and a reward paid of eight dollars, viz., five dollars to the person who apprehends him, and three dollars to the chief of the district in which he may be apprehended, on his delivery to the proper officer of the vessel. No master shall refuse to receive such deserter, under a penalty of twenty-five dollars. Deserters taken after the vessel has sailed shall be delivered up to the consul, to be dealt with as he may think fit. Any person who entices another to desert, secretes a deserter, or in any way assists him, shall be subject to a penalty of five dollars or one month's hard labour on the public roads.

9th. No master shall land a passenger without permission of the government, under a penalty of twenty-five dollars, and no individual shall be permitted to land or reside in Samoa, without special permission of the government. Any one so landing shall be compelled to leave by the first opportunity.

10th. If a sick person be left on shore from any vessel for the recovery of his health, he shall be placed under charge of the consul, who shall be responsible for his sick expenses, and will send him away by the first opportunity after his recovery.

11th. Any seaman remaining on shore after nine o'clock at night, shall be made a prisoner of, until the next morning, when he shall be sent on board, and shall pay a fine of five dollars.

12th. All fines to be paid in specie or its equivalent; or be com muted by the government, at the rate of one month's hard labour on the public roads for five dollars.

13th. Should the master of any vessel refuse to comply with any of these regulations, a statement of the case shall be furnished to the consul of the nation to which he belongs, and redress sought from thence.

14th. All magistrates or chiefs of districts, where vessels or boats may visit, shall enforce the rules and regulations relative to the landing of foreigners, and apprehension of deserters, or pay such a fine as the Malo shall impose.

15th. For carrying into effect the foregoing rules and regulations, the chiefs and governors of the respective districts shall be accountable, and elect one of their number to act as a magistrate or judge, to execute the laws.

16th. These regulations shall be printed, promulgated, and a copy furnished to the master of each vessel visiting these islands.

Done in council at the port of Apia, in the island of Upolu, this 5th day of November, A. D. 1839.

Malietoa,	his	×	mark.
Tai-ma-le-lagi,	"	×	"
Maletau,	"	×	"
Pea,	"	×	"
Tooa,	"	×	"
Mole,	"	×	"
Sanga,	"	×	"

Witnesses.

Charles Wilkes,
Commanding Exploring Expedition.

J. C. Williams,
U. S. Consul.

W. C. Cunningham,
H. B. M. Consul.

November 5th, 1839.

The foregoing commercial rules and regulations, having been signed by the chiefs in my presence, and submitted to me, I consider them just and proper, and shall forward to the American government a copy of the same, for the information of all masters of vessels visiting the Samoan or Navigator Group of Islands.

(Signed) Charles Wilkes,
Commanding Exploring Expedition,
United States of America.

U. S. Ship Vincennes,
Harbour of Apia, Island of Upolu,
Samoan, or Navigator Group,
November 6th, 1839.

X.

NUMBER OF PERSONS COMMITTED FOR TRIAL FROM 1831 TO 1837.

CRIMES.	1831.		1832.		1833.		1834.		1835.		1836.		1837.	
	M.	F.	M.	F.	M.	F.	M.	F.	M.	F.	M.	F.	M.	F.
Murder,	18	—	15	1	18	3	31	5	36	4	29	1	27	—
Attempt to murder, shooting at, stabbing, cutting and maiming, with intent,	12	—	15	1	18	3	31	5	18	—	30	2	26	4
Manslaughter,	2	—	4	—	2	—	5	—	5	—	13	—	4	—
Rape,	5	—	3	—	10	—	9	—	24	—	15	—	9	—
Arson,	5	—	6	—	1	—	1	—	7	—	—	—	11	2
Forgery,	4	—	6	—	9	—	16	—	21	—	1	15	1	22
Highway robbery, bushranging at large, with fire-arms,	37	—	46	—	82	—	158	3	111	5	78	1	77	2
Cattle, horse, and sheep-stealing,	54	2	42	—	25	—	114	1	118	1	114	2	134	—
Burglary,	37	—	21	—	30	—	41	—	33	—	42	3	50	—
Perjury,	7	—	10	—	12	2	19	3	8	2	12	3	16	5
Piracy and revolt,	—	—	34	—	—	—	6	—	—	—	4	—	11	—
Larceny, knowingly receiving stolen property,	271	34	269	54	253	44	367	58	484	89	562	115	450	89
Misdemeanour and assault,	69	9	84	11	94	19	95	23	91	20	106	11	112	10

	1836.		1837.	
	MALES.	FEM.	MALES.	FEM.
Sentence of death passed on prisoners, ...	32	—	23	—
Transportation, ...	254	42	164	16
Labour and imprisonment, ...	299	22	315	35
Executed, ...	26	—	12	—

X.—Continued.

ISSUES TRIED BEFORE COURTS OF QUARTER SESSIONS.

Abstract of Returns of the number of Criminal Issues and Prisoners tried before the several Courts of Quarter Sessions throughout the Colony, distinguishing those tried by Civil, and by Military Juries—laid before the Legislative Council on the 11th September, 1839.

WHEN AND WHERE TRIED.	BY CIVIL JURIES.			BY MILITARY JURIES.		
	CASES TRIED.	PERSONS TRIED.	CON-VICTED.	CASES TRIED.	PERSONS TRIED.	CON-VICTED.
1838.						
County of Cumberland:—						
Sydney	100	119	68	143	158	114
Campbelltown	11	18	12	7	9	5
Paramatta	12	15	12	14	18	14
Windsor	42	55	24	15	20	8
1st. Jan. 1837, to 15th Aug. 1839.						
Bathurst	—	16	12	—	48	29
1st. Jan. 1837, to 15th Aug. 1839.						
Maitland	38	—	—	41	—	—

CRIMINAL ISSUES TRIED BEFORE THE SUPREME COURT.

A Return of the number of Criminal Issues and Prisoners tried before the Supreme Court of New South Wales, from the 1st day of January, 1837, to the 15th day of August, 1839; distinguishing those tried by Civil, and by Military Juries.

YEAR.	NO. OF CASES TRIED.		NO. OF PRISONERS TRIED.		NO. OF PRISONERS CONVICTED.	
	BY A CIVIL JURY.	BY A MILITARY JURY.	BY A CIVIL JURY.	BY A MILITARY JURY.	BY A CIVIL JURY.	BY A MILITARY JURY.
1837	136	130	223	175	99	97
1838	108	136	168	181	96	118
1839	57	46	83	76	46	60
TOTALS	301	312	474	432	241	275

XI.

U. S. Ship Vincennes,
Sydney Cove, Dec. 11th, 1839.

Gentlemen,—

To prevent any misunderstanding relative to the conversation with you on the 30th ultimo, I would state that it was not my intention to make any proposition to you to leave the squadron, during our cruise south, but to receive any suggestions that you might offer, after being informed of our cruise, that might be considered more advantageous to your departments than accompanying the squadron in the Antarctic cruise.

It is my intention to sail from this harbour on the 18th instant, with all the vessels of the squadron for the Antarctic Ocean.

The vessels will be ordered to call at New Zealand in the spring, and may be expected there in March or April, if they are not detained by ice.

I request that such of you as may wish orders to join the squadron at New Zealand under these circumstances, will inform me by letter.

I am, respectfully, your obedient servant,

(Signed) Charles Wilkes,
Commanding Exploring Expedition.

To the Gentlemen of the Scientific Corps.

U. S. Ship Vincennes,
December 19th, 1839.

Gentlemen,—

Agreeably to your application, you will remain on shore, and join the squadron on the return of the Expedition, from the Antarctic Ocean, at the Bay of Islands, by the 1st of March next.

You will, of course, employ yourselves as may be most conducive to the interest of the Expedition, in New South Wales and New Zealand, and will keep a minute journal, making all possible collections, and a full report to me in writing, of your observations and duties.

Vouchers for passages and expenses on scientific excursions on public duty, will be taken when practicable for my approval.

I am, very respectfully, your obedient servant,

(Signed) Charles Wilkes,
Commanding Exploring Expedition.

To Messrs. Peale,
Rich,
Pickering,
Hale,
Dana,
Agate,
Drayton,
Scientific Corps.

Sydney, December 2d, 1839.

SIR,—

The undersigned, naval officers and naturalists, attached to the Expedition under your command, are prompted by sincere zeal in the public service, and a most disinterested desire to promote the important objects of the cruise, to submit to your indulgent consideration the following proposition:

We are induced to think that an English brig lately arrived in this harbour, and, in our judgment, very well adapted for such a service, may be purchased at a reasonable rate; and that, while the rest of the squadron proceeds on the southern tour, we may be advantageously employed in the survey and examination of many interesting and important islands. We, therefore, respectfully suggest the employment of this or a similar vessel, and volunteer our best endeavours to perform to the spirit such duties as you may conclude to assign to us during your absence in the south; and we beg you to accept our assurance, that, in case of your consent, no exertion shall be spared to justify this proposal, and recommend our temporary service to your highest consideration. Very respectfully.

WM. M. WALKER,*
JAMES ALDEN,
Lieutenants.
J. C. PALMER,
Acting-Surgeon.
WM. REYNOLDS,
SIMON F. BLUNT,
Passed Midshipmen.

T. R. PEALE,
WM. RICH,
JAMES D. DANA,
H. HALE,
ALFRED T. AGATE,
Scientific Corps

NOTE.—No notice was taken of this letter: it being one of the official communications of the cruise, has caused me to insert it.

CHARLES WILKES.

U. S. Ship Vincennes,
December 2d, 1839.

IN declining to sign the accompanying paper I do not wish to have it understood that I disapprove of the proposition in the main. If the commander of the Expedition should find that it better suited his convenience to charter a vessel to convey us to a rendezvous, and for other purposes, I should not conceive that I had a voice in the matter, but a direct proposition having already been made (with which I have

* Mr. Walker begs leave to disavow any thought of promotion on the projected service, and, in anticipation, begs respectfully, but most earnestly, to decline any acting appointment.

every reason to be satisfied), perhaps some expression of opinion may be looked for on my part.

I confess then freely that I have been actuated somewhat by selfish considerations. A visit to New Zealand is a point which I have more at heart than any other in these seas, and I conceive the time allotted by no means too much to be employed between Australia, New Zealand, and the Tonga Islands. Perhaps, if I had every thing at my disposal, I should feel tempted to touch at some southern point of New Zealand, before proceeding to the Bay of Islands; but there would be danger in this, both by land and water, besides the risk of losing the main points of observation. Farther than this, I should not think of attempting. I should therefore deprecate any change of plan, which would lead to a change of destination, but of course should not expect my individual voice to weigh against the general good of the Expedition.

CHARLES PICKERING.

To CHARLES WILKES, ESQ.,
Commanding Exploring Expedition.

U. S. Ship Vincennes,
Sydney Cove, Dec. 23d, 1839

SIR,—

You will proceed to New Zealand by the first opportunity, and attend to the following duties.

1st. You will make hourly observations of the tides.

2d. Observations of the temperature, at eight, twelve, and six; winds, weather, &c.

3d. Collect all specimens possible for the scientific corps, and other departments of the Expedition.

4th. Keep a journal, and note all the information you can obtain relative to New Zealand, and make a report in writing to me.

You will take with you a tide-staff, three thermometers, watch, stationary, &c.

I am, respectfully, your obedient servant,
(Signed) CHARLES WILKES,
Commanding Exploring Expedition.

MR. J W. WALDRON,
U. S. Brig Porpoise.

XII.

RETURN OF THE SCHOOLS IN NEW SOUTH WALES IN THE YEAR 1838

DESCRIPTION.	WHERE SITUATED.	NO. OF SCHOOLS.	NO. OF SCHOLARS ATTENDING. MALE.	FEM.	TOTAL.	EXPENSE PAID BY THE GOVERNMENT IN 1838. £. s. d.	REMARKS.
INSTITUTIONS AND SCHOOLS, SUPPORTED WHOLLY, OR IN PART, BY GOVERNMENT.							
Church of England.							
Male Orphan Institution..	Liverpool........	1	147	—	147	3,090 15 8½	Supported wholly by Government.
Female Orphan Institution	Paramatta.......	1	—	141	141	3,708 5 5	Supported chiefly by Government; some slight portion of the expense is paid by the parents of the children.
Primary and Infant Schools	Sydney	6	567	383	950	2,580 0 3¾	
	Country Districts	29	682	548	1,230		
		37	1,396	1,072	2,468	9,379 1 5	
Roman Catholic.							
Orphan Institution....	Waverly Crescent, near Sydney ..	1	40	22	62	800 0 0	Supported wholly by Government.
Primary and Infant.....	Sydney	5	239	225	464	769 13 11	Supported chiefly by Government; some slight portion of the expense is paid by the parents of the children.
	Country Districts	5	234	157	391		
Primary and Infant.....	Sydney	1	50	70	120	126 7 7½	These schools are supported by Government, and by voluntary contributions, the principle being, that in aid of each school, sums are paid from the Colonial Treasury, equal to those raised by private subscriptions.
	Country Districts	4					
		16	563	474	1,037	1,696 1 6½	
Presbyterian.							
Primary and Infant.....	Sydney	10	200	150	350	1,121 6 6	
	Country Districts	7	Estimated Nos.				
Australian School Society.							
Primary and Infant.......	Sydney	2	46	30	76	178 0 11	
Wesleyan.							
Sunday Schools	Sydney	4	*			50 15 8	
II. INSTITUTIONS AND SCHOOLS, SUPPORTED WHOLLY BY PRIVATE INDIVIDUALS.							
Colleges.							
King's School	Paramatta	1	105	—	105	—	The expenses are defrayed by the subscribers to these institutions, and by the fees paid by the parents or guardians of the students.
Sydney College...........	Sydney	1	125	—	125	—	
Australian College.....	Sydney	1	70	—	70	—	
		3	300	—	300	—	
Private Schools.							
Classical, Elementary, &c.	Sydney	30	390	530	920	—	
	Country Districts	37	442	444	886	—	
		67	832	974	1,806	—	
TOTALS		142	3,337	2,700	6,037	£12,425 6 0¼	

* Number of Children attending these Schools, 312, of which 157 are Males, and 155 Females.

XIII.

RETURN OF IMMIGRANTS AND CONVICTS ARRIVED, AND OF BIRTHS AND DEATHS, IN THE COLONY OF NEW SOUTH WALES, FROM THE YEAR 1837 TO 1840, INCLUSIVELY.

YEAR.	IMMIGRANTS.				CONVICTS.			BIRTHS.			GENERAL TOTAL.	DEATHS.				
												Male.		Female.		
	Men.	Women.	Children.	Total.	Male.	Female.	Total.	Male.	Female.	Total.		Men.	Children.	Women.	Children.	Total.
1837	1,769	1,138	1,368	4,275	2,892	533	3,425	1,159	1,111	2,270	9,970	937	280	303	279	1,799
1838	3,631	2,132	3,077	8,840	2,740	333	3,073	1,450	1,386	2,836	14,749	985	407	383	329	2,104
1839	5,843	3,719	3,796	13,358	1,565	728	2,293	1,678	1,626	3,304	18,955	1,109	500	449	423	2,481
1840	5,159	3,457	2,036	10,652	2,113	461	2,574	2,119	2,114	4,233	17,459	1,035	482	440	425	2,382
TOTALS	16,402	10,446	10,277	37,125	9,310	2,055	11,365	6,406	6,237	12,643	61,133	4,066	1,669	1,575	1,456	8 766

RECAPITULATION.

	Men.	Women.	Children.	Total.
INCREASE BY				
Immigrants . .	16,402	10,446	10,277	37,125
Convicts	9,310	2,055	—	11,365
Births	—	—	12,643	12,643
	25,712	12,501	22,920	61,133
DECREASE BY				
Deaths	4,066	1,575	3,125	8,766
	21,646	10,926	19,795	52,367
ADD				
Population of 1836 .	48,375	14,550	14,171	77,096
Population of 1840 .	70,021	25,476	33,966	129,463

ABSTRACT OF THE AVERAGE ASSETS AND LIABILITIES, AND OF THE CAPITAL AND PROFITS, OF THE SEVERAL BANKS OF THE COLONY OF NEW SOUTH WALES, FOR THE QUARTERS ENDING 31ST MARCH, 30TH JUNE, 30TH SEPTEMBER, AND 31ST DECEMBER, 1841.

PERIOD.	LIABILITIES.					ASSETS.						CAPITAL AND PROFITS.			
	Notes in Circulation.	Bills in Circulation.	Balances due to other Banks.	Deposits.	Total Liabilities.	Coin.	Landed Property.	Notes and Bills of other Banks.	Balances due from other Banks.	Notes and Bills discounted, and all other debts due to the Bank.	Total Assets.	Capital paid up.	Rate per Annum of last Dividend.	Amount of Dividend.	Amount of reserved Profits after paying Dividend.
Quarters ending	£ s. d.	£ s. d.	£ s. d.	£ s. d.	£ s. d.	£ s. d.	£ s. d.	£ s. d.	£ s. d.	£ s. d.	£ s. d.	£ s. d.		£ s. d.	£ s. d.
31st March, 1841	222,802 1 8	49,278 8 2	9,654 3 7	1,123,636 14 1	1,405,371 7 6	342,137 11 5	24,328 1 7	9,709 1 5	9,789 17 9	2,615,299 5 3	3,001,323 17 5	1,790,156 0 0	11½ per cent.	86,749 12 4	86,252 1 3
30th June, 1841	223,608 2 3	53,510 4 2	15,268 3 1	1,100,737 14 2	1,393,124 3 8	384,087 19 0	25,554 9 2	7,196 0 4	16,881 18 10	2,616,488 10 9	3,050,208 18 1	1,876,322 10 0	11½ per cent.	93,211 15 0	94,594 5 3
30th Sept., 1841	213,061 12 3	50,356 9 8	3,990 3 2¼	1,069,495 3 0½	1,336,906 8 11	388,570 1 8¾	27,655 2 2½	7,503 17 9¾	14,174 16 3	2,588,629 10 10¾	3,026,533 8 10¾	1,893,538 10 0	11½ per cent	95,217 15 1	109,528 4 11
31st Dec. 1841	200,328 8 11	64,508 7 11	1,660 1 5	947,350 11 8	1,213,847 9 11	404,059 4 0	29,170 9 5	6,683 5 1	12,080 11 7	2,508,397 7 6	2,960,399 17 7	1,995,094, 0 0	11½ per cent.	94,392 18 5	178,629 12 9
General Average for 1841	214,950 1 3¼	54,413 7 5¼	7,643 2 9¾	1,060,305 15 11	1,337,312 7 6	379,713 14 0¼	26,694 5 7	7,773 1 2	13,231 16 1¼	2,582,203 13 8	3,009,616 10 6	1,888,777 15 0	11½ per cent	92,393 0 4	102,251 3 6½

X V.

RETURN OF THE VALUE OF IMPORTS, FROM THE YEAR 1826 TO 1840, INCLUSIVELY.

YEAR.	From Great Britain.	From British Colonies.	From South Sea Islands.	From New Zealand.	From Fisheries.	From United States.	From Foreign States.	Total.	VALUE, After deducting Imports from New Zealand and the Fisheries.
	£	£	£	£	£	£	£	£	£
1826	280,000	30,000	—	—	—	—	50,000	360,000	—
1827	253,975	63,220	—	—	—	—	45,129	362,324	—
1828	399,892	125,862	—	—	—	—	44,246	570,000	—
1829	423,463	135,486	—	—	—	—	42,055	601,004	—
1830	268,935	60,356	—	—	—	—	91,189	420,480	—
1831	241,989	68,804	—	—	—	—	179,359	490,152	—
1832	409,344	47,895	—	—	—	—	147,381	604,620	—
1833	434,220	61,662	—	—	—	—	218,090	713,972	—
1834	669,663	124,570	—	—	—	—	197,757	991,990	—
1835	707,133	144,824	1,420	35,542	141,823	13,902	70,161	1,114,805	937,400
1836	794,422	220,254	1,972	32,155	103,575	22,739	62,289	1,237,406	1,101,676
1837	807,264	257,427	1,764	42,886	80,441	9,777	97,932	1,297,491	1,174,164
1838	1,102,127	255,975	5,548	53,943	71,506	8,066	82,112	1,579,277	1,453,828
1839	1,251,969	504,828	3,863	71,709	186,212	23,093	194,697	2,236,371	1,978,450
1840	2,200,305	376,954	1,348	54,192	104,895	24,164	252,331	3,014,189	2,855,102

RETURN OF THE VALUE OF EXPORTS, FROM THE YEAR 1826 TO 1840, INCLUSIVELY.

YEAR.	To Great Britain.	To British Colonies.	To South Sea Islands.	To New Zealand.	To Fisheries.	To United States.	To Foreign States.	TOTAL VALUE.
	£	£	£	£	£	£	£	£
1826	101,314	1,735	—	—	—	—	3,551	106,600
1827	70,507	4,926	—	—	—	—	881	76,314
1828	84,008	4,845	—	—	—	—	1,197	90,050
1829	146,283	12,692	—	—	—	—	2,741	161,716
1830	120,559	15,597	—	—	—	—	23,503	159,659
1831	211,138	60,354	—	—	—	—	52,676	324,168
1832	252,106	63,934	—	—	—	—	68,304	384,344
1833	269,508	67,344	—	—	—	—	57,949	394,801
1834	400,738	128,211	—	—	—	—	58,691	587,640
1835	496,345	83,108	2,696	39,984	38,445	18,594	3,011	682,193
1836	513,976	136,596	9,628	36,184	35,918	13,697	2,625	748,624
1837	518,951	118,447	485	39,528	54,434	10,617	17,592	760,054
1838	583,154	113,716	7,137	46,924	33,988	11,324	6,525	802,768
1839	597,100	194,684	1,347	95,173	34,729	18,568	7,175	948,776
1840	792,494	304,724	6,621	215,486	27,864	27,885	24,618	1,399,692

XVI.

RETURN OF LIVE-STOCK IMPORTED INTO NEW SOUTH WALES, IN EACH YEAR, FROM 1828 TO 1840.

YEAR.	DESCRIPTION OF STOCK.					
	Horses.	Horses and Mules.	Horned Cattle.	Sheep and Goats.	Sheep and Hogs.	Sheep.
	Number.	Number.	Number.	Number.	Number.	Number
1828	—	132	—	—	—	3,443
1829	—	16	—	—	—	2,215
1830	—	12	—	—	—	10
1831	—	Not stated	—	—	—	66
1832	—	—	—	36	—	—
1833	—	—	—	—	—	—
1834	6	—	—	—	62	—
1835	11	—	—	—	137	—
1836	8	—	4	—	449	—
1837	92	—	97	—	307	55,208
1838	185	—	74	—	192	9,822
1839	652	—	135	—	359	17,567
1840	1,008	—	244	—	252 Hogs.	19,958

RETURN OF VESSELS BUILT AND REGISTERED IN THE COLONY OF NEW SOUTH WALES, FROM THE YEAR 1822 TO 1840, INCLUSIVELY.

YEAR.	VESSELS BUILT.		VESSELS REGISTERED.	
	Number.	Tons.	Number.	Tons.
1822	3	163	3	163
1823	3	182	3	182
1824	5	157	5	157
1825	2	119	2	119
1826	12	654	19	1,634
1827	9	434	19	1,732
1828	6	162	13	478
1829	7	462	5	428
1830	3	78	25	1,777
1831	5	112	38	3,224
1832	5	220	21	2,143
1833	6	393	29	2,655
1834	9	376	19	1,852
1835	7	303	21	2,267
1836	9	301	39	4,560
1837	17	760	36	3,602
1838	20	808	41	6,229
1839	11	763	75	10,668
1840	17	1196	94	12,153

XVII.

RETURN OF THE NUMBER AND TONNAGE OF VESSELS ENTERED INWARDS, FROM THE YEAR 1826 TO 1840, INCLUSIVELY.

YEAR.	From Great Britain.		From British Colonies.		From South Sea Islands.		From New Zealand.		From Fisheries.		From United States.		From Foreign States.		TOTALS.	
	No.	Tons.	No.	Tons.	No.	Tons.	No.	Tons.	No.	Tons.	No.	Tons.	No.	Tons.	No.	Tons.
1826	33	11,848	23	3,969	—	—	—	—	—	—	—	—	6	1,361	62	17,178
1827	50	19,097	42	5,376	—	—	—	—	—	—	—	—	11	2,035	103	26,508
1828	59	20,585	65	8,789	—	—	—	—	—	—	—	—	13	3,185	137	32,559
1829	62	21,963	46	7,078	—	—	—	—	—	—	—	—	50	8,301	158	37,342
1830	41	14,400	45	7,221	—	—	—	—	—	—	—	—	71	9,604	157	31,225
1831	39	13,778	49	10,043	—	—	—	—	—	—	—	—	67	10,179	155	34,000
1832	56	18,588	76	13,122	—	—	—	—	—	—	—	—	57	9,640	189	41,350
1833	53	19,352	99	19,149	—	—	—	—	—	—	—	—	58	11,663	210	50,164
1834	58	20,906	112	23,730	—	—	—	—	—	—	—	—	75	13,896	245	58,532
1835	47	17,530	132	28,507	12	2,282	—	—	23	5,899	6	1,400	40	7,401	260	63,019
1836	60	23,610	124	25,861	4	546	41	5,430	25	6,031	3	975	12	2,962	269	65,415
1837	56	21,816	233	33,751	5	581	36	5,480	48	13,004	5	1,220	17	4,262	400	80,114
1838	102	41,848	241	34,469	6	616	38	4,291	31	7,928	1	274	0	2,351	428	91,777
1839	137	58,123	290	45,928	7	836	51	8,368	36	9,321	4	1,177	38	11,721	563	135,474
1840	190	80,806	347	53,625	6	750	68	13,123	27	8,087	8	2,520	63	20,047	709	178,958

XVII.—Continued.

RETURN OF THE NUMBER AND TONNAGE OF VESSELS ENTERED OUTWARDS, FROM THE YEAR 1826 TO 1840, INCLUSIVELY.

YEAR.	To Great Britain.		To British Colonies.		To South Sea Islands.		To New Zealand.		To Fisheries.		To United States.		To Foreign States.		TOTALS.	
	No.	Tons.	No.	Tons.	No.	Tons.	No.	Tons.	No.	Tons.	No.	Tons.	No.	Tons.	No.	Tons.
1826	26	7,063	23	6,198	—	—	—	—	—	—	—	—	11	3,759	60	17,020
1827	12	4,021	40	6,678	—	—	—	—	—	—	—	—	11	3,802	63	14,501
1828	15	4,565	38	8,913	—	—	—	—	—	—	—	—	16	6,708	69	20,186
1829	21	6,243	75	15,522	—	—	—	—	—	—	—	—	72	15,821	168	37,586
1830	12	4,551	55	12,263	—	—	—	—	—	—	—	—	80	12,008	147	28,822
1831	18	5,863	57	12,440	—	—	—	—	—	—	—	—	90	16,949	165	35,252
1832	25	8,190	81	15,122	—	—	—	—	—	—	—	—	88	19,545	194	42,857
1833	23	7,739	99	21,006	—	—	—	—	—	—	—	—	91	20,957	213	49,702
1834	27	8,639	88	16,005	—	—	—	—	—	—	—	—	105	28,729	220	53,373
1835	31	11,261	90	15,821	—	—	—	—	—	—	—	—	148	39,882	269	66,964
1836	32	9,759	106	22,895	—	—	36	4,709	36	8,484	—	—	54	16,987	264	62,834
1837	43	13,398	230	34,295	—	—	45	6,721	42	10,344	—	—	42	13,262	402	78,020
1838	35	12,367	216	28,494	5	362	39	5,358	40	10,385	1	289	73	35,749	409	93,004
1839	39	13,886	303	52,749	7	1,216	81	13,581	28	7,718	2	621	88	35,005	548	124,776
1840	54	18,774	350	66,131	6	737	83	17,111	22	6,199	4	950	146	53,802	665	163,704

XVIII.

RETURN OF WOOL EXPORTED FROM THE COLONY OF NEW SOUTH WALES, FROM 1822 TO 1840.

YEAR.	QUANTITY.	VALUE, As entered in the Returns of Exports.	YEAR.	QUANTITY.	VALUE, As entered in the Returns of Exports.
	lbs.	£		lbs.	£
1822	172,880	Not known	1832	1,515,156	73,559
1823	198,240	—	1833	1,734,203	103,692
1824	275,560	—	1834	2,246,933	213,628
1825	411,600	—	1835	3,893,927	299,587
1826	552,960	48,384	1836	3,693,241	369,324
1827	407,116	24,306	1837	4,448,796	332,166
1828	834,343	40,851	1838	5,749,376	405,977
1829	1,005,333	63,555	1839	7,213,584	442,504
1830	899,750	34,907	1840	8,610,775	566,112
1831	1,401,284	75,979			

RETURN OF AMOUNT OF AUCTION DUTY, AT $1\frac{1}{2}$ PER CENT., PAID INTO THE COLONIAL TREASURY, FROM 1824 TO 1840, INCLUSIVELY.

YEAR.	AMOUNT OF DUTY.			YEAR.	AMOUNT OF DUTY.		
	£	s.	d.		£	s.	d.
1824	328	3	11	1833	1,540	2	8
1825	555	3	1	1834	2,327	6	10
1826	576	7	$1\frac{1}{2}$	1835	3,135	16	2
1827	682	18	$11\frac{3}{4}$	1836	4,697	11	5
1828	1,325	10	$7\frac{1}{2}$	1837	4,820	3	11
1829	1,228	7	$1\frac{3}{4}$	1838	6,137	10	1
1830	1,423	18	$3\frac{3}{4}$	1839	7,700	16	5
1831	1,353	7	$4\frac{1}{4}$	1840	18,701	2	10
1832	1,415	15	$10\frac{1}{4}$				
TOTAL	£8,889	12	$4\frac{3}{4}$	TOTAL	£49,060	10	4

XIX.

RETURN OF THE AMOUNTS RECEIVED FROM THE SALE OF CROWN LANDS, FROM 1824 TO 1840, INCLUSIVELY.

YEAR.	AMOUNT.			YEAR.	AMOUNT.		
	£	s.	d.		£	s.	d.
1824	279	17	9½	1833	24,956	1	1
1825	5,548	14	10	1834	41,844	9	1
1826	2,596	2	6	1835	80,784	14	6
1827	2,274	11	7	1836	126,458	16	0
1828	5,004	13	2	1837	120,427	0	5
1829	2,710	15	0	1838	116,324	18	11
1830	943	5	10	1839	152,962	16	4
1831	2,597	1	10	1840	316,626	7	5
1832	12,509	13	10				
TOTAL	34,465	2	4½	TOTAL	980,385	3	9

XX.

ESTIMATED QUANTITY OF LAND IN CULTIVATION, EXCLUSIVE OF GARDENS AND ORCHARDS, ON 31ST DECEMBER, 1840.

CROPS.

WHEAT.	MAIZE.	BARLEY.	OATS.	RYE.	MILLET.	POTATOES.	TOBACCO.	SOWN GRASSES.
Acres.	Acres.	Acres.	Acres.	Acres.	Acres.	Acres.	Acres.	Acres.
74,133	24,966	5,144	5,453	609	115	2,594	381	12,721

PRODUCE.

WHEAT.	MAIZE.	BARLEY.	OATS.	RYE.	MILLET.	POTATOES.		TOBACCO.		SOWN GRASSES (HAY.)
Bushels.	Bushels.	Bushels.	Bushels.	Bushels.	Bushels.	Tons.	Cwt.	Tons.	Cwt.	Tons.
1,116,814	777,947	105,389	66,020	8,863	3,338	11,050	15	215	—	21,329

XXI.

U. S. Ship Peacock,
Sydney Cove, New South Wales,
December 21st, 1839.

SIR,—

Having thoroughly examined this ship at different times, and particularly at this place, I would respectfully submit the following report of her condition. First, that the sheer-streak is quite rotten in many places, as well as the gun and berth-deck water-ways; and from the frequent calking the gun and spar-decks have undergone, they have become much worn, and quite leaky, particularly the spar-deck; also, the stanchions supporting the bulwark on the spar-deck are very much decayed, and with the exception of three or four of them, are unsafe, and not able to support the rail and boats attached to it, under any thing more than ordinary circumstances.

I am, sir, yours respectfully,
(Signed) JONAS DIBBLE,
Carpenter.

WILLIAM L. HUDSON, ESQ.,
Commanding U. S. Ship Peacock.

U. S. Ship Peacock,
Sydney, New South Wales,
December 22d, 1839.

SIR,—

I have delayed reporting in writing the defective state of the Peacock until the present time, with a full knowledge it would have been utterly impossible to have completed the necessary repairs at this port in sufficient time for our Antarctic cruise. I feel, too, that the government and the whole country are anticipating results from that quarter, and are under the full belief that all the ships composing this squadron were thoroughly overhauled, and amply prepared to encounter every kind of weather. I have no wish at present to undeceive them, but feel it my duty to state to you on the present occasion, that the Peacock's sheer-streak, to which the channels are bolted and ports hung, is perfectly decayed, fore and aft, and that all the stanchions of the upper-deck bulwarks, are either rotten, or in an advanced state of decay. Against these defects, however, I feel it my duty to contend, without anticipating any thing but favourable results, but at the same time prepared for the worst that may occur.

Yours, respectfully,
(Signed) WM. L. HUDSON.

CAPTAIN CHARLES WILKES,
Commanding Exploring Expedition.

2 N

XXII.

RETURN OF TIMBER EXPORTED, FROM THE YEAR 1830 TO 1840, INCLUSIVELY.

YEAR.	Cedar.	Blue Gum, Pine, and other Timber.	Treenails.	VALUE.
	Quantity.	Quantity.	Number.	
				£
1830	368,830 Superficial Feet.	179,403 Superficial Feet.	23,959	5,218
1831	580,393	416,857	24,316	8,401
1832	418,930	233,653	186,831	6,132
1833	1,086,437	147,170	328,503	13,153
1834	899,492	30,065	212,467	7,941
1835	907,921	145,628	178,969	10,489
1836	1,409,467	106 Logs 3,778	35,094	14,611
1837	116,828	18,828	62,989	14,463
1838	699,066	9,000	73,450	6,382
1839	729,001	823 Deals, 15 Logs	40,588	8,815
1840	1,250,786	151,500 Superficial Feet.	4,350	20,971

RETURN OF OIL, ETC., EXPORTED, FROM THE YEAR 1830 TO 1840, INCLUSIVELY.

YEAR.	Sperm Whale.	Black Whale.	Whalebone.		Seal Skins.	VALUE.
	Tuns.	Tuns.	Tons.	Cwt.	No.	
						£
1830	983	98	9	16	9,720	59,471
1831	1,571	505	28	5	4,424	95,969
1832	2,491	695	43	6	1,415	147,409
1833	3,048	418	—	—	1,890	146,855
1834	2,760	975	43	15	890	157,334
1835	2,898	1,159	112	—	641	180,349
1836	1,682	1,149	79	—	386	140,220
1837	2,559	1,565	77	8	107	183,122
1838	1,891	3,055	174	—	3 Cases	197,644
1839	1,578	1,229	134	14	7 Cases	172,315
1840	1,854	4,297	250	—	474	224,144

XXIII.

STATEMENT OF THE INTRODUCTION AND PROGRESS OF THE BREED OF FINE WOOLLED SHEEP IN NEW SOUTH WALES, DELIVERED AT THE RIGHT HON LORD HOBART'S OFFICE, 26TH JULY, 1803.

The samples of wool brought from New South Wales having excited the particular attention of the merchants and principal English manufacturers, Captain M'Arthur considers it his duty respectfully to represent to His Majesty's ministers, that he has found, from an experience of many years, the climate of New South Wales is peculiarly adapted to the increase of fine-woolled sheep; and that, from the unlimited extent of luxuriant pastures with which that country abounds, millions of those valuable animals may be raised in a few years, with but little other expense than the hire of a few shepherds.

The specimens of wool that Captain M'Arthur has with him, have been inspected by the best judges of wool in this kingdom; and they are of opinion that it possesses a softness superior to many of the wools of Spain; and that it certainly is equal, in every valuable property, to the very best that is to be obtained from thence.

The sheep producing this fine wool are of the Spanish kind, sent originally from Holland to the Cape of Good Hope, and taken from thence to Port Jackson.

Captain M'Arthur being persuaded that the propagation of those animals would be of the utmost consequence to this country, procured, in 1797, three rams and five ewes; and he has since had the satisfaction to see them rapidly increase, their fleeces augment in weight, and the wool very visibly improve in quality. When Captain M'Arthur left Port Jackson in 1801, the heaviest fleece that had then been shorn weighed only three pounds and a half; but he has received reports of 1802, from which he learns that the fleeces of his sheep were increased to five pounds each;* and that the wool is finer and softer than the wool of the preceding year. The fleece of one of the sheep originally imported from the Cape of Good Hope, has been valued here at four shillings and sixpence per pound; and a fleece of the same kind bred in New South Wales is estimated at six shillings a pound.

Being once in possession of this valuable breed, and having ascertained that they improved in that climate, he became anxious to ex-

* In the grease. The average weight of the fleeces of fine-woolled sheep in New South Wales, washed, is two and a half pounds.

tend them as much as possible; he therefore crossed all the mixed-breed ewes of which his flocks were composed, with Spanish rams. The lambs produced from this cross were much improved; but when they were again crossed, the change far exceeded his most sanguine expectations. In four crosses, he is of opinion, no distinction will be perceptible between the pure and the mixed breed. As a proof of the extraordinary and rapid improvement of his flocks, Captain M'Arthur has exhibited the fleece of a coarse-woolled ewe, that has been valued at ninepence a pound; and the fleece of her lamb, begotten by a Spanish ram, which is allowed to be worth three shillings a pound.

Captain M'Arthur has now about four thousand sheep, amongst which there are no rams but of the Spanish breed. He calculates that they will, with proper care, double themselves every two years and a half; and that in twenty years they will be so increased as to produce as much fine wool as is now imported from Spain and other countries, at an annual expense of £1,800,000 sterling. To make the principle perfectly plain upon which Captain M'Arthur founds this expectation, he begs to state, that half his flock has been raised from thirty ewes purchased in 1793, out of a ship from India, and from about eight or ten Spanish and Irish sheep purchased since. The other half of his flock were obtained in 1801, by purchases from an officer who had raised them in the same time, and from about the same number of ewes that Captain M'Arthur commenced with. This statement proves that the sheep have hitherto multiplied more rapidly than it is calculated they will do in future; but this is attributed to the first ewes being of a more prolific kind than the Spanish sheep are found to be; for since Captain M'Arthur has directed his attention to that breed he has observed the ewes do not so often produce double lambs.

As a further confirmation of the principle of increase that Captain M'Arthur has endeavoured to establish, and which he is positive time will prove to be correct, he would refer to the general returns transmitted from New South Wales. In 1796, (since when not one hundred sheep have been imported,) one thousand five hundred and thirty-one were returned as the public and private stock of the colony. In 1801, six thousand seven hundred and fifty-seven were returned; and although between those periods all the males have been killed as soon as they became fit, yet there is a surplus over the calculation of six hundred and thirty-three.

Captain M'Arthur is so convinced of the practicability of supplying this country with any quantity of fine wool it may require, that he is earnestly solicitous to prosecute this, as it appears to him, important object; and on his return to New South Wales, to devote his whole attention to accelerate its complete attainment. All the risk attendant on the undertaking he will cheerfully bear; he will require no pecuniary aid, and all the encouragement he humbly solicits, is the protection of government, permission to occupy a sufficient tract of unoccupied lands to feed his flocks, and the indulgence of selecting from amongst the convicts, such men for shepherds, as may, from their previous occupations, know something of the business.

(Signed) JOHN M'ARTHUR.

London, 26th July, 1803.

XXIV.

U. S. Flag-Ship Vincennes,
New Zealand, Bay of Islands, April 5th, 1840.

MY DEAR SIR,—

I need not tell you how much I feel interested in your cruise. From the interest you took in the outfit of our expedition, I am sure you well know the interest it excites, and how much this feeling is heightened by a knowledge on my part of what you have undertaken, and have to go through. This prompts me to a desire to be useful to you if possible, and to give you my experience of the last season among the ice, whither you are bound.

Your cruise will be an arduous one, no matter how you may be enlightened on your course; but you have so much knowledge of the ice, and the manner of treating it, that it appears almost presumptuous in me to sit down to give you any hints relative to it. But, believing as I do, that the ice of the Antarctic is of a totally different character from that of the Arctic, I venture to offer you a few hints that may be useful to you in your undertaking; and although my instructions are binding upon me relative to discoveries, I am nevertheless aware that I am acting as my government would order, if they could have anticipated the case, knowing how deeply it feels the liberal assistance and great interest evinced by all the societies and distinguished men of Great Britain, to promote and aid this, our first undertaking in the great cause of science and usefulness; and I must add the pleasure it gives to me personally, to be able to return, though in a small degree, the great obligation I myself feel under to you, and many others, the promoters of your undertaking.

WINDS.—The winds for the first fortnight of our time, to the eastward of longitude 140° E., were from the northward and westward, light generally, accompanied occasionally with clear weather for hours, and again with dense fogs of short duration, with a long swell from the same quarter.

After passing longitude 140° E., or to the westward of it, we experienced fine weather, with southeast winds and occasional snow-squalls, lasting but ten or fifteen minutes, and a dry healthy atmosphere.

The barometer, during our stay on the coast, was always indicative of wind by its depression; and was a true guide. Its mean standing was 28 in. The temperature surprised me: we seldom, if ever, had it above 30°, even in the sun at mid-day, and I do not think that three times it was found above 35°.

Gales come on very suddenly, and are always attended with snow, sleet, and thick fogs, rendering it extremely hazardous; for one must be found, when they do come, more or less surrounded with ice-islands. They sometimes last for thirty-six hours. After they set in, you may calculate that they will blow strong for at least half that time. The nearer you are to the land, the more violent they are, though not of such long duration. Fine weather usually precedes them, and we found them to happen and the weather to be more changeable near the full and change, although I am no believer in the lunar influences upon the weather.

Currents.—During the whole of our stay along the icy coast, we found no perceptible current by the reckoning and current log. During a gale of wind I was induced to believe that some existed, from the short sea that was formed, thinking there was more than was to be expected. *Tides* on such an extent of coast there undoubtedly must be, but of little strength, or we should have perceived them.

In many of the icy bays we were stationary for a sufficient time to perceive them if they had been of any magnitude, and where the current was repeatedly tried.

The winds have their effect upon the loose drift-ice, or that which is detached from the icy barrier. Owing to a change of wind from southeast to north, with a fresh breeze, the Peacock became embayed, and the ice forced in upon her, which brought about the accident. The northerly winds are always accompanied with a heavy swell, and her escape is attributable to a rare exercise of good seamanship and perseverance. If Captain Hudson's ship had been as strong as adamant itself, he is of opinion she would have been ground to atoms by a longer exposure; her stem was abraded to within an inch and a half of the wood-ends.

There are places in which the barrier is within the floe-ice several miles. I enclose you the mean temperature during the summer months.

You will see there is but little chance of the ice melting or disappearing, as from accounts frequently takes place in the Arctic Ocean. Your time, being unlimited, will allow you to wait some days in a situation to make experiments.

I frequently found myself so closely beset that I thought it next to impossible to escape, and if the wind had not been extremely constant in its direction, I should have been shut up or much injured; as it was, I escaped with scarcely a scratch, although we took some heavy thumps.

The charts will show you the tracks and state of the ice. It was constructed as I went on, and the ice-islands laid down by carefully-kept diagrams by the officer of the deck during his watch. This I found gave me more confidence in proceeding, and facilities in case of having to return.

Magnetic Pole.—I consider we have approached very near to the pole. Our dip was 87° 30′ S., and the compasses on the ice very sluggish; this was in longitude 147° 30′ E., and latitude 67° 04′ S. Our variation, as accurately as it could be observed on the ice, we made 12° 30′ E. It was difficult to get a good observation, on account of the sluggishness of our compasses. About one hundred miles to the westward, we crossed the magnetic meridian.

The pole, without giving you accurate deductions, I think my observations will place in about latitude 70° S., and longitude 140° E.

On the meridian of 140° E., you will find a small bay, partly formed by ice-islands and rocks, which I have named Piner's Bay, and I think among the rocks you may find a snug little harbour. I was driven out of the bay by a gale of wind; sounding about one and a half miles from the shore in thirty fathoms. The icebergs being aground, form good shelters; but I was too much exposed to venture to remain, and my object was to trace the land and the icy barrier, which I have done, as you will see it laid down on my chart.

We had delightful and clear weather ten days or a fortnight along the coast, with the wind at from southeast to south-southwest; the two latter points particularly. The drift-ice is in large pieces, so large as to give a ship an awkward thump; but when I found it tolerably open I have run through it to get to clear water, and in hopes of making the land, but our progress was soon stopped by the firm barrier, impenetrable, through which there is no passing.

I am of opinion that there is little movement of the ice during the season. Strong gales may change its position a trifle, but I think not materially.

The only prospect of nearing the land is through a sea well studded with large icebergs, nearly thirty or forty miles in width; and I generally found that we got nearer to the shore in those places than elsewhere. One thing I must tell you, as respects filling your water: you will sometimes find a pond of delicious water on the top of an old iceberg, frozen over, but on cutting through it you will see a supply sufficient for a navy. It will save you fuel, and discomfort and cold to you, your vessels, and their crews.

I was very fortunate in the weather the latter part of the time; and indeed altogether I was scarcely a day without some observation, (except during the gales, of which we had three, occupying about eight days,) and generally half a dozen.

My time for six weeks was passed on deck, and having all daylight, I of course had constant employment, and with the many assistants, I could make rapid progress; and you will find that no opportunity ought to be lost in this navigation, if one is to do any thing. One's ship is in constant danger, and the Vincennes, a first-class sloop of seven hundred and eighty tons, it requires all the foresight and activity one is possessed of to look-out for her.

I consider that I have had a most providential escape; and if this ship had not been enabled to "do every thing but talk," I should not have been where I now am; but she had inspired me with so much confidence, among the coral reefs last summer, that I could put full faith in her doing her duty. I must refer you to the chart, on which I have noted remarks, variations, &c.

I should have mentioned, that in 1838 and 1839 I went south in the brig Porpoise, in order to trace Palmer's Land on its eastern side, (but too late for any trial to reach high latitudes,) and hoping that the lateness of the season would enable me to run some distance along it. I got within three miles of the coast, and saw it trending to the south-southeast about thirty miles; but it was so blocked up with ice as to render it impossible to get through. I have little doubt myself, in favourable seasons, Weddell's track may be followed, notwithstanding what the Frenchman may say, there being no land to which the ice is attached; and that the ice in those parts changes very much, the currents being exceedingly strong, as I myself witnessed. I could not afford the time to be frozen up, as my other duties were and are paramount to passing the winter in such a situation. But you are differently situated, and I should advise you, by all means, to try to penetrate between longitude 35° and 45° W.

I am, &c.,

Charles Wilkes,

Commanding Exploring Expedition

To Captain James C. Ross,

Commanding H. B. M. Ships Erebus and Terror.

XXV.

U. S. Ship Vincennes,
Sydney Cove, Dec. 23d, 1839.

Sir,—

The following instructions will claim your particular attention during our contemplated Antarctic cruise.

1st. It cannot be too strongly impressed upon you the necessity to use every means in your power to avoid a separation, as the lives of those entrusted to your particular care, and those comprising the squadron, may be jeoparded by it.

2d. The most careful attention must be given by you to the health and comfort of your crew, and the most economical care and expenditure of your stores and provisions. The greatest attention to the cleanliness and airing of the vessels, and the drying of the clothes and bedding to avoid all dampness and foul air, are likewise essential.

In the event of your being frozen up or detained by the ice, your responsibilities will be great, and every precaution must be taken to insure your safety. You may rest assured that every exertion will be made to relieve you at the earliest possible day.

I now give you a short sketch of what I deem to be the principal object of our cruise towards the Antarctic Circle: first, to reach as high a southern latitude as can be attained, without hazarding the safety of your vessel and crew; and last, to make all the observations that my former instructions have pointed out, including the dip and intensity.

In the event of parting company, you will rendezvous, first, at Macquarie Island, if it should occur before reaching its latitude, where you will remain forty-eight hours, off and on; thence proceed to the Emerald Isle, waiting on and off thirty-six hours, and from thence to the south as far as the ice will permit, and continue cruising along its borders, between the longitude of 160° E. and 105° E., until the 1st of March, at which period it is deemed the season will become too far advanced to proceed south with any chance of success.

On your return, you will proceed to the Bay of Islands, New Zealand, where you will await my arrival, or find orders with the American consul, resident there.

If the Lord Auckland Group should lay in your path on your return, you will find a secure harbour in that of Sarah's Bosom, in latitude 50° 38′ S., longitude 166° 16′ E.

On your return, you will vary your route as much as possible, or as the weather and state of your vessel will permit.

In case you should not be able to reach the Bay of Islands, prior to the 25th of March, you will thence make for the island of Tongataboo, Friendly Group, where you will find me or orders directing your further movements. (If no intelligence before the 15th of April, you will then proceed to the Feejee Group of Islands, Harbour of Takanova, or Sandalwood Bay, and there continue to employ yourself making the various surveys directed in the copy of the instructions.)

Leave letters where you may stop, and place signal No. 2 on a conspicuous spot; No. 1 will represent the Vincennes, No. 3 the Porpoise, and No. 4 the Flying-Fish.

In the event of your penetrating the ice, with the prospect of a clear sea before you, you will steer to the westward, bearing in mind that the only prospect of again clearing it is on the route you first followed, or that supposed to have been taken by Weddell, between the longitude of 35° and 49° W.

On clearing it, you will proceed to carry out the foregoing instructions, to make the appointed rendezvous with the least possible delay.

(It is my intention, if I am not detained by ice, to reach the harbour of Sandalwood Bay, in the Feejee Group, in April, and after examining those islands, to proceed to the Sandwich Islands for provisions early in July. I state this that you may govern yourself accordingly, in the event of your not being able to meet me or the Peacock at any of the rendezvous pointed out. Should you fall in with the Peacock, Captain Hudson will give you the necessary instructions.)

I am, very respectfully,

Your obedient servant,

(Signed) Charles Wilkes,

Commanding Exploring Expedition.

The foregoing instructions were sent to Captain Hudson. To Lieutenants-Commandant Ringgold and Pinkney, without the clause in parenthesis.

U. S. Ship Vincennes,

At Sea, December 29th, 1839.

Sir,—

The following observations and duties will be attended to on board the Flying-Fish, under your command, during her present cruise to the Antarctic Ocean:

1st. You will keep a daily journal of every occurrence of interest.

2d. Your route will be daily laid down upon the skeleton chart;

also, the position of all land, islands of ice, &c., which you may fall in with. Astronomical bearings will be taken when the weather will permit.

3d. You will note in your journal the variation of the compass, daily, sketches of refractions, and appearances of ice-islands and formations.

4th. You will obtain, if possible, and preserve any stones, specimens of earth, &c., from the ice, and note the appearances of any halos, aurora australis, &c.

5th. The observations, &c., required in the General Order of the 25th August, 1838, will be attended to, when practicable, excepting those of the barometer and hygrometer, and the observations of the masthead.

I am, very respectfully,

Your obedient servant,

(Signed) CHARLES WILKES,

Commanding Exploring Expedition.

To LIEUT. COM. PINKNEY,

U. S. Schooner Flying-Fish.

XXVI.

BAROMETRICAL OBSERVATIONS DURING THE GALES OF WIND EXPERIENCED BY U. S. SHIP VINCENNES, JAN. 28TH, 29TH, 30TH, 31ST, AND FEB. 1ST AND 2D.

DATE.	HOUR.		BAROMETER.	REMARKS.
1840.	A. M.	P. M.		
JAN. 28TH,	3		29·50 in.	Light snow; S. E. by S. fresh.
"	9		29·40	Wind fresh from S. E. by S., cloudy.
"		3	29·24	Wind quite fresh S. E. by S., light snow.
"			29·10	Wind blowing a gale from S. E., snow.
"		10 30	28·90	
29TH,	1 30		28·79	
"	2 30		28·68	
"	3		28·68	
"	4		28·59	
"	4 45		28·57	
"	6		28·57	Weather was pleasant the remainder of 29th.
"	9		28·68	
"	10		28·67	
"	11		28·78	
30TH,	9		29·16	
"	11		29·10	At 3 A. M. of the 30th, the Barometer rose to 29·16 in., and remained at that point several hours.
"	12		29·04	
"		1	29·00	
"		2	28·98	
"		3	28·91	
"		4	28·90	
"		5	28·90	
"		6	28·88	
"		7	28·82	
"		8	28·80	
"		9	28·80	
"		10	28·76	
"		11	28·76	
"		12	28·74	

XXVI.—Continued.

Date.	Hour.	Barometer.	Hour.	Barometer.
1840.	A. M.		P. M.	
Jan. 31st,	1	28·72 in.	1	28·60 in.
"	2	28·72	2	28·60
"	3	28·73	3	28·60
"	4	28·70	4	28·60
"	5	28·70	5	28·60
"	6		6	28·60
"	7 30	28·68	7	28·60
"	8		8	28·66
"	9	28·62	9	28·70
"	10	28·60	10	28·70
"	11	28·60	11	28·70
"	12	28·60	12	28·70
Feb. 1st,	2	28·72		
"	3	28·73	3	28·90
"	4	28·73		
"	5	28·73		
"	9	28·76	9	29·00
Feb. 2d,	3	29·12	2	29·31
"	9	29·25	3	29·31
"	11	29·30	7	29·35
"	12	29·32	9	29·37
"			10 30	29·300
"			12	29·210

XXVII.

U. S. Ship Vincennes,
At Sea, January 31st, 1840.

Sir,—

It becomes my duty, in consequence of the report of the assistant-surgeons of this ship (a copy of which is enclosed), relative to the health and condition of the crew of this ship, to restore you to duty for their benefit, and to obtain all the medical advice in my power to enable me to carry out the instructions of the government.

You will, after due examination and consideration, relative to the health and condition of the crew, report to me in writing, your opinion in regard to the same.

I am, very respectfully,
Your obedient servant,
(Signed) Charles Wilkes,
Commanding Exploring Expedition.

Dr. Edward Gilchrist,
Acting Surgeon, Vincennes.

U. S. Ship Vincennes,
At Sea, January 31st, 1840.

SIR,—

I have received your letter of this date, restoring me to duty, enclosing a report of the assistant-surgeons of this ship, and directing me after due examination and consultation relative to the health and condition of the crew, to report to you in writing my opinion in regard to the same.

In obedience to that order, I respectfully report that, in my opinion, the health of the crew is materially affected by the severe fatigue, want of sleep, and exposure to the weather, to which they have lately been subjected; that a continuance of these hardships, even for a very short period, will entirely disqualify a great number of men for their duty; and that the necessary attention to the health of the crew, and their future efficiency and usefulness, demand the immediate return of the ship to a mild climate.

I am, very respectfully,

EDWARD GILCHRIST,
Acting Surgeon.

CHARLES WILKES, ESQ.,
Commanding Exploring Expedition.

U. S. Ship Vincennes,
At Sea, January 31st, 1840.

SIR,—

In answer to your letter of this date, enclosing the report of the medical officers on the present state of the health of the crew, and requesting the opinion of the ward-room officers, as to the expediency of pushing farther south under the present circumstances, we would state, that, in our opinion, it would be as well to hold on until to-morrow at meridian, in order, should the weather then prove more favourable, to attempt making the recently discovered land at another point; but that on a material increase of the sick-list, or continuance of the present bad weather beyond the above period, we should run to the northward. We are, very respectfully,

Your obedient servants,

(Signed) OVERTON CARR,
First Lieutenant.
A. LUDLOW CASE,
JOSEPH A. UNDERWOOD,
Lieutenants.
EDMUND H. DE HAVEN,
SAMUEL R. KNOX,
Acting Masters.

TO CHARLES WILKES, ESQ.,
Commanding Exploring Expedition.

U. S. Ship Vincennes,
At Sea, January 31st, 1840.

SIR,—

In answer to your communication of to-day, addressed to the wara-room officers, calling upon them for their opinion with regard to the practicability of our prosecuting the desirable researches in these latitudes, I am of opinion, with due regard to the report of the medical officers, which you have submitted for our perusal, that it is very desirable to ascertain the extent of the recently discovered land, by another attempt to the westward; provided, of course, this object can be attained without further endangering the health of the crew.

Very respectfully,
Your obedient servant,
(Signed) JAMES ALDEN.

CHARLES WILKES, ESQ.,
Commanding Exploring Expedition.

U. S. Ship Vincennes,
At Sea, January 31st, 1840.

SIR,—

In answer to your letter, with the report of the medical officers enclosed, requesting the opinion of the ward-room officers of this ship. as to the expediency under the present circumstances of prosecuting our discoveries south at this time, I beg leave to state, that in my opinion the report of all the medical officers of the ship, as to the reduced condition of the crew, would be a sufficient reason for putting back; and, in addition, I would state other reasons which occur to me.

We have been almost surrounded with drift-ice and ice-islands for the last twenty-three days, and coasting along the barrier of field-ice, which has rendered it impossible to penetrate further south in this vicinity; and, although gratifying it would be to land upon the Antarctic Continent, I am not aware that any advantages to be derived from it would be commensurate for the dangers it would be necessary to incur; and if the discovery of new land in these regions is important, I consider it equally so that every precaution be taken to communicate the same to others.

Farther, in my opinion, the continuance of the severe gale which we have encountered for the last five days, accompanied with sleet, hail, and snow, and the necessity of keeping all hands on the watch, owing to the thick weather, &c., and difficulty of navigating among ice, makes it not only most dangerous, but, if necessary to be continued in, will render the watch officers and crew unfit for the arduous duties now and hereafter required of them, at a time when the ship and rigging are enveloped in ice.

In fact, I am of opinion, that the future operations should not be hazarded by encountering dangers and risks to be run, under all the circumstances, greater than I have ever encountered in the course of seven years' sea service.

In expressing, however, finally and respectfully, my opinions, I most cheerfully yield to those of more experience and skill, if they do not concur with mine.

I am, sir, very respectfully,
Your obedient servant,
(Signed) R. R. WALDRON,
Purser U. S. Navy.

CHARLES WILKES, ESQ.,
Commanding Exploring Expedition.

U. S. Ship Vincennes,
At Sea, January 31st, 1840.

SIR,—

In reply to your communication of this date, addressed to the ward-room officers of this ship, I would state, that I think that it would be advisable to remain in this vicinity at least two days longer, and if possible, get farther information respecting the recently discovered land. At the expiration of the above mentioned time, if the stormy weather we have experienced should continue, or the number of sick be increased, I think it would be expedient to bear up to the northward. In expressing this opinion, I am sensible of the hardship and danger to which the officers and men are exposed, but I am also prepared to share the same with them in any manner you may dictate.

Respectfully, yours,
(Signed) JARED ELLIOTT.

CHARLES WILKES, ESQ.,
Commanding Exploring Expedition.

XXVIII.

BAROMETER AND DAILY MEAN OF TEMPERATURE OF AIR AND WATER, DURING THE CRUISE OF THE U. S. SHIP VINCENNES, NEAR THE ANTARCTIC CIRCLE, IN JANUARY AND FEBRUARY, 1840.

DATE.	BAROM.	AIR.	WATER.	DATE.	BAROM.	AIR.	WATER.
JAN. 11	29·36 in.	32·45°	31·29°	FEB. 1	28·85 in.	30·75°	29·50°
12	28·97	32·45	30·00	2	29·26	32·45	29.41
13	28·87	32·45	30·45	3	28·74	32·04	31·33
14	29·17	32·37	31·74	4	28·50	32·80	32·75
15	28·87	32·95	31·16	5	29·23	32·58	32·62
16	28·68	33·95	30·50	6	29·21	31·17	32·75
17	28·84	30·95	31·20	7	29·00	31·00	31·90
18	28·87	32·52	31·75	8	29·14	32·16	31·75
19	28·77	32·48	31·26	9	29·19	32·18	31·50
20	28·91	31·70	32·04	10	29·08	31·83	31·75
21	29·02	34·56	31·09	11	29·08	30·12	30·00
22	29·03	25·18	30·63	12	29·15	29·00	30·03
23	29·04	26·16	30·45	13	29·08	27·75	30·20
24	29·15	26·37	30·75	14	29·15	25·29	29·90
25	29·22	23·04	28·45	15	29·28	27·00	30·80
26	29·06	25·69	29·00	16	29·33	26·58	30·20
27	29·29	26·40	28·91	17	29·16	28·79	30·00
28	29·31	25·91	28·87	18	28·91	28·58	30·00
29	28·88	28·75	29·00	19	28·76	30·12	30·75
30	29·00	26·04	28·75	20	28·97	28·00	30·62
31	28·66	29·00	29·00	21	29·06	29·08	31·70
				22	28·89	32·22	32·50
				23		35·25	35·45
				24		36·08	34·82

XXIX.

[Copy.]

U. S. Ship Peacock,
Sydney, New South Wales,
March 3d, 1840.

SIR,—

I have the honour to report the arrival of the Peacock at this port, for the purpose of making such few repairs as have become necessary, preparatory to the further prosecution of the objects of the Expedition, and avail myself of the occasion to say, that in our recent adventures south, we fell in with a barrier of ice in the latitude of 65° S., and longitude 159° E., and had followed its trendings as far as 67° S., longitude 150° E.

On Sunday, January 19th, while standing into a bay of ice, in latitude 66° 31′ S., and longitude 153° 40′ E., we made (what we believed to be) land to the southward and westward.

It was seen towering above and beyond some large icebergs, that were from one hundred to one hundred and fifty feet in height. We endeavoured to work up for this land, which presented the appearance of an immense mass of snow, apparently forming a vast amphitheatre, with two distinct ridges or elevations throughout its extent. After working up until midnight through detached portions of ice, we reached the barrier at the head of the bay, and were compelled to give up any further attempt to near it, (what we believed to be land,) and passed out of the bay again, which was some twenty miles in extent, through drift-ice, into a more open space for pursuing our course to the southward and westward along the barrier.

On the 23d of January we made, beyond the barrier, which was thickly studded with bergs and islands of ice, (what we believed it to be,) high land, at least so far as terra firma can be distinguished where every thing is covered with snow, and worked into a bay for a nearer and more minute examination. The sea-water had been discoloured for some days, but no bottom obtained by soundings; in the bay, however, it changed to a dark dull green, and gave every indication that we were on soundings, and not far from land.

The result confirmed the appearances: we obtained bottom in three hundred and twenty fathoms, of slate-coloured mud, and the lead brought up with it a piece of stone, about an inch in length, of nearly the same colour, while the lower part of the lead showed a fresh and deep indentation, as though it had struck on a rock. Dip observations were made on the ice with Robinson's and Lloyd's needles; the former gave 86·10°, the latter 86·23°.

While ascertaining the dip, a large king-penguin was captured on the ice, and brought to the ship; to add to our collections, in his stomach were found thirty-two pebbles of various sizes, which appeared to have been very recently obtained, and afforded additional evidence of our immediate proximity to land.

While further pursuing the object of our search in this vicinity, on the morning of the 24th, and endeavouring to clear some ice ahead of us, the ship made a sternboard, and came in contact with a large piece of ice, which carried away one of the wheel-ropes, wrenched the neck of the rudder, and rendered it useless.

We immediately commenced working ship with the sails and ice-anchors into a more open sea. In this we were successful for a time, until an increase of wind, and a change in its direction, brought in upon us masses of ice for miles in extent, which completely beset the ship, finished the work of destruction on our rudder, and forced us into the immediate vicinity of an ice-island some seven or eight miles in extent, with an elevation equalling our topgallant-masthead, and its upper portion inclining towards the ship. In this situation we furled all but the fore-and-aft sails, and hung by our ice-anchors. Fortunately, between us and a portion of this island, lay a larger piece of ice, one end of which held us by the counter, until forced beyond it by the pressing masses of ice outside, which started our anchors, and set us stern on to the island, carrying away our spanker-boom and stern-davits, and forcing the starboard quarter-deck bulwarks in end some three or four inches, jamming a signal-gun hard and fast in the gang-way, and breaking off all the bulwark stanchions on that side of the quarter-deck. We took this occasion to cant her, with the jib, into a narrow channel alongside the island, and with the help of other sails, passed by a portion of it without further injury to our spars, until an opportunity presented of forcing her into a small opening in the ice, with the head towards the sea.

Our rudder, which we unshipped and got in upon deck while wedged in the ice, came in over the side in two pieces, the head and neck entirely broken off, with the two midship pintles, and we shortly afterwards found the upper and lower braces gone from the stern-post.

Towards midnight the sea was increasing, accompanied with snow, with every indication of a gale from seaward; and the ice, with which we were continually in contact, or actually jammed, more formidable in character, rapidly accumulating outside of us, and forming a compact mass. I found, as we were nearing the open sea, that we had been carried so far to leeward by the ice, as to be in great danger of taking up our last residence in the barrier, amongst bergs and islands of ice. There was, therefore, no choice left but to force her out, or grind and thump the ship to pieces in the attempt.

Aided by a kind Providence, we reached an open space on the morning of the 25th, after having beat off the gripe of the ship, &c., and at meridian the carpenters had so far secured our rudder that it was again shipped, in the two remaining braces left on the stern-post.

We were yet surrounded by ice and icebergs, in a bay some thirty miles in extent, from which no outlet could be seen from the masthead. At midnight, however, we found a passage, about half a mile in width, between some bergs and field-ice.

On the morning of the 26th, having reached a partially clear sea, and thoroughly turned over in my mind the state of the ship, with the head of the rudder gone, hanging by two braces only, and in such a state we could hardly hope to have it answer its purpose, through the boisterous weather with which we should have to contend before reaching the nearest port, and its utter unfitness for further cruising amongst icebergs and near ice, through the foggy, thick weather, and frequent snows to which those latitudes are subject, and when rapid evolutions are often necessary, in which the rudder must perform its part,—with the ship considerably strained, her starboard spar-deck bulwarks gone as far as the gangway, the gripe off and stern mutilated, and the further fact before me, that the other vessels of the squadron were ranging over the same longitude, with directions to leave on the 1st of March for surveying operations in the north; that the ship's bottom would have to be examined, and repairs made before leaving another port, (which would occupy, with all the facilities this quarter of the world affords, at least four weeks,) during which time the services of this ship would be lost in surveying the Feejee, &c., I determined to proceed at once to Sydney, expedite as much as possible the repairs of the ship, and be ready at the earliest moment to co-operate with the rest of the squadron.

The Vincennes was seen by us in the distance on the 19th, and the brig Porpoise on the 23d of January.

On the night of the 7th, and morning of the 8th of February, we had frequent and unusually brilliant displays of the aurora australis, one of which made its first appearance in the southwest portion of the horizon, but soon diffused its beams of light from east to west, throwing them up to a concentrated point in the zenith, where they were attended with continued quick flashes, resembling heat lightning, and extending over about a third part of the heavens. The rays or beams of light composing this magnificent spectacle, varied in colour from a light orange to tints of pale red, assuming in their changes hues I should in vain attempt to describe.

During intervals of the brightest flashes in the zenith, however, they lost their distinctive outlines, and mingled in the glow of bright twilight which nearly overspread the heavens.

This exhibition was to us so perfectly unique and strongly marked in character, as to excite the attention of those on board most indifferent to such phenomena, and called forth from all, exclamations of surprise and pleasure.

The ship's compasses were minutely examined on this occasion, but exhibited no symptoms of being affected by the presence of the aurora. The motion of the ship, however, from the effect of the sea at the time, would have rendered any change imperceptible, if the disturbing cause had not produced an oscillation of the needle beyond four or five degrees.

During the aurora, a single squall of light hail passed over the ship.

After a rough and boisterous passage north, we anchored within the Heads at Sydney, on the night of the 21st of February.

The officers and crew have all enjoyed good health, indeed, we have been for some days past without a man on the sick-list; and it affords me great pleasure again to bear testimony to the zeal and efficiency of the officers and men in the performance of their various duties.

I feel quite confident we shall have completed our repairs, and be ready to leave Sydney, in about three weeks.

I am, sir, most respectfully,
Your obedient servant,
(Signed) WM. L. HUDSON,
Commander

To J. K. PAULDING, Esq.,
Secretary of the Navy.

U. S. Ship Peacock,
Sydney, New South Wales,
March 12th, 1840.

SIR,—

The foregoing report was prepared for the Honourable Secretary of the Navy, immediately after my arrival in port, but no conveyance having offered, I take leave to address it to you as commander of the squadron; availing myself of the occasion to add, that we fell in with the first ice, on our passage south, in the latitude of 61° 32′ S., longitude 161° E., and made the barrier on the 15th of January, in the latitude of 65° 53′ S., longitude 159° E., and followed its trendings as far as 67° S., and 150° E.

Our compasses were at the time exceedingly sluggish, and gave no evidence of the ship changing her position, unless kept in continual agitation by shaking.

Immediately after my arrival at this port, I commenced the repairs of the Peacock, which are now so far advanced that I shall be ready for such duties as you may assign me by the 25th of the present month.

The Peacock, as you have seen, has been considerably strained; and we have found on examination while repairing, that after her gripe was beaten off, the ice had chafed the stem to within one inch and a half of the wood-ends of the planking.

I trust my efforts to carry out your orders, and the course I have pursued in returning to Sydney, and expediting the repairs of the ship, have been such as to meet your approbation, as well as that of the Honourable Secretary of the Navy.

I am, sir, yours, respectfully,

(Signed) WILLIAM L. HUDSON,
Commander.

CHARLES WILKES, ESQ.,
Commanding Exploring Expedition.

XXX.

[Copy.]

U. S. Brig Porpoise,
Bay of Islands, New Zealand.
March 31st, 1840.

SIR,—

I have great gratification in reporting my arrival at this place on the night of the 26th instant, all in good health; and respectfully report the following results, as occurring since the period of separating from you.

The 12th of January was consumed in diligent search and endeavours to regain: failing to do so, I proceeded westerly. At 10 P. M., the day following, I entered an inlet formed by the barrier, for the purpose of making a close examination and experimenting on dip.

On a near approach to the margin, numbers of Phocæ proboscidæ were seen reposing: I succeeded in taking a pair, the skins of which were subsequently placed on board the Peacock.

Very lofty ridges of ice, and the loom usual over high land, were visible along the southern horizon, over the barrier. The compass at the time being very sluggish, showing the brig's head to the north, when her head was correctly south.

From appearances to the southward, with the numerous Phocæ proboscidæ, I was strongly impressed with the belief of the close approach to land.

On the 15th the water was much discoloured, casting in one hundred and fifty fathoms without success. Dense fogs and snow prevented further examination. I fell in and communicated with the Peacock on the evening of the 15th, having her in sight on the 21st and 22d.

On the evening of the 16th, strong appearances of land again arose, in corroboration of which I insert an extract from my journal. as well as the remarks from the log-book.

EXTRACT FROM JOURNAL.

"At 6^h 30^m P. M. I went aloft to take a look, the weather being clear, horizon good, and clouds lofty. I heard the noise of a penguin; soon after, one was seen very near the brig, with a large seal to windward. After reaching masthead, I saw over the field of ice, an object, large, dark, and rounding, resembling a mountain in the distance. The icebergs all were bright and brilliant, and in great contrast.

"I watched for an hour to see if the sun in his decline would change the colour of the object by a difference of rays: it remained the same, with a white cloud above, similar to those generally hovering over high land; at sunset it remained the same. I took the bearing accurately, intending to examine it closely as soon as I got a breeze. I am strongly of the opinion it is an island, surrounded by immense fields of ice now in sight."

EXTRACT FROM LOG.

"At 7 P. M. discovered what was supposed to be an island, bearing south-by-east,—a great deal of field-ice in sight.

(Signed) "J. H. NORTH."

17th, the indications were again noticed, corroborating those of the day preceding. From the 19th to the 21st, I was not in sight of the barrier, owing to adverse winds and thick weather. On the afternoon of the last date, I closed in again. On the 22d, 4 A. M., appearances of land again to the southward and eastward, at the time passing an iceberg with dark veins and dusty appearances, exciting again confident hopes of soon making positive discoveries.

On the 23d, I attained the parallel of 66° 49′ S., 151° 24′ E., by observation, having reached the southern extreme of an extensive gulf, studded with islands of ice, and far in the distance to the southward high and lofty bergs were identified and entangled in the main barrier, rendering our advancement one step further impossible. I put about, tried dip, and retraced my steps, exchanging colours with Peacock at 2^h 30^m P. M., then on her way in.

After separating from you on the 27th, I proceeded westward, keeping the barrier close to, reluctantly meeting insuperable obstacles in every effort to pass the Antarctic Circle.

On the 28th, I experienced a heavy gale from southeast of thirty-six hours, with snow and dense fog, rendering my situation dangerous in the extreme, from the vast numbers of icebergs, and quantities of floe and drift-ice, surrounding the brig at the time.

On the afternoon of the 30th, at 3^h 45^m, a ship was discovered ahead; at 3^h 50^m, another appeared in company; being in latitude 64° 52′ 30″ S., and longitude 135° 27′ E. I determined to speak them: supposing them to be the Vincennes and Peacock; at 4^h made them out to be standing to the northward, under easy sail, and discovered them to be strangers; at 4^h 30^m hoisted our colours, knowing that an expedition under Captain Ross was expected in these seas, I took them to be his ships, and stood ready to cheer the discoverer of the North Magnetic Pole. At 4^h 50^m, having gained considerably upon them, and being within I suppose a mile and a half, the strangers showed French colours, the leeward and sternmost displaying a broad pennant; and concluded they were the French discovery ships under Captain D'Urville.

Desirous of speaking, and exchanging the usual and customary compliments incident to naval life, I closed with them, designing to pass within hail under the flag-ship's stern. When within short musket-shot, my intentions too evident to excite a doubt, so far from a reciprocity being evinced, I saw, with surprise, sail made on board the flag-ship. Without a moment's delay, I hauled down my colours and bore upon my course.

On the morning of the 31st, at 8 A. M., I found myself completely embayed in an immense gulf, with a field of table-ice one hundred and fifty feet high, bearing to the northward, and from east to west so far as eye could discern. After consuming the day in trending to windward, I passed out along its eastern margin without accident.

In my progress and examinations, I obtained frequent specimens of sandstone, granite, and red clay, from the field and floe-ice.

I gained the meridian of 105° E., on the 12th of February, latitude 64° 54′ S.; the weather was at intervals misty, affording little opportunity for observation; many strong indications of land presented themselves. The barrier assumed a dark discoloured appearance, with numerous stratified veins of earth and rocks, and with lofty and conical peaks, remotely placed along its southern portion; the impression of land, surrounded and covered by field-ice, was often strongly urged. Penguins and seals were seen, and in my anxiety to land and convince my mind, I was embayed in a narrow and dangerous inlet, which, with the aid of a strong southeast wind, was cleared in safety ere night closed in.

The wind easterly and the weather becoming clear, the occasion seemed so auspicious, that I was induced to extend my researches a day or two, believing it would meet your approbation.

As I advanced westward, the marks of the approach to land were becoming too plain to admit a doubt. The constant and increasing noise of penguins and seals, the dark and discoloured aspect of the ocean, with frequent huge masses of black frozen earth identified therewith, strongly impressed me with the belief that a positive result would arise in the event of a possibility to advance a few miles further south.

On the afternoon of the 13th, I landed and extracted from an immense mass of black earth identified with the barrier, some hundreds of yards back from the margin, specimens of rock corresponding to those previously obtained. At sunset of the 13th, one hundred and fifty-one icebergs, all assuming a discoloured and earthy appearance, were counted from deck.

At 6 A. M. of the 14th, the wind having set in from the northwest, I stood back along the barrier on my return, hauling in to the southward, and making it within the meridians of 120° and 130° E., being a portion which escaped my personal observation, arising from inclement weather.

I had reached the meridian of 100° 07′ 40″ E., latitude 64° 15′ S. I attempted to close with the barrier on the 24th, in order to procure a supply of ice, being reduced to a half-gallon allowance of water; after several unsuccessful attempts, owing to rough and foggy weather, I concluded to proceed north, in conformity with my instructions, and at 2 P. M. bore up in a gale from northwest; the wind prevailing heavy from that quarter gradually carried me near the Lord Auckland Isles, passing the last iceberg in latitude in 55° 24′ S., longitude 148° 37′ E.

I availed myself of your suggestion,—put in and anchored on the night of the 7th of March in the harbour of Sarah's Bosom. I procured a supply of wood and water, and sailed again on the afternoon of the 10th. The sketch of the island in my possession, though not entirely correct, is sufficiently so to guide a vessel safely to the anchorage. I proceeded to this place, passing to the eastward of New Zealand,—strong northeast winds prolonging the passage.

I spoke the American whale-ship Mary and Martha, Coffin, master, of Plymouth, Massachusetts, off Cook's Straits, on the night of the 17th instant: twenty-seven months out; twenty-two hundred barrels of oil; reported at least one hundred ships engaged on the east coast in the fisheries.

The weather during the cruise has been attended with great variety and sudden transitions. The great anxiety I felt to attain a high southern parallel, and obtain convincing proofs of the existence of land from the indications presented, added to the ardour of the officers and crew, often involved us in situations, alike interesting, critical, and dangerous, attributing our escapes without injury to the too plain guidance of the watchful hand of Providence.

Among the most pleasing reflections are those of the perfect exemption from sickness and disease,—not a serious case occurring during the whole period, and not a symptom of incipient scurvy. I have avoided all unnecessary exposure, affording every convenience and comfort to the crew, ever keeping in mind, and rigidly adhering to, your sanitary regulations.

I cannot sufficiently express the satisfaction I feel in reporting the very exemplary conduct of the crew; a universal desire to perform their several duties was evinced, from the eldest to the youngest. I beg leave to recommend them in the strongest terms to your notice and consideration.

To the officers I return my thanks: they were ever attentive and unremitting in their duties, greatly contributing to the gratifying and safe termination of the cruise.

I feel great pleasure in speaking in high terms of them, and feel assured they will receive from you the merit which they deserve.

The observations resulting from the cruise, together with the currents, soundings, &c., are minutely and correctly placed upon the chart accompanying, which indicates the track of our researches along the Antarctic Circle.

I have the honour to be,
Very respectfully,
Your obedient servant,
(Signed) CADWALADER RINGGOLD,
Lieut. Com. U. S. Navy.

To CHARLES WILKES, ESQ.,
Commanding Exploring Expedition.

XXXI.

[Copy.]

SIR,—

In reply to your letter of yesterday, I have to inform you that Captain William Hobson, R. N., arrived here on the 29th January, ult., in H. B. M. ship Herald, and that on the following day the two proclamations now enclosed were made. During the next week meetings with some of the chiefs were held by Captain Hobson, when the treaty (of which I have forwarded you a copy) was signed by a few chiefs. Subsequently Captain Hobson and suite visited Hokianga and the Thames, and obtained a few signatures at either place; hitherto these are the only proceedings which have taken place relative to the cession of any rights, by the chiefs of New Zealand, to the British crown.

Referring to the above, the other apparent measures taken by Captain Hobson to establish the British authority here are, the holding a court of sessions at Kororarika, which is in active operation, having a strong police force under its control; the formation of a General Post for New Zealand, and the appointment of various government officers for New Zealand, by His Excellency the Governor of New South Wales.

It is, however, to be remarked, that no laws relative to the mode or form of government intended to be pursued in this colony, have as yet been published.

I have the honour to be, &c.,
(Signed) James R. Clendon,
U. S. Consul.

PROCLAMATION.

Victoria, Queen of England, with her affectionate remembrance to the chiefs and tribes of New Zealand, desires to point out to them their chieftainships in these lands; and that they may keep in peace and live in comfort, thinks it right to send an English chief to advise with the natives of New Zealand, that they may accept the government of the Queen over all their land and islands. Because there will be thousands of the Queen's subjects to reside in the lands, and they are coming.

The Queen is desirous of establishing a government, that all the evils now upon the natives from the English living in idleness and lawlessness may be removed.

Now the Queen is pleased to send me, William Hobson, Captain, Royal Navy, as governor of all the islands of New Zealand, which will at another time be given to the Queen.

The Queen says to the collection of the tribes of New Zealand, and all other tribes of New Zealand, these are the laws that we have spoken of.

First. That the chiefs at the assembly, and those that were not at the assembly, hereby give up entirely to the Queen for ever the government of all their land.

Secondly. The Queen of England agrees and consents to secure to all the tribes, chiefs, and all men in New Zealand, and the head chiefs, all their rights in their lands, villages, and other property. But the chiefs are to give to the Queen the right of purchasing all the lands that the owners are willing to sell, at the price they choose to put on it, and the Queen says she will pay for it herself.

Thirdly. This is the consent to the government of the Queen. The Queen will protect all the natives of New Zealand, and secure to them all the rights and privileges of the people of England.

(Signed) William Hobson,
Consul and Lieutenant-Governor.

We the chiefs at the collection of the tribes of New Zealand, assembled at Waitanga, are the chiefs of New Zealand, and see the truth of these words and accept them, and therefore we put our names and marks thereto.

Done at Waitanga on the 6th day of February, in the year of our Lord one thousand eight hundred and forty.

TRANSLATION OF THE TREATY.

Her Majesty, Victoria, Queen of England, in her gracious consideration for the chiefs and people of New Zealand, and her desire to preserve to them their lands and to maintain peace and order amongst them, has been pleased to appoint an officer to treat with them for the cession of their country, and of the islands adjacent.

The Queen seeing that many of Her Majesty's subjects have already settled in this country, and are constantly arriving, and that it is desirable for the protection of the natives to establish a government amongst them.

Her Majesty has accordingly been pleased to appoint me, William Hobson, a Captain in the Royal Navy, to be governor of such parts of New Zealand as may be now or hereafter ceded to Her Majesty, and proposes to the chiefs of the confederation of the united tribes of New Zealand, and the other chiefs, to agree to the following articles:

Art. I. The chiefs of the confederation of the united tribes, and the other chiefs who have not joined the confederation, cede to the Queen of England for ever the entire sovereignty of the country.

Art. II. The Queen of England confirms and guarantees to the chiefs and tribes, and to all the people of New Zealand, the possession of their lands, dwellings, and all their property. But the chiefs of the confederation and the other chiefs grant to the Queen the exclusive right of purchasing such lands as the proprietors thereof may be disposed to sell, at such prices as shall be agreed upon between them and the persons appointed by the Queen to purchase from them.

Art. III. In return for the cession of the sovereignty to the Queen, the people of New Zealand shall be protected by the Queen of England, and the rights and privileges of British subjects shall be granted to them.

(Signed) William Hobson,
Consul and Lieutenant-Governor.

Now we the chiefs of the confederation of the united tribes of New Zealand, being assembled at Waitanga, and we the other chiefs of New Zealand having understood the meaning of these articles, accept of them and agree to them all. In witness whereof our names or marks are affixed.

Done at Waitanga the 6th day of February, in the year of our Lord one thousand eight hundred and forty.

Here follow signatures of chiefs.

END OF THE SECOND VOLUME.

1340412

Made in the USA